www.cengage.com

cengage.com is the World Wide Web site for Cengage Publishing and is your direct source to dozens of online resources.

At *cengage.com* you can find out about supplements, demonstration software, and student resources. You can also send e-mail to many of our authors and preview new publications and exciting new technologies.

cengage.com
Changing the way the world learns®

FOURTH EDITION

Generalist
Case Management

A Method of
Human Service
Delivery

Marianne Woodside
Tricia McClam

University of Tennessee

BROOKS/COLE
CENGAGE Learning®

Australia • Brazil • Japan • Korea • Mexico • Singapore • Spain • United Kingdom • United States

Generalist Case Management: A Method of Human Service Delivery, **Fourth Edition**
Marianne Woodside,
Tricia McClam

Publisher: Jon-David Hague

Assistant Editor: Naomi Dreyer

Editorial Assistant: Coco Bator

Media Editor: Elizabeth Momb

Senior Brand Manager:
Elisabeth Rhoden

Market Development Manager:
Kara Kindstrom

Manufacturing Planner:
Judy Inouye

Rights Acquisitions Specialist:
Roberta Broyer

Copy Editor: Laurene Sorensen

Cover Designer:
Kathleen Cunningham

Cover Image: © iStockphoto
.com/Dan Tero

Art and Cover Direction:
Carolyn Deacy, MPS Limited

Production Management:
Teresa Christie, MPS Limited

Composition: MPS Limited

For product information and technology assistance, contact us at **Cengage Learning Customer & Sales Support, 1-800-354-9706**

For permission to use material from this text or product, submit all requests online at **www.cengage.com/permissions** Further permissions questions can be e-mailed to **permissionrequest@cengage.com**

Library of Congress Control Number: 2012952304

Student Edition:

ISBN-13: 978-1-285-17322-1
ISBN-10: 1-285-17322-8

Brooks/Cole
20 Davis Drive
Belmont, CA 94002-3098
USA

Cengage Learning is a leading provider of customized learning solutions with office locations around the globe, including Singapore, the United Kingdom, Australia, Mexico, Brazil, and Japan. Locate your local office at **www.cengage.com/global**

Cengage Learning products are represented in Canada by Nelson Education, Ltd.

To learn more about **Brooks/Cole**, visit **www.cengage.com/brooks/cole**

Purchase any of our products at your local college store or at our preferred online store **www.CengageBrain.com**

Printed in the United States of America
1 2 3 4 5 6 7 17 16 15 14 13

About the Authors

Together, Tricia McClam and Marianne Woodside have over fifty years of experience in human service education, as well as many years working as practitioners in education and vocational rehabilitation. Currently they are professors in the Department of Educational Psychology and Counseling in the College of Education, Health, and Human Sciences at the University of Tennessee. They are committed to research in teaching and learning in the human services and are the authors of several other texts including *Introduction to Human Services; An Introduction to Human Services: Cases and Applications; The Helping Process: Assessment to Termination;* and *Interviewing: What Students Want to Know.*

Brief Contents

Contents

2   Historical Perspectives on Case Management   38

• • • • • • •

3   Models of Case Management   70

• • • • • • •

5   The Assessment Phase of Case Management   142

• • • • • • •

6   Effective Intake Interviewing Skills   172

• • • • • • •

7   Service Delivery Planning   208

• • • • • • •

8   Building a Case File   244

• • • • • • •

9 Service Coordination   298

• • • • • • •

10   Working within the Organizational Context   330

• • • • • • •

11   Thriving and Surviving as a Case Manager   364

• • • • • • •

Preface

For us, the purpose of writing textbooks is to share with our students and colleagues what we have learned about the profession of human service delivery during our years of teaching and working in the field. This philosophy guided the preparation of this fourth edition of *Generalist Case Management: A Method of Human Service Delivery*. Primary informants for this edition were both educators and human service professionals, especially those working as case managers or care coordinators. Through our associations with colleagues in professional organizations and educators who used the third edition, we learned about current trends, challenges, and new knowledge and skills necessary for effective case management. Interviews we conducted with case managers across the United States for the past fifteen years enabled us to capture their voices as they described the realities of service delivery. We believe this adds a real-world perspective to the text.

Change occurs rapidly these days, and case management as a service delivery strategy is no exception. Factors affecting case management today include the economic downturn, federal legislation, emerging client groups, technology, shifting demographics, funding challenges, new service delivery models, increasing multicultural and ethnic perspectives, and ethical and legal dilemmas. The fourth edition of *Generalist Case Management* reflects these changes. There are updated references and examples, a focus on strengths-based case management, a review of technological advances, and an emphasis on collaboration that embraces family, friends, and clients as case managers. We integrated into our discussion of the delivery of case management services the demands related to current economic and political conditions and context. We included discussions of new trends and challenges in case management. A new emphasis on diversity in its broadest sense—ethnic, religious, gender, and lifestyle—pervades the text. Finally, we strengthen the voice of the case manager in each chapter, providing the reader with a realistic picture of the day-to-day work.

The concept of case management is dynamic. Just as the process has changed during the last decade, so it will continue to evolve during the twenty-first century. Many factors will influence human service delivery in the future: economic instability, the managed care environment, technology, the scarcity of resources, demands for accountability, the changing political climate, and the influence of diversity. In this text, we defined and described case management as it is practiced today, but with an eye to the future.

Goals

In this text we explored professional issues and skills related to case management and described the most up-to-date aspects of case management. In short, our goals for this text were fourfold: to define case management, to describe many of the responsibilities that case managers assume, to discuss and illustrate the many skills that case managers need, and to describe the context in which case management occurs. Underlying these goals are the human service values and principles that guide them.

The first four chapters of *Generalist Case Management: A Method of Human Service Delivery* focus on defining case management. Chapter 1 begins by defining case management, outlining the process and components of the case management process, and articulating the principles and goals that guide the work. Roy Roger Johnson's case illustrates the three phases of case management: assessment, planning, and implementation. Chapter 2 expands the definition of case management by reviewing its history. The case of Sam, who was institutionalized early in childhood, illustrates how the changing definition of case management has been reflected in the care of clients. First-person accounts of clients in the early days, as well as excerpts from relevant legislation, enliven the history. Managed care, which has a strong influence on human service delivery today, is defined and discussed in terms of its effects on the case management process. In Chapter 3, we focus on three models of case management and specific roles and responsibilities that case managers assume as they work in agencies with clients. Vignettes and cases illustrate both models and roles and responsibilities. In addition, salient aspects of multicultural case management, such as identity, power, and advocacy, provide the groundwork for understanding how this perspective influences all case management. Finally, a discussion of ethical and legal perspectives follows in Chapter 4, addressing specific issues and challenges relevant to this specific method of helping. Issues include confidentiality, autonomy, working with violent clients, the duty to warn, and the question of when to break the rules.

Chapters 5 through 9 describe in detail the phases of the case management process. In Chapter 5, we begin to trace the case management process from the intake interview to termination. This chapter explores the assessment process in case management including the interview process, types of interviews, issues related to confidentiality, and application and evaluating an application for services. Guidelines for documentation conclude the chapter. In Chapter 6 we provide an in-depth view of the intake interview, the skills needed, and how to adapt the interviewing process to special population. We also include information related to attitudes, characteristics, and skills of interviewers and pitfalls to avoid while interviewing.

Planning, the second major phase of the case management process, is introduced in Chapter 7. Students learn useful information about how to formulate goals and objectives, how to find resources, and how to gather additional information beyond the intake interview process. Tests and their appropriate uses are discussed. Building on the planning process, Chapter 8 describes the case file and explains its multiple components, such as physical examinations, psychological evaluations, social histories, and testing. All of this information is useful to build a comprehensive view of the client.

Chapter 9 describes the third phase of case management, that of service coordination. The chapter focuses on the case manager's interaction with other colleagues. A discussion of service coordination explores the process, including referrals and effective communication with other professionals. Advocacy, a major responsibility of the case manager, is discussed in depth. This chapter also examines how to work effectively as a team member and as a team leader.

Chapters 10 and 11 conclude the text with an examination of the context in which the work of the case manager occurs. In Chapter 10, we introduce concepts such as organizational structure and climate, budgeting, and the commitment to quality care. Chapter 11 concludes the text with a discussion of how to thrive and survive in the "real world" of case management and human services. This chapter provides students with a realistic picture of what it is like to be a case manager, including introducing the knowledge and skills that case managers see as critical. In addition, the chapter introduces challenges and tools that work to support quality client care, such as using time management and asserting oneself, and the importance of professional self-care strategies including supervision.

Features

The fourth edition of *Generalist Case Management: A Method of Human Service Delivery* incorporates many aspects present in the third edition and introduces new features designed to provide a realistic and current view of case management and maintain student interest.

CASE MANAGEMENT AS A PROCESS

Each chapter of the text builds upon the next. Understanding case management and the roles and responsibilities of the case manager becomes a dynamic process as students learn to define the concepts, understand the process from assessment to termination, and study the context in which the work takes place.

PRACTITIONER QUOTES

Throughout the text, quotes from interviews with case managers illustrate aspects of the case management process and ways chapter concepts occur within service delivery context. The practitioners represented in this text reaffirm the use of case management in a variety of settings (e.g., education, vocational rehabilitation, child and family services, mental health, corrections, substance abuse) and various populations (e.g., aging, veterans, homeless, children and youth, mentally ill).

VIGNETTES AND CASE EXAMPLES

Each chapter includes multiple vignettes and case examples to expand student understanding of the concepts introduced. Chapters 1 and 8 follow the case of

Roy Johnson, a client receiving vocational rehabilitation services, from his application of services through service coordination. We read Paulette Maloney's case manager case notes about Rosa Knight, a 23-year-old single mother, who applies for services at Paulette's agency. In Chapter 9, we meet Rube Manning, an adult parolee attempting to integrate into society after release from prison. In Chapter 10, Carlotta Sanchez, who works for the Sexual Assault Crisis Center in a city of 400,000, has just begun her responsibilities as a case manager for the agency.

STUDENT INSTRUCTIONAL SUPPORT

In each chapter, organizational and study materials frame the content. At the beginning of each chapter, we list objectives. At the conclusion of each chapter, students may review summaries of the chapter and answer questions to review their learning.

New Features

ETHICAL AND LEGAL PERSPECTIVES

In this edition of the text we moved Ethical and Legal Perspectives to an earlier chapter in the text. Ethical and Legal Perspectives now becomes one aspect of defining case management. This revision allows students to read about commitments to ethics earlier in their study and reinforces the values related to case management described in Chapter 1.

MULTICULTURAL PERSPECTIVES

The fourth edition of the text reinforces today's need to consider each case management encounter as multicultural. Each chapter provides guidelines for performing case management within an increasingly diverse society. For example, Chapter 3 introduces a new section on multicultural perspectives, linking these perspectives to identity, power, and advocacy. This means it is essential to attend to the ethnic and cultural dimensions of who the client is, the experiences the client brings, and the direction the client wishes the case management process to take. We include standards for cultural competence of case managers in this chapter. In Chapter 5, we include ways for clients to evaluate the cultural competence and justice of the case manager. Topics ranging from a culturally sensitive medical examination to considerations of culture when taking a social history provide specific ways the case manager may assume a multicultural stance.

DEEPENING YOUR KNOWLEDGE: CASE STUDY

Each chapter includes a case study designed to help students apply concepts to practice. At the end of each case study there are questions to guide student learning. We believe the case and questions may also provide the basis for a classroom activity. In Chapters 1, 2, 3, 4, 5, 6, 7, 9, and 11, the case study concludes the chapter and helps the student review her or his understanding of the concepts introduced. In Chapters 8 and 10 the case study is embedded into the whole of the chapter, providing a continuous integration of content and practice.

VOICES FROM THE FIELD: RESEARCH AND PRACTICE

This new section links the student with current case management practices in the government, agencies, and private practice. The purpose of Voices From the Field is to expose students to how text material translates into practice in the real world. For example, in Chapter 4, addressing potential violence in the workplace, we include the Occupational Safety and Health Administration (OSHA) *Guidelines for Preventing Workplace Violence for Health Care and Social Service Workers* (2004). Federal government guidelines for promoting quality health care, stemming from the Affordable Health Care Act, illustrate the commitment to quality discussed in Chapter 10.

WANT MORE INFORMATION?

The Internet is a source of information with which students are familiar. This section targets one concept in each chapter and provides ways that the student may further investigate the most current practices in case management. For example, Chapter 3's Want to Know More? feature asks students to search on the terms "case management roles," "case management responsibilities," and "case management jobs." Chapter Eleven's Want to Know More? section focuses on natural disasters and how government agencies and not-for-profit agencies are responding to expanding human service needs.

THRIVING AND SURVIVING AS A CASE MANAGER

In response to reviewers and our own sense of telling the story of the case manager, we introduce in-depth quotes from case managers interviewed in 2012. We wanted to emphasize the passion and commitment these professionals have toward their clients and their work. These professionals also talk about the issues and challenges they face in their day-to-day work.

STUDENT WORKBOOK FOR GENERALIST CASE MANAGEMENT: A METHOD OF HUMAN SERVICE DELIVERY

We updated the *Student Workbook for Generalist Case Management: A Method of Human Service Delivery,* 4th edition, which students and instructors may use to accompany this text. Our purpose in writing this student workbook was to provide opportunities for our students to think about, practice, apply, and reflect on some of the many skills that are integral to successful case management. The workbook format allows students to record answers, complete exercises, and develop plans.

All chapters begin with some type of pretest (answers provided) to reinforce the main ideas of *Generalist Case Management.* A chapter summary follows the pretest. Exercises relevant to each chapter provide practice opportunities that relate to various case management skills. Among them are writing plans and case notes, examining tests, applying time management strategies, and using case management models. Cultural sensitivity and current topics in human services are integrated into the chapters. Next is a section that explores a concept, idea, client group, or issue in more depth. Topics explored in more detail in other chapters include children as a client group; PACT as a case management model; advocacy; ethical issues that may arise when working with individuals with HIV/AIDS; and vicarious trauma. Additional exercises focus on a combination of textbook concepts, case studies, and new information from this section. All chapters conclude with a focus on self-assessment.

Conclusion

We hope that you and your students benefit from *Generalist Case Management*, 4th edition. It was a pleasure to update. We learned so much about case management as it is practiced today and pass this new understanding on to you.

Acknowledgments

Many people contribute to an undertaking such as this text, and we would be remiss if we failed to acknowledge them. Our colleagues in the National Organization for Human Service and the Council for Standards in Human Service Education have encouraged and supported our efforts to investigate case management—offering suggestions, reviewing materials, and sending information. Ray Vaughn also has our gratitude for sharing his perspectives on case management. Chris Morgan prepared many of the Deepening Your Knowledge: Case Study sections. For the new section Thriving and Surviving as a Case Manager, Katie, Ellen, Sara, and Jessica provided their expertise related to case management and their work. Everett Painter and Elizabeth McClanahan Davison assisted with the Instructor's Manual.

The case managers that we interviewed over the past fifteen years made many contributions to this book. They shared their time, experiences, successes, and failures to enlighten us about the complexities of case management. It is their words that give this text a firm grounding in reality. Among their contributions are definitions of case management, perspectives on the components of the process, and evidence of the trends and challenges that the future holds. Most of all, we thank them for helping us understand the dynamics of the rich and varied process of case management.

Throughout our careers we have valued the review process. The comments and suggestions of the copyeditor, Laurene Sorenson, and the proofreader, Maura Woodside were critical to the development of this text. As they read the printed version, we hope they will be able to see how their unique contributions have improved the text.

Of course, our friends at Cengage Learning deserve our thanks. The expertise and assistance of Amelia Blevins was central to the project. Amelia Blevins and Jon-David Hague.

Last, but not least, we thank our families for their support during this effort. We have spouses who encourage our writing and support us in our academic endeavors.

As the field of human services continues to grow and develop, we look forward to hearing from you. We hope you will share with us your observations and experiences with case management in the field, as well as your reactions to this text. Please send us your comments.

Marianne Woodside
Tricia McClam

. . . .

CHAPTER ONE

Introduction to Case Management

Surviving and Thriving as a Case Manager

Ellen

The agency I work for is located in the northwestern United States. We serve all age ranges. It is a community mental health center. The center has several different campuses across the county. I believe they serve around 18,000 people: children, adults and older adults. And the programs that they offer are quite extensive. They have counseling services, forensic services, housing and rehabilitation, case management, intensive case management, and then different psycho-educational sorts of things they do as a group. I had two positions within the agency. It is not unusual to stay in an agency and assume a new position.

At first I worked for a program that provided extended support and we provided intensive case management to adults and older adults who were chronically mentally ill. So I worked with a lot of folks who had psychotic disorders and anxiety and depression that were living mostly in adult family homes in the community, which are small residential facilities. They have twenty-four–hour care within the homes and so my role as a case manager was to go to those homes a few times a week to do just case management things. The case manager's job is to make sure clients are thriving in their environment, and everyone is safe and healthy.

I worked in that position for about two years and I carried a caseload of between 20 and 30 people at any given time. We spent a lot of time traveling between houses. And then with the shifts in the budget, I transferred to a different position. I worked in one of the adult community support clinics in the south side of the county. At that particular clinic I was a case manager. Most of our clients would come to us. These clients were more capable of managing public transportation in order to make it to appointments, but they were still very much mentally ill. They had other marginalizing sorts of issues: housing issues, financial issues.

> —Permission granted from Ellen Carruth, 2012, text from unpublished interview

In this agency we focus on meeting the needs of individuals and their families. The individuals, our clients, have difficult medical diagnoses and our goal is to allow them to live in their homes. In additional, all of our clients have other needs, reflecting social, educational, financial, and other family concerns. Meeting these multiple needs requires service coordination. We provide services that meet the specific needs of each client. And we involve the client and the families in service delivery. Coordination and integration support the management process. Sometimes professionals working in mental health and developmental disabilities do not understand how to work together to serve a single client. We provide the bridge.

—Case manager, children's services, New York, NY

The agency I work for helps adolescent females. It would be difficult to describe the average client. Our clients come from various economic circumstances and they present very different issues. For some, they have resided in state custody for a number of years and they need short-term housing. For others, they are in crisis and parents or guardians either asked for help or the court referred them. Some are homeless. The girls can stay at this facility for as long as fourteen to thirty days, depending upon who provides the payment. In this agency we offer an array of services including individual and family therapy, psycho-educational groups, and a mental health assessment. Of course, we give the girls food, shelter, and clothing.

—Case manager, youth shelter, St. Louis, MO

Intensive Case Management Program is the name and focus of our program. Our commitment is to meet the long-term needs of the persistently mentally ill. These clients will always need focused help, so when we enroll clients we take the long view. We do everything we can to help these folks. One of our goals is to normalize their experiences; we try to give them a life in the community. We also hope to reduce the stigma in the community. Some of the services we provide include daily living skills training, transportation, health needs, and medications.

—Case manager, mental health comprehensive care services, Knoxville, TN

The preceding quotations represent the words of case managers involved in the delivery of human services. This chapter introduces you to the subject and presents a model of case management that guides many helping professionals who work in human service delivery. Focus your reading and study on the following objectives.

CASE MANAGEMENT DEFINED
- Describe the context in which human service delivery occurs today.
- Differentiate between traditional case management and case management today.

THE PROCESS OF CASE MANAGEMENT
- List the three phases of case management.
- Identify the two activities of the assessment phase.
- Illustrate the role of data gathering in assessment and planning.
- Describe the helper's role in service coordination.

THREE COMPONENTS OF CASE MANAGEMENT
- Define case review and list its benefits.
- Describe why there is the need for documentation and report writing in case management.
- Trace the client's participation in the three phases of case management.

PRINCIPLES AND GOALS OF CASE MANAGEMENT
- List the principles and goals that guide the case management process.
- Describe how each principle influences the delivery of services.

Case Management Defined

The world in which case managers function is changing rapidly. Because of client tracking systems, the electronic transfer of records, dual-diagnosis clients, limited resources, and rapid communication capabilities, current service delivery is vastly different from that of a few years ago. One result is that the time between policy development and implementation is much shorter. Another is that many human service agencies and organizations have chosen to limit the services they provide. More and more, case managers need skills in teamwork, networking, referral, and coordination in order to obtain the services clients need. All this takes place in a constellation of service providers that continues to grow and change.

Service delivery is affected by the current economic downturn and negative economic climate resulting in an expanding number of individuals, families, and communities needing help and support to meet basic needs. Rising unemployment and underemployment, loss of homes to foreclosures, rising health care costs, and increasing costs of postsecondary education, to name a few, are consequences of the late-2000s financial crisis in the United States (Altman, 2009). In addition, many individuals and communities are dealing with the aftermath of increasing weather-related crises.

Changing demographics and multicultural perspectives present additional challenges to delivering case management services to clients in need. A report by Kotkin (2012) for the Smithsonian Institute described significant shifts to youthful and diverse. Accordingly, although there is an increase in the number of those over the age of sixty-five living in the United States, the number of those between fifteen and sixty-four is projected to increase by 42 percent. In addition, by 2050, it is projected that the minority population will increase from 30 percent to 50 percent of the population. The growth of the U.S. minorities reflects an increase in mixed-race individuals, Latinos, Asians, and immigrants from diverse countries and backgrounds. Immigration remains a significant aspect of the changing demographics globally. Experts anticipate that approximately one million people a year will come to the United States (University of Southern California, 2011), and that these immigrants will represent predominantly an educated and skilled workforce contributing to public and private businesses.

In addition, the current political climate brings the role of government under close scrutiny, especially with regard to human services. How involved should government be in meeting human needs? What is its role? What is the proper relationship between state and federal governments? As these questions are examined and debated, case managers sometimes find themselves working under a cloud of uncertainty that influences the work they do, their professional identity, and their professional development.

The quotations that introduced this chapter share a common theme: All three situations require providing and coordinating services for the individuals and families served. Our first case manager directs an agency that provides intensive case management to children and families with complex medical problems. In this agency, the case management process begins as early as the diagnosis of a medical problem and can be terminated once clients are back home and able to manage their own care. An assessment, planning, and coordination process supports clients. There is a continuous evaluation of both client needs and the effectiveness of the care provided. Because the ultimate goal is for the family to manage its own case, all plans and services focus on and build on family strengths.

The services provided by the youth shelter are different. Its primary responsibility is to provide housing, assessment, and counseling for two weeks; the staff then makes appropriate referrals. Although contact is short term, the girls receive intensive physical and psychological care, participate in determining their own treatment plan, and receive shelter and nutritious food. The treatment plan is based on their needs, strengths, and interests. Accountability means developing plans based on the girls' priorities as established on the day they arrive.

The third case manager works in an agency that provides long-term managed care for people with mental illness. Rarely do they close a case. People with severe mental illness who reside in the community require service coordination that is long term, closely monitored, and supportive. The agency's commitment to these clients is to assess their needs periodically and adjust plans and provide services accordingly. Often this agency is the only lifeline for these adults. Because the agency maintains a long-term relationship with clients, its staff develops ways to update assessments and service plans.

These diverse examples illustrate service delivery today. As you can see, the care varies from agency to agency, from helper to helper, and from client to client. One element each example has in common is the use of case management to coordinate and deliver services, moving an individual through the service delivery process from intake to closure.

Traditional Case Management

To define case management, it is helpful to look at the ways in which case management was traditionally regarded. In mental health service delivery in the 1970s, case management was a necessary component of service delivery because clients with complex needs required multiple services. Case management was a process linking clients to services that began with assessment and continued through intervention. In the 1980s, there was a shift in the focus of case management. Many professionals and clients objected to the use of the word *manage* because it connotes control. This language did not seem to reflect a commitment to client involvement or empowerment. Terms such as *service coordination* and *care coordination* were considered to indicate more completely these new goals of case management. Many believed that the term *service coordination* more accurately represented the primary work of the case management process—linking the client to services and monitoring progress. Jackson, Finkler, and Robinson (1992) describe the development of the term *care coordination* during their work with Project Continuity, which facilitated care for infants and toddlers who required repeated hospitalization and who qualified for intervention under the Education for All Handicapped Children Act (1975) and its 1986 amendment, the Preschool Infant/Toddler Program (PL 99-47; parts B and H).

> Over the course of this project, the term *care coordination* evolved from what is popularly described as case management. Staff expressed dissatisfaction with the case management term because they did not feel families should be viewed as cases needing to be managed. Therefore, the project changed the description to care coordinator, which reflects the role as coordinator of care services for the child and family (p. 224).

Case Management Today

Since the late 1990s and early 2000s, many effective case managers have assumed the dual role of linking and monitoring services and providing direct services. In many instances, this dual role is called *intensive case management* and reflects the time and financial resources committed to the client. The trend of the dual role continues today. The principles of integration of services, continuity of care, equal access to services and advocacy, quality care, and client empowerment, described later in this chapter, guide the case management service delivery.

Today, case management characterizes an accepted way of providing human services to clients and their families. For example, the certification of the Human Services Board-Certified Practitioner includes demonstrated competence in case management, professional practice and ethics as one of the four knowledge and skills assessment components (Center for Credentialing and Education, 2011). In addition, the National Association of Social Workers offers BSW social worker case managers the Certified Social Work Case Manager credential (National Association of Social Workers, 2011). In the area of substance abuse, the Substance Abuse and Mental Health Services Administration (SAMHSA), specifically the Center for Substance Abuse Treatment (CSAT), provides a comprehensive guide for the case manager and case management function (Siegal, 1998). Many states developed their own case management certification on the roles, responsibilities, and competencies and skills outlined in the SAMHSA Treatment Improvement Protocols for addiction-related and other human service professionals (Kansas Association of Addiction Professionals, 2011; Oklahoma Behavioral Health Case Management Certification, 2012). Multicultural concerns are embedded in each of these efforts to professionalize case management. There is an emphasis on understanding multicultural competencies required of case managers and addressing issues of advocacy and social justice. It is interesting to note that the certifications include attention to ethnic and cultural aspects of providing services.

We conducted numerous interviews with service providers who are performing the role of case manager, and some indicated a preference for terms other than *case management* and *case manager* in describing their jobs and job titles. Three primary objections to these terms surfaced. One is that the practitioners

Want More Information?

The Internet provides in-depth resources related to the study of case management. Search the terms listed below to read more about how professional organizations and the federal and state government describe case management.

- Human Services Board Certified Practitioner
- Social Work Case Manager
- Commission for Case Management Certification
- SAMHSA Treatment Improvement Protocols

find it objectionable to think of clients as "cases." A second relates to the resentment clients may feel at being managed. Third, these helpers believe that they do more than case management. Many of the helpers interviewed did refer to themselves as case managers, but not necessarily in the traditional sense of the term.

What has emerged today is a broader perspective on service delivery, one that encompasses traditional case management as well as case management with a broader focus. In some situations, it includes case management with a new focus. **Case management** is a creative and collaborative process, involving skills in assessment, communication, coordination, consulting, teaching, modeling, and advocacy that aim to enhance the optimum social functioning of the client served and positive outcomes for the agency (Commission for Case Manager Certification, 2009). Note that it includes the dual role of coordinating and providing direct service. The goal of **case managers** is to help those who need assistance to manage their own lives and to support them when expertise is needed or a crisis occurs. These professionals gather information, make assessments, and monitor services. They find themselves working with other professionals, arranging for services from other agencies, serving as advocates for their clients, and monitoring resource allocation and quality assurance. They also provide direct services. Social justice as a consideration for client rights and equality, as well as respect for the client's culture, guide this work.

The evidence is clear that case management is more a part of service delivery than ever before. In fact, case management is defined and mandated through federal legislation, has become part of the services offered by insurance companies, and is now accepted by helping professionals as a way to serve long-term clients who have multiple problems.

The diversity of professionals with case management responsibilities is reflected in the many job titles they have: case manager, intensive case manager, service coordinator, counselor, social worker, service provider, care coordinator, caseworker, and liaison worker. In some cases, these professionals provide services themselves; in others, they coordinate services or manage them. Increasingly, they are assuming new responsibilities, such as cost containment and budget management. There is little agreement about what to call those they serve, but most frequently they talk about *clients*, *individuals*, or *participants*.

The diversity of job titles, the range of individuals and groups served, and the variety of job responsibilities are all indications that service delivery is changing. This text explores case management as a complex, evolving, and diverse process. As you study this text, you will review traditional case management, learn about the new ways in which case management is being applied, and explore the new roles and responsibilities given to helpers.

One of the important ways of learning about case management is through the voices of helping professionals themselves, as in the many concrete examples in this book. As you read, note their different job titles, roles, responsibilities, service delivery methods, and terminology. The examples that illustrate concepts and principles generally use the terminology of the particular setting involved. When a case or example does not define the terminology, the term *case management* will be used to mean the responsibilities of both service provision (e.g., counseling) and service coordination (e.g., arranging for services from others).

The term also refers to the management skills needed to move a case from intake to closure. In referring to the service provider, the term *case manager* will mean the professional who performs the tasks of case management.

The section that follows introduces the process of case management and its three phases. The case of Roy Johnson illustrates each phase.

The Process of Case Management

The three phases of case management are assessment, planning, and implementation. **(See Figure 1.1.)** Each phase will be discussed in detail in later chapters. Human service delivery has become increasingly complex in terms of the number of organizations involved, government regulations, policy guidelines, accountability, and clients with multiple problems. Therefore, the case manager needs an extensive repertoire of knowledge, skills, techniques, and strategies.

Let's see how these phases occur in three different settings. Steve is a case manager at an agency that works with children and families or guardians of children served by the juvenile court system. He maintains a caseload of young people being sent to correctional facilities. For him, assessment is complex and multifaceted. He describes it this way.

> First, I try to collect lots of information about the child I am helping. I will call schools, doctors, psychologists, and other professionals who worked with the child. If I think it is needed, I will also request additional testing or examinations. This might include a mental health status exam, psychological exam, intake interview, or environmental scan of the child's home. This I would do in a home visit and I might talk with other family members or neighbors.

Maria is a director at a children's services center. She describes the process of planning how her staff will provide services to clients.

> When I work with clients, I follow certain steps. This is important for me and for the client. Many clients see so much that is wrong with their

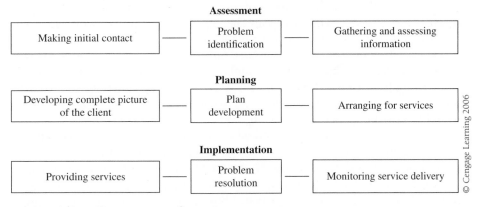

Figure 1.1   The process of case management

lives and they want to fix everything all at once. So, from my staff and me the clients hear the same phrase, "one step at a time." We help clients understand that change is difficult and a long-term process. For many new case managers, this is a difficult concept to put into practice. They want to save the day and make a difference. Their enthusiasm for helping leads to rushing in with grandiose ideas. These high expectations often cannot be met or cannot be met in a short period of time.

Fredrico, a social worker, serves as a case manager in the emergency room of a large metropolitan hospital. He provides frontline assessment and referral for treatment for emergency admissions from various sources. Finding a place for patients to stay represents the implementation phase.

So in my job, I see the patient initially. Usually patients come to the emergency room involuntarily; the police bring them in. Once this occurs, I fill out the necessary paperwork for the hospital to accept responsibility for patient commitment. Once the hospital is ready to dismiss the patient, I find the patient a place to stay. That is really challenging, especially when the patient is angry.

As you can see, the responsibilities at each phase vary, depending on the setting and the case manager's job description. It is important to understand that the three phases represent the flow of case management rather than rigidly defined steps to successful case closure. An activity that occurs in the first phase (e.g., the information gathering that Steve does) may also appear in the second or third phases, as in Maria's planning and the hospital social worker's (Fredrico's) referral. Other key components in effective case management appear throughout the process, including case review, report writing and documentation, and client participation. Ultimately, the goal of case management, stated earlier, is to empower clients to manage their own lives as well as they are able. The case of Roy Johnson illustrates how this happens.

Roy Johnson is a real person, but his name and other identifying information have been changed. The case as presented here is an accurate account of Roy's experience with the human service delivery system and the case management process. His case exemplifies the three phases of case management. The agency that served Roy uses the terms *counselor* and *client*. The following background information will help you follow his case through assessment, planning, and implementation.

Roy referred himself for services after suffering a back injury at work. He was twenty-nine years old and had been employed for five years as a plumber's assistant; he hurt his back lifting plumbing materials. After back surgery, he wanted help finding work. Although he had received a settlement, he knew that the money would not last long, especially since he had contracted to have a house built. He heard about the agency from a friend who knew someone who had received services there and was now working. The agency helps people with disabilities that limit the kind of work they can do. An important consideration in accepting a person for services at the agency is determining whether services will enable that person to return to work. The agency opened Roy's case; we will follow it to closure.

Assessment

The **assessment** phase of case management is the diagnostic study of the client and the client's environment. It involves initial contact with an applicant as well as gathering and assessing information. These two activities focus on evaluating the need or request for services, assessing their appropriateness, and determining eligibility for services. Until eligibility is established, the individual is considered an applicant. When eligibility criteria have been met, the appropriateness of service is determined, and the individual is accepted for service, he or she becomes a client. You will read more about assessment in Chapter 5.

THE INITIAL CONTACT

The initial contact is the starting point for gathering and assessing information about the applicant to establish eligibility and evaluate the need for services. In most organizations, the data gathered during the initial contact is basic and demographic: age, marital status, educational level, employment information, and the like. Other information may be obtained to provide detail about aspects of the client's life, such as medical evaluations, social histories, educational reports, and references from employers.

Roy was self-referred to the agency. He initiated contact by telephoning for an appointment. Fortunately, a counselor was able to see him that week, so he made an appointment for May 24 at 10:30. The agency sent him a brochure about its services and a confirmation of his appointment. When he arrived at the agency, Roy completed an application for services. (See Figure 1.2.) The agency believes the applicant should supply the information in this initial information gathering. He completed it without too much trouble, although he wasn't sure how to answer the question about where he had heard about the agency. He didn't know the name of his friend's friend. The receptionist helpfully told him to write in "self-referral." She suggested that he leave any questions blank if he wasn't sure about the response. She also asked him not to sign the application until he had met with a counselor. She stated that each counselor liked to explain the paragraph at the end of the application in order to make sure that applicants understood the implications of applying for services and the conditions that apply to the release of any client information. She would later transfer this information to the agency's client database.

Roy had brought a copy of a letter prepared by his orthopedic surgeon, Dr. Alderman, for his attorney a year earlier. (See Figure 1.3.) Dr. Alderman had expressed the opinion that Roy would be left 10% disabled as a result of the injury. Dr. Alderman was also careful to clarify that Roy's condition did not reflect a preexisting disability even though he had suffered back problems previously. Tom Chapman, the counselor who saw Roy, made a copy of the letter and returned Roy's copy to him.

APPLICATION FOR SERVICES

Part 1

A. Office No. ☐☐   B. Counselor No. ☐☐☐   C. Client SSN |1|2|3|-|4|5|-|6|7|8|9|

D. Review Date ☐☐-☐☐-☐☐   E. Referral Date ☐☐-☐☐-☐☐

F. Name___Johnson,___ _{Last}   ___Roy___ _{First}   ___Roger___ _{Middle}

G. Address___Rt. #51___ _{Street}   ___Centerville___ _{City}   ___County___ ☐☐ |T|N| _{State} |4|7|8|7|1| _{Zip}

H. Phone No.|7|0|7|-|5|8|7|-|7|0|8| Directions to home:_____

I. Birthdate |0|7|-|1|6|-|8|0| Age _29_ Place of birth: _Michigan_   _East Lansing_

J. Referral Source ☐☐ ___self___

K. Disability ☐☐☐ ___bad back___   L. Sex |M| M-Male F-Female

L. Cause of Disability: _accident_   Age at beginning of disability: _28_

M. How does disability limit activities? _walking, standing, pushing, pulling, lifting_

N. Other physical or mental problems: _none_

O. Have you previously received Services?   Yes____ No _✓_ State_____ Date_____

Part 2

A. Race |1| 1-White 2-Black 3-American Indian or Alaskan Native 4-Asian or Pacific Islander

B. Highest grade of school completed |1|2| (MR=99) Year____ Name & Address:_____

_____

Other Training:_____

C. Martial status |5| 1-M,2-W,3-Div,4-Sep,5-NM D. No. of Dep☐☐ E. Hisp. Origin: Yes☐No☑ F. Vet: Yes☐ No☑

G. Wk status ☐ H. Wkly Earn ☐☐☐ I. Hrs Worked ☐☐ J. Primary source of support ☐ K. Institution ☐☐

L. Public Assistance/Support:

SSI-Aged	Yes ☐	No ☐	Amount_____	SSDI	Yes ☐ No ☑	Amount_____
SSI-Blind	Yes ☐	No ☐	Amount_____	VA Disability	Yes ☐ No ☑	Amount_____
SSI-Disabled	Yes ☐	No ☐	Amount_____	Other Disability	Yes ☐ No ☑	Amount_____
AFDC	Yes ☐	No ☐	Amount_____	Other PA/PS	Yes ☐ No ☑	Amount_____
Gen. Ass't	Yes ☐	No ☐	Amount_____	Total monthly amount (Nearest $)☐☐☐	(none 000)	

M. Medical Insurance Coverage: Yes ☐ No ☐ Name & Type Coverage:_____

_____ Coverage ID No._____

N. Availability of Medical Insurance through Client's Employment for a Salary or Wages at Application: ☐
  0 - Insurance not available   1 - Insurance is available   2 - Client not working at application

Figure 1.2 Application for services

During the initial contact, the case manager determines who the applicant is, begins to establish a relationship, and takes care of such routine matters as filling out the initial intake form. An important part of getting to know the applicant is learning about the individual's previous experiences with helping professionals, his or her strengths, and his or her perception of the presenting problem; the referral source; and the applicant's expectations. As these matters are discussed, the case manager uses appropriate verbal and nonverbal communication skills to

Name & Address of Family Physician: _Dr. Alderman_

Name & Address of Physicians and Dates Seen for Disability: _Dr. Alderman   06/16/XX_

Date Last Hospitalized: _May XX_   Hospital: _Centerville_   Reason: _Surgery_

Do you wear any type artificial appliance? Yes_____ No _✓_ Type:_____

Family Members Living in Home:

NAME	AGE	RELATION	EDUCATION	JOB	MONTHLY WAGE
Separate family entity					

Past Work Record: [LIST LAST JOB FIRST]

EMPLOYER	ADDRESS	JOB TITLE	DATES	WEEKLY WAGE	WHY DID YOU LEAVE THIS JOB?
Memorial Hosp.	_Centerville_	_Plumber_	_2 mos. XX_	$ _7.50/Hr._	_hurt on job_
Rock City	_Mechanical_			$	
		Construction work all		$ _15/hr_	
				$	

Type or Area of Vocational Interest: _undecided_

List 2 persons (Other than listed in home) who would always know your address:

Name: _Terry Jones_ Address:_____ Phone: _987-6543_

Name: _Mae Johnson_ Address:_____ Phone: _123-4567_

APPLICATION FOR REHABILITATION SERVICES*

I hereby make application to receive services I may be eligible to receive. I understand that the information contained in my record will not be disclosed, other than in the administration of the program unless written consent is obtained from me or my parent/guardian. Also, I hereby request and authorize any person(s), agency or institution to release to the agency any medical, social, psychological, vocational and/or financial information they may have or may receive, pertaining to me.

Signature of Client _Roy R. Johnson_ _____ Date ☐☐-☐☐-☐☐

Signature of Parent/Guardian (if required)_____ Date_____

Signature of Counselor _Tom Chapman_ _____ Date _5/24/XX_

*To be explained at the time of initial interview   Review Date ☐☐-☐☐-☐☐

Figure 1.2   *(Continued)*

establish rapport with the applicant. (These skills will be discussed in Chapter 6.) The case manager demonstrates sensitivity to cultural considerations that influence applicant strengths, perceptions, and expectations and cultural considerations. (These skills will be discussed in Chapter 3.) Skillful use of interviewing techniques facilitates the gathering of information and puts the applicant at ease. The counselor makes the point at the conference that the client is considered an expert and that self-reported information is very important. By providing

ORTHOPEDIC ASSOCIATES

200 W. MAIN STREET
DOUGLAS, USA 12345-6789

June 16, XXXX

Mr. Jefferson Maupin
Attorney at Law
215 Fourth Street
Douglas, USA 12345-6788

Re: Roy Johnson

Dear Mr. Maupin:

HISTORY: Mr. Roy Johnson presented to my office on April 12, with a history of an injury to the lumbar spine, which occurred at work on March 15. He reported that he was lifting an object weighing approximately 60 lb. at the time of onset. Prior to that event, he had not experienced significant low back or lower extremity pain for the preceding four years. The patient does have a significant history, including previous lumbar spine surgery. The surgical procedure was performed in July XXXX, and included a bilateral L4 laminectomy and diskectomy with a left L5 foraminotomy. When the patient presented to my office on this occasion, he complained of right lower extremity pain rather than left lower extremity pain. My initial disposition was to refer the patient to physical therapy and ask him not to work for two weeks.

The patient returned to my office on April 25, at which time he continued to complain of severe low back and right lower extremity pain. On that occasion, an MRI study was ordered, which revealed evidence of a herniated disc at the L4 level primarily on the right consistent with the patient's right lower extremity pain.

Figure 1.3   Dr. Alderman's letter

information about routine matters, the case manager demystifies the process for the applicant and makes him or her more comfortable in the agency setting. Some of the routine matters addressed during the initial meeting are completing forms, gathering insurance information, outlining the purpose and services of the agency, giving assurances of confidentiality, and obtaining information releases. Of course all of this has to be complete with attention to multicultural aspects of the client's experiences. Documentation records the initial contact.

Mr. Johnson was admitted to the hospital for definitive surgery and, on May 4, he underwent a right L4 diskectomy with the operating microscope. The procedure was more difficult because of previous surgical scar tissue. At surgery, the patient had several free extruded disc fragments consistent with acute lumbar disc herniation.

Since surgery, the patient has returned to the office on several occasions for routine follow-up. He has also been attending physical therapy for a routine postoperative physical therapy program.

DISCUSSION: This patient sustained an acute lumbar disc herniation on March 15, as evidenced by acute onset of back and leg pain documented by MRI study and by positive surgical findings. Because the patient has a history of significant previous lumbar disc disease, it is my opinion that he will undergo a gradual recovery and that he will be left with a disability to the body as a whole as a result of the injury described above of 10 percent to the body as a whole. This disability rating does not reflect his preexisting disability. The disability rating is based on anatomical findings and is in accordance with the AMA guidelines.

Sincerely,

MF Alderman

Marvin F. Alderman, M.D.

MFA:bj

Figure 1.3 *(Continued)*

In the agency Roy went to, case managers fill out a counselor's page **(Figure 1.4)**, which describes the initial meeting, and a client master record **(Figure 1.5)**. The client master record provides basic information about the client, his or her sources of support, and his or her employment. Its format was designed so that data could easily be entered into the computer, thus simplifying the agency's recordkeeping. At this point, Roy was still considered an applicant for services in accordance with agency guidelines.

NAME: _____ Date _____

Mr. Johnson is a 29-year-old referral who has an orthopedic back problem and brought with him a doctor's report from the first accident as well as the second. The client has limitations in most of his daily living activities. The client has just settled with unemployment for $40,000 and is presently building a home, which will deplete this very soon. The client has a 12th-grade education and stated he had been through 2 college quarters. The client was pleasant and well mannered and answered the questions without any problems.

The rules, regulations, and time limits were explained and understood by the client as well as the order of selection. The client is seeking possible training, but really is undecided about what he can and cannot do. He would meet the economic guidelines for certain services at this time and, if the statement is correct, for all services, once his settlement is depleted.

The client was given a functional limitation sheet for Dr. Alderman, which he is to return to this office. We will sponsor a general medical and psychological evaluation with Barbara Hillman. We may put him in a vocational evaluation at the TVTC. We will place this case in status 02 as of this date.

TC:bj

Figure 1.4   Counselor's page

Although Dr. Alderman's letter provided helpful information about Roy's presenting problem, agency guidelines stated that all applicants must have a physical examination by a physician on the agency's approved list. Mr. Chapman also felt that a psychological evaluation would provide important information about Roy's mental capabilities. He discussed both of these

CLIENT MASTER RECORD

APPLICANT STATUS 02

A. Office No. ☐☐

B. Counselor No. ☑☑☑

C. Client No. ☑☑☑■☑☑■☑☑☑

D. Client Name Johnson, Roy R.
   LAST   FIRST   MIDDLE

E. Review Date: ☐☐■☐☐■☐☐

F. Date of: Status 02 ☑☑■☐☑■☒☒

P. Work Status
1. Wage or salaried worker-competitive labor market
2. Wage or salaried worker-sheltered workshop
3. Self-employed-except state agency managed business
4. State agency-managed business enterprise
5. Homemaker
6. Unpaid family worker
7. Not working-student
(8.) Not working-other
9. Trainee or worker (non-competitive labor marker)

G. Race
(1) White
2. Black
3. American Indian or Alaskan Native
4. Asian or Pacific Islander

Q. Hours Worked ☑☑

H. Highest Grade Of School Completed ☑☑
(M.R. = 99)

R. Institution
(00) Not in institution
01 - Public Mental Hospital
02 - Private Mental Hospital
03 - Psychiatric inpatient unit of General Hospital
04 - Community Mental Health Center-inpatient
05 - Public institution for the mentally retarded
06 - Private institution for the mentally retarded
07 - Alcoholism treatment center

I. Marital Status
1. Married   4. Separated
2. Widowed   (5) Never married
3. Divorced

08 - Drug abuse treatment center
09 - School and other institution for the blind
10 - School and other institution for the deaf

J. Number of Dependents ☑☑

11 - General Hospital
12 - Hospital or specialized facility for chronic illness
13 - Institution for aged
14 - Halfway House

K. Hispanic Origin   Yes ☐   No ☒

15 - Correctional Institution - adult
16 - Correctional Institution - juvenile
17 - Other institutions and special living arrangements
     including group home

L. Veteran   Yes ☐   No ☒

M. Occupation (Title) none

S. Public Assistance/Public Support
(1) Not on public assistance
1 - SSI - aged
2 - SSI - blind

Occupation Code ☒☑☑☑☑☑

N. Weekly Earnings (Nearest $) $☐☐☐None☒ooo

3 - SSI-disabled
4 - AFDC
5 - GA only
6 - SSDI
7 - Veterans' disability benefits
8 - Other disability benefits
9 - All other PA/PS payments

O. Primary Source of Support
A. Current earnings, interest, dividends, rent
B. Family and friends
C. Private relief agency
D. Public assistance, at least partly with Federal funds

T. Public Assistance   Monthly Amount
   (Nearest $) $☐☐☐ None☒ooo

E. Public assistance, without Federal funds (General Assistance only)
F. Public institution-tax supported
G. Workmen's compensation
H. Social Security Disability Insurance benefits
J. All other public sources
(K.) Annuity or other non-disability insurance benefits (private insurance)
L. All other sources of support

U. Weeks Unemployed ☑☑

Tom Chapman   6/25/xx
COUNSELOR'S SIGNATURE   DATE

Figure 1.5   Client Master Record

with Roy, who was eager to get started. As Roy prepared to leave, Mr. Chapman explained that it would take time to process the forms and review his application for services. He would be in touch with Roy very soon, explaining the next steps. (*See Figure 1.6.*)

July 20, XXXX
Roy R. Johnson
Rt. 51
Centerville, USA 12345

Dear Mr. Johnson:

We have scheduled the following appointment(s) for you: We have authorized a general physical for you with Dr. Jones, Suite 201, Physicians Office Building, 172 Lake Road. Please call 589-2111 to schedule the appointment. We have also authorized a psychological evaluation with Barbara Hillman. She will call you to schedule an appointment.

Please make every effort to keep the appointment(s) that we have scheduled. If, however, you will be unable to keep the appointment(s) please contact this office prior to the date of the appointment. Our phone number is 596-5120. Your cooperation is appreciated. If you have any questions, please feel free to contact us.

Sincerely,

Tom Chapman

Counselor

TC:bj

Figure 1.6   Tom Chapman's memo

GATHERING AND ASSESSING INFORMATION

If the applicant is accepted for services, the client and the case manager will become partners in reaching the goals that are established. Therefore, as they work through the initial information gathering and routine agency matters, it is important that they identify and clarify their respective roles, as well as their expectations for each other and the agency. From the first contact, client participation and service coordination are critical components in the success of the process. The case manager must make clear that the client is to be involved in all phases of the process. A skillful case manager makes sure that client involvement begins during the initial meeting. Client involvement helps support attention to individual client needs and demonstrates respect.

In Roy's case, the counselor reviewed the application with him. There were some blanks on the application, and they completed them together. Roy had not been sure how to respond to the questions about primary source of support and members of his household. As Roy elaborated on his family situation, the counselor completed these items. Roy felt positive about his interactions with Tom Chapman because Tom listened to what he said, accepted his explanations, and showed insight, empathy, and good humor.

In gathering data, the case manager must determine what types of information are needed to establish eligibility and to evaluate the need for services. Once the types of information are identified, the case manager decides on appropriate sources of information and data collection methods. His or her next task is making sense of the information that has been gathered. In these tasks, assessment is involved: The case manager addresses the relevance and validity of data and pieces together information about problem identification, eligibility for services, appropriateness of services, plan development, service provision, and outcomes evaluation. During this process the case manager checks and rechecks the accuracy of the data, continually asking, "Does the data provide a consistent picture of the client?"

Client participation continues to play an important role throughout the information-gathering and assessment activities. In many cases, the client is the primary source of information, providing historical data, perceptions about the presenting problem, strengths assessment, cultural considerations, and desired outcomes. The client also participates as an evaluator of information, agreeing with or challenging information from other sources. This participation establishes the atmosphere to foster future client empowerment.

The counselor needed other information before a certification of eligibility could be written. In addition to Dr. Alderman's letter, a general medical examination, and a psychological evaluation, the counselor requested a period of vocational evaluation at a regional center that assesses people's vocational capabilities, interests, and aptitudes. Tom Chapman had worked with all these professionals before, so he followed up the written reports he received with further conversations and consultations. Following a two-week period at the vocational center, the evaluators met with Roy and Mr. Chapman to discuss his performance and make recommendations for vocational objectives. After completing the report, Mr. Chapman and Roy met several times to review information, identify possibilities, and discuss the choices available to Roy.

> Mr. Chapman's knowledge of career counseling served him well as he and Roy discussed the future. Unfortunately, an unforeseen complication occurred, delaying the delivery of services. Tom Chapman changed districts, and another counselor, Susan Fields, assumed his caseload. Meanwhile, Roy moved to another town to attend school. Although he was still in the same state, Roy was now about 200 miles from his counselor. While Roy was attending his first semester at school in January, Ms. Fields completed a certificate of eligibility for him. (See Figure 1.7.) This meant that he was accepted as a client of the agency and could now receive services. In May, his case was transferred to another counselor (his third) in the town where he lived and attended school.

Planning

The second phase of case management is **planning**, which is the process of determining future service delivery in an organized way. When planning begins, the agency has usually accepted the applicant for services. The individual has met the eligibility criteria and is now a client of the agency. During this planning process, the counselor and the client turn their attention to developing a service plan and arranging for service delivery. Client participation continues to be important as desired outcomes are identified, services suggested, and the need for additional information determined. The actual plan addresses what services will be provided and how they will be arranged, what outcomes are expected, and how success will be evaluated. More information on planning is in Chapter 7.

A plan for services may call for the collection of additional information to round out the agency's knowledge of the client. Some case managers suggest that the service delivery process is like a jigsaw puzzle, with each piece of information providing another clue to the big picture. During this stage, the case manager may realize that a social history, a psychological evaluation, a medical evaluation, or educational information might provide the missing pieces. You will read more about this information in Chapter 8. The plan identifies what services are needed, who will provide them, and when they will be given. The case manager must then make the appropriate arrangements for the services.

During the assessment phase, Tom Chapman did a comprehensive job of gathering information about Roy. When Roy was accepted for services, the task facing him and his new counselor was to develop a plan for services. (**See Figure 1.8.**) Clarity and succinctness characterize the service plan, which the counselor and the client complete together, emphasizing the client's input in the process. The plan lists each objective, the services needed to reach that objective, and the method or methods of checking progress.

Suppose that Tom Chapman had believed that a psychological evaluation was unnecessary and had been able to establish eligibility solely on the basis of the medical and vocational evaluations. Susan Fields, the new counselor, might find that a psychological evaluation would be beneficial, especially since the agency was contemplating providing tuition and support for training. One

---

Contact Report and Case Memorandum

Name   Roy R. Johnson   Date of contact: 1/5/xx

District   31   Counselor: S. B. Fields

Status: Last report   02   This report   10   Date of change   1/5/xx

Certification of Eligibility

Disability: The client's primary disability is a back condition. The enclosed general medical exam and orthopedic report from Dr. Porter substantiate this. At present, there are no secondary disabling conditions identified.

Functional limitations/vocational handicap: The client has had long-term treatment for his back condition. He is able to lift only 20 lbs; is restricted to walking, standing, sitting, and occasionally bending; and has restrictions on his lower extremities. The client's vocational handicap restricts him from numerous job areas, such as construction work and truck driving.

Reasonable expectations: I feel that the client can benefit from an educational program before going to selective job placement.

Severely disabled: The client presently meets our guidelines for severe disability, as he will require multiple services over an extended period of time. He clearly has long-term impairment.

Priority category: The client is being assigned to Priority Category I.

___Susan Fields___   ___1/5/xx___

Counselor's signature   Date

bj

Figure 1.7   Certificate of eligibility

objective of the plan would then be to provide a psychological evaluation of the client. This is an example of continuing to gather data during the planning phase, as well as continuing to assess the reliability and validity of the data.

Roy's plan indicates that he is eligible for services and meets agency criteria. His program objective, business communications, was established as a result of evaluation services, counseling sessions with Mr. Chapman, and Roy's stated

---

SERVICE PLAN

1. **NAME** ___JOHNSON, ROY_____ **PROGRAM TYPE** ☒ INITIAL ☐ AMENDMENT

2. **YOU ARE ELIGIBLE FOR:** ☒ VOCATIONAL REHABILITATION SERVICES ☐ EXTENDED EVALUATION SERVICES ☐ PAST EMPLOYMENT SERVICES

   BECAUSE: ☒ A. YOU HAVE A PHYSICAL OR MENTAL DISABILITY WHICH CONSTITUTES A SUBSTANTIAL HANDICAP TO EMPLOYMENT AND:

   ☒ B. YOU CAN REASONABLY BE EXPECTED TO BENEFIT IN TERMS OF EMPLOYABILITY FROM SERVICES.

   ☐ C. IT CANNOT BE DETERMINED WHETHER OR NOT YOU CAN BENEFIT IN TERMS OF EMPLOYABILITY FROM
   REHABILITATION SERVICES.

   ☐ D. POST EMPLOYMENT SERVICES ARE NEEDED FOR YOU TO MAINTAIN EMPLOYMENT.

3. **PROGRAM OBJECTIVE:**___Business Communications_____**ANTICIPATED DATE OF ACHIEVEMENT: MONTH__1_YEAR_XX_**
   ESTIMATED DATES TO REACH OBJECTIVE & RECEIVE SERVICES
4. **INTERMEDIATE OBJECTIVE, SERVICES METHODS OF CHECKING PROGRESS.**

	RESPONSIBILITY	FROM	TO
OBJECTIVE To correct physical impairment so that client might reach vocational objective	Client	1/XX	1/XX
SERVICES Possible office visit with the doctor			

METHOD OF CHECKING PROGRESS___Medical information_

	RESPONSIBILITY	FROM	TO
OBJECTIVE To provide background information and educational skills so that client might reach vocational objective	VR	1/XX	1/XX
SERVICES __A. Tuition		1/XX	1/XX
B. Miscellaneous Educational Expenditures		1/XX	1/XX

METHOD OF CHECKING PROGRESS ___R-11, Grade Reports_

	RESPONSIBILITY	FROM	TO
OBJECTIVE To follow client's progress and develop plan amendment if needed so that client might reach objective	Client	1/XX	1/XX
SERVICES A. Possible RP-B		1/XX	1/XX
B. Client/Counselor Contacts		1/XX	1/XX

METHOD OF CHECKING PROGRESS R-11

5. **CLIENT OR FAMILY AND AGENCY RESPONSIBILITIES AND CONDITIONS:** I. Client is responsible to maintain contact with counselor twice each semester by mail, phone or in person. II. Client is responsible to furnish VR Counselor with a copy of grades at the end of each term. III. Client is responsible to maintain an average load of classes and average grades throughout his program. IV. Client is responsible to file for any similar benefits which might help him pay for his program. V. Client is responsible to furnish VR counselor with a resume and a list of potential employers to interview with during the first part of his senior year. VI. Client is responsible to notify counselor of any significant change of address, health, phones number of financial status.
6. **CLIENT'S VIEW OF PROGRAM** ___The client and I have discussed the services necessary to help him reach his vocational objective and we are in mutual agreement with his plan._

I HAVE PARTICIPATED IN THE DEVELOPMENT OF THIS PROGRAM AND I UNDERSTAND IT.
I UNDERSTAND AND ACCEPT THE STATEMENT OF UNDERSTANDING WHICH HAS BEEN EXPLAINED TO ME.

Roy Johnson	_5-6-XX_	_Susan Fields_	_5/6/XX_
Client's Signature	Date	Supervisor Signature	Date

Figure 1.8 Service plan

vocational interests. The three stated intermediate objectives would help Roy achieve the program objective. The plan also provides a place to identify the responsibilities of Roy and of the agency in carrying out the plan. Note that this agency takes very seriously the participation of the client in the development of the plan, even asking that the client sign it, as well as the counselor. There is one additional part of the plan to be signed by the client. Because one of Roy's objectives is to acquire training, which involves a significant expenditure on the agency's part, Roy has also signed a Student Letter of Understanding that further describes his responsibilities. In a sense, this letter is a contract between the student and the agency; the counselor signs it as the agency's representative. The letter also states that this agreement is valid as determined by federal and state regulations.

Once the plan is completed, the counselor begins to arrange for the provision of services. He or she must review the established network of service providers. Experienced case managers know who provides what services and who does the best work. Nonetheless, they should continue to develop their networks. For beginning helpers, the challenge is to develop their own networks by identifying their own resources and building their own files of contacts, agencies, and services. Chapter 7 provides information about developing, maintaining, and evaluating a network of community resources.

Implementation

The third phase of case management is **implementation**, when the service plan is carried out and evaluated. It starts when service delivery begins, and the case manager's task becomes either providing services or overseeing service delivery and assessing the quality of the service delivery. He or she addresses the questions of who provides each service, how to monitor implementation, how to work with other professionals, and how to evaluate outcomes. This phase is discussed further in Chapter 10.

In general, the approval of a supervisor may be needed before services can be delivered, particularly when funds will be expended. Many agencies have a cap (a fee limit) for particular services. In addition, a written rationale is often required to justify the service and the funds. As resources become increasingly limited, agencies redouble their efforts to contain the costs of service delivery. In Roy's case, the agency's commitment to pay his training tuition represented a significant expenditure. Susan Fields submitted the plan and a written rationale to the agency's statewide central office for approval.

Who provides services to clients? The answer to this question often depends on the nature of the agency. Some are full-service operations that offer a client whatever services are needed in house, as described in Chapter 3. As a rule, however, the client does not receive all services from a single worker or agency. It is usually necessary for him or her to go to other agencies or organizations for needed services. This makes it essential for the case manager to possess referral skills, knowledge of the client's capabilities, and information about community resources.

No doubt you remember that Roy's first counselor, Tom Chapman, arranged for a psychological evaluation. Many agencies like Tom's have so many clients needing psychological evaluations that they hire a staff psychologist to do in-house evaluations of applicants and clients. Other agencies simply contract with individuals—in this case, licensed psychological examiners or licensed psychologists—or with other agencies to provide the service. Whatever the situation, the counselor's skills in referral and in framing the evaluation request help determine the quality of the resulting evaluation.

Another task of the case manager at this stage is to monitor services as they are delivered. This is important in several respects: for client satisfaction, for the effectiveness of service delivery, and for the development of the case manager's network. Monitoring is doubly important because of the personnel changes that constantly occur in human service agencies. Moreover, there may be a need to revise the plan as problems arise and situations change. Cultural considerations play a part in monitoring services. Clients need to receive services within a culturally sensitive context. They need to feel respected and heard and believe the services represent the values they hold.

The implementation phase also involves working closely with other professionals, whether they are employees of the same agency or another organization. A case manager who knows how to work successfully with other professionals is in a better position to make referrals that are beneficial to the client. These skills also contribute to effective communication among professionals about policy limitations and procedures that govern service delivery, the development of new services, and expansion of the service delivery network.

Perhaps there is no other point in service delivery at which the need for flexibility is so pronounced. For example, during the implementation stage it often becomes necessary to revise the service plan, which must be regarded as a dynamic document to be changed as necessary to improve service delivery to the client. Changes in the presenting problem or in the client's life circumstances, or the development or discovery of other problems, may make plan modification necessary. Such developments may also call for additional data gathering.

In his second semester at school, Roy heard about a course of study that prepared individuals to be interpreters for the deaf. This intrigued him, because he was already proficient in sign language. His mother was severely hearing impaired, and as a child, Roy signed before he talked. He also thought back to the evaluation staff meeting, at which the team discussed the possibility of making interpreter certification a vocational objective for him. Roy liked the interpreting program and the instructors, so he applied to the program. The change in vocational objective made it necessary to modify his plan. His counselor (by now, his fourth) revised the plan at the next annual review to include his new vocational objective of educational interpreting.

 # Three Components of Case Management

Case review, report writing and documentation, and client participation appear in all three phases of case management; they are discussed in detail in later chapters. Here we introduce the concepts by examining how each applies to Roy's case.

Case review is the periodic examination of a client's case. It may occur in meetings between the case manager and the client, between the case manager and a supervisor, or in an interdisciplinary group of helpers, called a *staffing conference* or *case conference*. A case review may occur at any point in the case management process, but it is most common whenever an assessment of the case takes place. Case review is an integral part of the accountability structure of an organization; its objective is to ensure effective service delivery to the client and to maintain standards of quality care and case management.

Roy's case was reviewed in several ways. Each time a new counselor assumed the case (unfortunately, this was often), a review was held. There were also reviews on the occasion of the two counselor contacts Roy had per semester. At the end of each semester, his grades were checked—also part of the case review. The staffing related to Roy's vocational evaluation is an example of case review by a team. In this case, the client was an active participant in the case review. Roy also participated in developing the service plan, which involved a review of the information gathered, the eligibility criteria, and the setting of objectives. The agency serving Roy implemented the important component of case review in various ways at different times throughout the process.

An important part of case review is the documentation of the case. **Documentation** is the written record of the work with the client, including the initial intake, assessment of information, planning, implementation, evaluation, and termination of the case. It also includes written reports, forms, letters, and other material that furnishes additional information and evidence about the client. The particular form of documentation used depends on the nature of the agency, the services offered, the length of the program, and the providers. A **record** is any information relating to a client's case, including history, observations, examinations, diagnoses, consultations, and financial and social information. This also includes agency documents such as application forms, financial assessments, outcome assessments, case review notes, referrals, confidentiality and HIPAA documents, and transition or termination documents. The case manager's professional expertise must include documenting events appropriately and in a timely manner and preparing reports and summaries concisely but comprehensively.

Roy's file includes many different types of documentation. In this chapter, the written record includes computer forms, applications for services, counselors' notes, medical evaluations, reports, and letters. Other documentation (not shown here) in Roy's file is a psychological evaluation, a vocational evaluation, specialized medical reports, and medical updates. In Roy's case, all this documentation turned out to be indispensable because he worked with five different counselors. In addition to Roy's documentation, strengths assessment and cultural assessments were included in his file. Both of these assessments help case managers provide Roy with services that reflected his unique individual needs. For continuity of service, good case documentation is essential.

Client participation means the client takes an active part in the case management process, thereby making service delivery more responsive to client needs and enhancing its effectiveness. In some cases a partnership is formed between the case manager and the client; an important result of this partnership is client empowerment. One of the many factors involved in forming a partnership with the client is clear communication or two-way communication, bearing in mind cultural dimensions of communication. The case manager must explain to the client the goals, purposes, and roles of the case manager as defined by the agency. The case manager encourages the client to define his or her goals, priorities, interests, strengths, and desired outcomes. At this point the client also commits to assuming responsibility within the case management process. As client participation continues and the partnership develops, it is helpful for the case manager to have knowledge of subcultures, deviant groups, reference groups, and ethnic minorities so as to communicate effectively with the client about roles and responsibilities. Other factors can affect client involvement, including the timing, setting, and structure of the helping process. Minimizing interruptions, inconveniences, and distractions always enhances client participation.

Encouraging client participation has identifiable components. The first is the initial contact between the client and the case manager. It is easier to involve clients who initiate the contact for help, as Roy did, because they usually have a clearer idea of what the problem is and are motivated to do something about it. In Roy's case, the clarification of roles and responsibilities occurred at three points in the assessment phase. Roy and his counselor were able to talk about the agency and the services available, and the counselor encouraged Roy to talk about his goals, motivations, strengths, and interests.

When Roy completed his application, the counselor reviewed it with him, especially the statement at the bottom of the second page. On signing such a statement, the client voluntarily places himself or herself in the care of the agency. With this agreement come roles and responsibilities for both the client and the counselor, which the counselor reviews at that point. A second opportunity to clarify roles and responsibilities comes with the completion of a service plan. Both the client and the counselor sign the service plan, which designates the responsibility for each task and the time frame for completion of each service.

The middle phase of case management is devoted to identifying problem areas, reaffirming client strengths, and developing and implementing a plan of services. It was during this phase that Roy decided to change from business communications to interpreting, which required an amendment to the plan. Roy also needed help paying for a tutor in a science course that was particularly difficult for him. Client participation during this phase ensures that the client's perceptions of the problem and its potential resolution are taken into account.

The final phase of client participation comes at the termination of the case. At this time, the client and the case manager together review the problem, the goals, the service plan, the delivery of services, and the outcomes. They may also discuss their roles in the process. Thus, in terms of client participation, termination means more than just closing the case. It is an assessment of the client's progress toward self-sufficiency, the ultimate goal of client empowerment. Self-sufficiency is defined differently for each client.

Now that you have some sense of the flow of the process and its component parts, we will review the responsibilities of managing cases in all three phases.

This section introduces the principles and goals that guide service delivery and discusses how they influence the work of case management.

 # Principles and Goals of Case Management

The guiding principles and goals of case management have emerged through the work of early pioneers in helping, through federal legislation, and through current practice. These include integration of services, continuity of care, equal access to services/advocacy, quality care, client empowerment, self-determination, and evaluation. The subsections that follow discuss these principles and their relevance to case managers.

Integration of Services

Case managers must be committed to a holistic view of the individual receiving help. This means they recognize that they work with the **whole person**, acknowledging the many human dimensions that are considered in service delivery: social, psychological, medical, financial, educational, and vocational. Most likely, the client has problems in more than one of these areas.

Multiproblem clients are best served by integrating the needed services. This means developing and implementing a plan that brings together a variety of services to help a client. Integrated service delivery means each service enhances and supports the other efforts and is a guiding principle of case management. Many people enter the system with multiple needs or needs that change over a long period. To address multiple needs, case managers integrate the work of many agencies and professionals. With service integration, there is less chance of fragmentation and duplication. Integrated services also facilitate effective priority setting and encourage positive interaction among the services provided. One case manager at a full-service high school describes how she integrates services for her students.

> I love my job! And I believe my job is really important to the educational mission of our high school and to the success of our students. First, I am responsible for documenting all of the services we have available for students here at school, day and evening and documenting all of the services the community provides. I established a coordinating team made up of school members, parents, students, and members of community organizations. Organizations such as the Board of Education, the Department of Health and Wellness (includes Behavioral Health), the Families First program, and the Department of Human Services (Foster Care, Children and Youth Services, and Child Protective Services) all come together to help provide services to students and their families here at the high school.
>
> (Director of Full-Service Schools, Los Angeles, CA)

Voices from the Field: Research and Practice

Buck and Alexander (2006) reported findings from a qualitative research project in which consumers with serious mental illness were interviewed. They explored how these consumers talked about their intensive case management experiences. The authors described three themes from the interviews that provide insights into the consumer experience: getting services; being social; being there for me. For those interviewed the help they receive from their case managers related to service provision. They ranked this service provision as most important. The responses indicated consumer appreciation of the concrete and tangible benefits they received, including housing and employment. According to those interviewed, case managers helped consumers with financial matters and budgeting. Quotes from these consumers provided insight into the specific tasks integrated into the case management role and relationship.

> "She helped me get a Social Security card and an ACCESS card. She also helped me get into a group home. And now I can get my medicines for free."
>
> "This past Saturday, when I called him he got me into the shelter. I had to leave the boarding house because they were increasing the rent and I couldn't afford it."
>
> "He took me to the eye doctor for an eye exam, I was fearful that the Luvox was affecting my eyesight."
>
> "She helped me when she thought I could get a job. She had confidence that I could do it."
>
> "He was most helpful with budgeting my money, trying to save and pay my bills. He did it in the beginning, and that was one of my goals to do that. Now it comes to me easy. I've been doing it on my own for four or five months now." (p. 475)

A second theme, "getting rides," emerged as salient for these consumers with serious mental illness. Many expressed gratitude since they did not have access to transportation; they did not drive, did not have a car, or did not live on the route of a bus, metro, or train. Quotes from the participants about the transportation for them indicated its importance.

> "When he took me to court. I had no way to get there. If not, I would have gone to jail."
>
> "He took me to the doctor and gives me rides when I need them."
>
> "He takes me to the mall, he gives me rides. He is generous. He's a nice person." (p. 476)

Finally, a third theme consumers used to describe the relationship with their case managers was "being social." Focused on what type of help they wanted from their case managers, they indicated the consumers wanted to spend "me time" with them. Participating in events outside their home was a priority. Quotes exemplified the type of social outings the consumers wished for.

Finally, participants just wanted the case managers to be there for them. Their constancy and consistency provided support they needed and wanted. And consumers described their frustration when case managers were not there for them.

"I'd like him to take me out to eat to get a milkshake or something."

"Movies, or a baseball game or football game."

"When I picked up a drink and I was having a meeting with the people here who were on my back. I was afraid they would throw me out. She comforted me and explained that they probably wouldn't throw me out but they would need to give me notice."

"She was there for me, not for her job or the system, but because she wanted to be."

"I wish she would carry through with what she says. She told me she'd help with Christmas presents and never did."

"I need to get my driver's license. It expired. I've asked her a few times to help me call Harrisburg to find out what I have to do. She hasn't gotten back to me or acted on it. It's important." (pp. 476, 477)

Summarized from Buck, P. W., & Alexander, L. B. (2006). Neglected voices: Consumers with serious mental illness speak about intensive case management. *Administration and Policy in Mental Health Services Research*, 33(4), 470–481, DOI: 10.1007/s10488-005-0021-3. Retrieved from http://www.springerlink.com/content/f12727m05710123g/fulltext.pdf

Continuity of Care

Continuity of care has two meanings. First, continuity means that services are provided to the client uninterruptedly, from the first phone call to termination and often beyond. An individual's needs are addressed, if possible, even before he or she makes the first visit to the agency. For example, the case manager may phone before the initial visit or the intake interview, asking if there are any questions, or he or she might mail a welcoming letter with information about the agency. Continuity of care extends beyond termination, through a transition period when services are no longer needed but the client needs minimal support or just the knowledge that short-term assistance or advice is available if needed.

Second, the term *continuity of care* refers to the comprehensiveness of the care provided. This aspect of continuity involves therapeutic intervention along with support in the environment, maintaining a relationship with the client's family and significant others, crisis intervention, and social networking beyond the mere linking of services. This term is also used in reference to managed care, denoting the intensity level of the services provided. For example, a troubled teen might receive in-home counseling, after-school clinical treatment, or commitment to a residential facility. Each represents a continuum of more intensive intervention. Levels of treatment may interfere with family values or cultural norms; these must be factored in to planned services.

Lois, director and case manager at a mental health agency providing intensive case management services in Los Angeles, provides the following illustration of continuity of care.

A serious issue our clients face occurs once they leave the hospital setting. At times they leave the hospital after their situation stabilizes.

This usually occurs once they take the correct medicine on a regular schedule. Then the clients come to our agency looking for care coordination and follow-up care. Earlier in 2000, the first follow-up was with the hospital. That did not work; between the dismissal and the follow-up, client mental health deteriorated. A majority of the clients did not even make it to that first appointment. Others forget to take their medication and then do not refill the prescription. In our work, we help clients immediately after they leave the hospital. In many cases, we meet the client at the hospital during their formal dismissal.

In the situation described, case managers are able to provide a continuity of care that extends beyond hospitalization. As they continue to work with the client, it becomes necessary to closely monitor their follow-up appointments and their medication prescriptions.

Equal Access to Services/Advocacy

Equal access to services/advocacy means that everyone in need of assistance has the same opportunity to approach, apply for, and use case management services without regard to ethnicity, race, religion, sexual orientation, socioeconomic class, or disability. Attention is also given to developing ways to extend access to services, such as fee waivers, transportation, and outreach efforts. Affordability is directly linked to the issues of eligibility and access. To ensure access to services, eligibility must be defined so as to include those who lack traditional economic, social, and political access. There must also be a plan for individuals who are on the margins of eligibility. Fredrico, a case manager at a metropolitan hospital we introduced earlier, discusses a problem he encountered.

We had a patient come to the emergency with serious medical injuries. He was very aggressive and hostile to the hospital staff and me. The police brought him; he had been in a brawl in a downtown bar. He was transitioning from male to female. Also an individual recently came to the emergency room because of a suicide attempt. He was also a cross-dresser, going to become a female. He had advanced HIV infection. He could not comprehend he was HIV positive. I had to find him a place to live after dismissal from the hospital. Many shelters don't want to admit people who are HIV positive and his transition from male to female increased the difficulty of placing him in either an all-men or all-women facility. I had difficulty finding him a place to stay.

This case also illustrates the advocacy role case managers assume as part of their commitment to equal access. Advocacy is the act or process of representing the interests of the client and teaching the client to advocate for him or herself. It may involve speaking or writing in defense of a person or cause. In representing the interests of clients, case managers often assume a dual role, representing both the institution and the client. Advocacy is one way to assure equal access to needed services. It is discussed more fully in Chapter 10.

Quality Care

Quality care is providing superior services to all clients and implies a commitment that respects the rights of the client and demands accountability on the part of human service professionals. The watchwords are *effectiveness* and *efficient care*; both are key considerations in service delivery. Quality assurance is a component of case management; the term signifies professional excellence, high standards of care, and continuous improvement (Mullahy, 2010).

Workers at youth services in St. Louis point out that success is defined in different ways, depending on the client and the goals of the service.

> Primary work in our agency is outreach. Our success varies depending upon the youth and his or her needs. We measure success on an individual basis, since we develop care plans for each individual. Needs differ, care plans differ. For some youth, we have simple goals (although not necessarily simple to achieve), such as finding housing or shelter. There are also homeless youth who want to remain on the streets. These youth, we help them find food and medical care and social support. Some youth want temporary support, a meal, a phone call, or help during the cold weather. Since the needs and desires of these youth differ, the level of success varies.
>
> (Outreach worker, youth services, St. Louis, MO)

Effectiveness means getting results. In this age of scarce resources, it is important that the available resources are used wisely, that is, *efficiently*. In service delivery, the productive use of resources involves determining the outcomes desired and developing a plan to achieve those outcomes. Efficiency is measured in terms of the resources required, the time expended, the cost of services, and the outcomes achieved. The plan is constantly monitored, necessary adjustments are made, and appropriate justification is presented when additional resources are needed. The case manager must maintain efficiency because of the complexity of many of the problems he or she faces. It is essential that professionals work together to deliver quality services in an efficient manner.

Client Empowerment

Client empowerment within the case management process means respecting clients as individuals, building on their strengths and interests, learning about their cultural context, placing them in a partnership role, and moving them toward self-sufficiency. Respect for the client stems from the long-standing belief that all individuals, regardless of their needs or disabilities, have integrity and worth. This belief guides the case manager to place the client in a central role in the helping process. Ensuring the client's full participation during the service delivery process means involving him or her in every step, including identifying the problem, gathering information, establishing goals, planning, implementing the plan, and evaluating the plan and outcomes.

Client empowerment within the process expands respect for the client to include teaching basic service coordination skills and, as time and skill development allow, encouraging the client to manage his or her own case. The goal is to develop self-sufficiency so that the client can manage his or her own life without depending on the human service delivery system.

In many agencies, self-sufficiency is a primary goal, as Helen, a case manager in a family advocacy agency says.

[We are advocates] for the participant . . . and self-sufficiency is critical in our work. We teach clients how to plan and how to set goals and then plan a way to meet them. A reason for the focus on self-sufficiency is the limited amount of time we can provide services. We want clients to be able to make it on their own once our services are no longer available. Many of the clients begin the service delivery process because they believe they cannot make it on their own. We try to instill confidence, build skills, and teach them how to negotiate the system. We want them to leave because they know they can make it on their own.

Another way of showing respect for the client is to treat him or her as a customer. This means asking clients about their needs, providing services that match those needs, offering clients an opportunity to evaluate the services they receive, and making changes to improve services based on their feedback. The children's services agency in New York City is an agency that has worked hard to treat its clients as customers. It was a struggle for the agency to make this transition.

Children services focuses on family empowerment. So we forge excellent and ongoing relationships with our parents. Both the children and the parents are our clients. We believe that parents should drive the decision-making process. We give parents as much information as they need to make choices and determine outcomes. In other words, we help parents choose the best interventions for their children. We talk to parents using terms they understand. In terms of the language and forms we use, we make them very user friendly. We have such admiration for our parents. They become our customers. The agency began to help the parents of young children with complex medical needs. We continually seek input from our parents so that we may improve our services. Finally, a key aspect stated before is that we form strong bonds and partnerships with our parents.

Client empowerment also means providing services based on client strengths. This focus begins as early as the initial assessment phase, when the case manager and the client begin to identify positive client characteristics on which to build a treatment plan and continues through implementation.

Self-Determination

Building upon client empowerment is the commitment to **self-determination**. In service delivery, self-determination means allowing clients, whenever possible, to choose their goals and interventions. This commitment to the client engenders an attitude of respect to the client and to the client's need for services. Practically speaking for self-determination to occur, the client and family needs information about all of the possible services and interventions available. Self-determination, supported by cultural considerations means this information, conveyed in a format and language the client understands is essential. Critical information includes knowledge about available resources, financial constraints, and services that are not available. During the process, clients may also need to consider and weigh their options, so teaching clients a decision-making or problem-solving process may also be necessary.

There are legal dimensions of the commitment to self-determination. In some instances the client is not capable of determining his or her needs or assessing which services might be best. In other instances, clients may be very young or mentally incompetent. Care taken to implement self-determination in whatever way possible while engaging assistance from guardians or the state. Barbara, a case manager for an intensive case management program for the persistently mentally ill, describes her challenge with self-determination.

> Sometimes my clients literally cannot think for themselves. This usually occurs when they stop taking their medication or when they need a change in medication. Just last week I visited one of my clients. I could not reach him by phone. I changed my schedule so I could drive by to see him that day. He did not answer the door, but the door was cracked. He was sitting on the couch. Just sitting on the couch. He could not talk and did not acknowledge my presence. I called the emergency medical staff and they sent an ambulance. Calling emergency medical was my idea, not his.

During the process of case management service delivery, optimal cooperation occurs among the client, the family, and the case manager. Depending upon which model of case management is used (see Chapter 3), the needs of the client, and the resources available, assigning responsibilities for decision making to the client and choosing interventions that place responsibilities upon the client, family, and multiple professionals is ideal. (Chapter 4 addresses some ethical issues related to self-determination.)

Evaluation

As a critical part of case management, **evaluation** is the assessment of the process, the outcomes, and the quality of the process. Evaluation takes place throughout the case management process. The focus is on relevance for the client, client progress and satisfaction, integration of services, quality of services, and outcomes. Both professionals and clients are involved in the evaluation process. One agency approaches evaluation this way:

> We evaluate all of our work with clients. Each month we review a client case file. Every six months we include the client and the family in the review. At either review we can change the plan, as client needs change.
>   (Case manager, training center, Parkston, SD)

Deepening Your Knowledge: Case Study

Tom is a social worker in Brooklyn, New York. He works for an agency that assists recently widowed mothers with finding housing and other services to meet their needs and the needs of their families. In the following case, he works with Deborah. After reviewing her application, Tom knows that Deborah is a mother of three children (ages two, four, and seven) and recently (four months ago) lost her husband to a loading dock accident.

Tom meets with Deborah for an initial interview and conducts an assessment (diagnostic study of the client and client's environment). Tom keeps in mind that he must determine her eligibility in order for Deborah to change from an applicant to a client. Tom learns from Deborah that she is a twenty-seven-year-old African-American widow. Her education level is a high school diploma, she has been a stay-at-home mom for her three kids for the past eight years, and she worked for two years in a grocery store before marrying her husband and having children.

Her relevant medical history reveals that she has Type II diabetes, is slightly obese, and has high blood pressure. Her social information reveals that she does not spend much time outside of her apartment other than taking the children to run errands or to the park. She used to be involved with her church, but hasn't gone since a few months before her husband's accident. She isn't even sure if the members of the church realize he has passed away. She does have some extended family (cousins and mother) who live in the Bronx. Her mother has not been involved in her life since she was seventeen; she talks to her cousins on the phone rarely and sees them once every few years.

During the initial interview, Deborah becomes very emotional and discusses how she feels hopeless about the death of her husband, her lack of money, her health conditions, and her lack of social connection with anyone outside of her children. She does not know how to navigate paying bills, acquiring health insurance, or many of the other demanding tasks that she is now faced with. She says that her landlord, with whom she has a limited but positive relationship with, told her, during a talk about the upcoming rent, about the help Tom's agency may be able to provide. This landlord offered to give her two months to figure things out before she would take a more aggressive approach to payment.

Morgan, C. (2012). Unpublished manuscript, Knoxville, Tennessee. Used with permission.

Discussion Questions

1. What issues would you focus on when helping Deborah set goals for her situation?
2. What roles might Tom need to assume with respect to service coordination for Deborah?
3. Describe how the principles of case management should drive Tom's decision making when working with Deborah.

Chapter Summary

Managing client services is an exciting and challenging responsibility for helping professionals who work as case managers. Today case management is integrated into a variety of human service delivery systems. To assist clients with multiple problems, case managers must know the process of case management and be able to use it. The process can be adapted to many different settings, for work with a variety of populations. There are multicultural considerations associated with all phases of the case management process. The coordination and the provision of direct services are guided by common principles and goals, committing the agency to give the client quality services and ensuring that he or she participates in the case management process.

The three phases of case management—assessment, planning, and implementation—each represent specific responsibilities assumed by the case manager. The process of case management is nonlinear; for example, a case manager may make assessments early on and return to conduct assessment during the planning and implementation work with the client. Three components of the case management process appear in all three phases of case management: case review, report writing and documentation, and client participation. Note that the first two components also include interaction with and participation by the client. These components require ongoing evaluation and written documentation of the case activities. The process of case management described here is based on important principles and goals of case management. These include integration of services, continuity of care, equal access to services/advocacy, quality care, client empowerment, self-determination, and evaluation. The purpose of these principles and goals are to provide the most effective and efficient services to the client.

 # Chapter Review

● Key Terms ....

Case management   Record
Case manager   Whole person
Assessment   Continuity of care
Planning   Equal access to services/advocacy
Implementation   Quality care
Case review   Client empowerment
Self-determination   Evaluation
Documentation

● Reviewing the Chapter ....

1. Describe the context in which case management services are delivered today.
2. How has the definition of case management changed over the years?
3. What is the expanded definition of case management?

4. Distinguish between the terms *applicant* and *client*.
5. What should be accomplished during the assessment phase?
6. What occurs during the initial contact between the case manager and the individual seeking services?
7. Describe the routine matters that are discussed during the initial contact.
8. Identify the types of information that are gathered during the initial interview.
9. Which factors determine what information has been obtained by the end of the initial interview?
10. Using the case of Roy Johnson, discuss the advantages of a partnership between the case manager and the client.
11. Describe the case manager's activities during the planning phase.
12. What questions guide implementation?
13. Why is flexibility so important during the implementation phase?
14. Define case *review*.
15. List the three keys to successful case review.
16. Why is documentation important in service coordination?
17. How can the case manager promote client participation?
18. How will a client's resistance affect his or her participation in the service coordination process?

Questions for Discussion ....

1. Why do you think that case management as a method of service delivery is changing?
2. From your own work and study of human services, what evidence do you have of the importance of assessment and planning?
3. If you were a case manager, what three principles would guide your work? Provide a rationale for your choices.
4. What do you think Roy Johnson would say about his experience with the case management process?

References ....

Altman, R. C. (2009). The Great Crash: A geopolitical setback for the West. *Foreign Affairs*, January/February 2009. Retrieved from http://www.foreignaffairs.com /articles/63714/roger-c-altman/the-great-crash-2008

Center for Credentialing and Certification (2011). Human services board-practitioner application packet. Retrieved from http://www.cce-global.org/credentials-offered /hsbcp

Commission for Case Manager Certification (2009). Care coordination: Case managers "connect the dots" in new delivery models. *Issue Briefs*, 2, 1. Retrieved from http://www.ccmcertification.org/node/760

Jackson, B., Finkler, D., & Robinson, C. (1992). A case management system for infants with chronic illnesses and developmental disabilities. *Children's Health Care*, 21(4), 224–232.

Kansas Association of Addiction Professionals (2011). *KAAP certification commission: Person-centered case manager.* Retrieved from http://www.ksaap.org/casemanager .shtml

Kotkin, J. (2010, July–August). The changing demographics of America. *Smithsonian.* Retrieved from http://www.smithsonianmag.com/specialsections/40th-anniversary /The-Changing-Demographics-of-America.html

Mullahy, C. M. (2010). *The case manager's handbook* (4th ed.). Sudbury, MD: Jones and Bartlett.

National Association of Social Workers (2011). The certified social work case manager. Retrieved from http://www.socialworkers.org/credentials/specialty/c-swcm.asp

Oklahoma: Behavioral Health Case Management Certification (2012). *Oklahoma Department of Mental Health and Substance Abuse.* Retrieved from http://www.ok.gov /odmhsas/Consumer_Services/Behavioral_Health_Case_Management/Behavioral _Health_Case_Management_Certification_Steps/index.html

Siegal, H. A. (1998). *TIP 27: Comprehensive case management for substance abuse.* Retrieved from http://www.ncbi.nlm.nih.gov/books/NBK14516/

United Nations. (2006). Demographic yearbook. Retrieved from http://unstats.un.org /unsd/demographic/products/dyb/dybcens.htm#MIGR

University of Southern California. (2011). USC analysis projects growth and change of U.S. foreign born population and their children through 2040. Retrieved from http ://www.usc.edu/uscnews/newsroom/news_release.php?id=2511

CHAPTER TWO

Historical Perspectives on Case Management

Surviving and Thriving as a Case Manager

Sara

The agency I work for started out just working with pregnant teenage women. It has expanded greatly within the time that I have been there, and its purpose is prevention. We are trying to get out in the community and help families in crisis. We work primarily with at-risk adolescents, so most of our work is with substance abuse and trauma.

We have two residential programs—a level two and a level three—as well as an intensive outpatient program that serves the level-two clients' therapeutic needs as well as the community. We hire case managers to support the Department of Human Services. We even have a continuum of foster care and foster parents, so we go from foster homes all the way up through level three.

Actually I'm a therapist now but I'm a case manager as well. So I can tell what that work looks like and how it has changed. When I was just a case manager, I had to maintain contact with the clients and maintain the caseload. It ranged at times but it was typically about sixteen to twenty. I did carry thirty-two for a little while, but I needed to complete two face-to-face contacts with my clients each month, and the contact was more like weekly. It was primarily crisis intervention and I was on call, so anytime anything happened with a client they would call me, twenty-four/seven. If they were unable to de-escalate a kid I was called, and if I couldn't do it over the phone then I had to go in.

But on top of the face-to-face contact I was in charge of these kids' lives, providing them with all their services and making sure that they were getting everything that they needed. All of the kids I worked with were in state custody so I worked in collaboration with the Department of Human Services. I was facilitating the child and family team meetings, going to court, writing court reports, and driving the kids' treatment. I was maintaining contact with their families, getting families involved in the treatment.

—Permission granted from Sara Bergeron, 2012, text from unpublished interview.

*F*amilies we work with, they have lots of needs and many of the families get really lost in the system. Some never had services before. Our job is to provide casework . . . this means helping them pay for services such as utilities . . . we also work with them in their homes. Going there lets them know we will extend ourselves for them. It is respectful . . . and we help our clients help themselves by advocating for them.

 —Case manager, family services, New York, NY

*W*e absolutely refuse to work against other agencies . . . we build a cooperative environment . . . we believe there are enough clients for everyone . . . our case managers work with intake and establishing eligibility. That means they are gatekeepers . . . we have a certain district that we serve . . . a good number of our clients come from the hospital.

 —Director and case manager, counseling center, Tucson, AZ

*O*ur services began with the apartments in the northeast part of the city. We needed housing for clients coming out of psychiatric hospitals. The clients had a difficult time finding a stable place to live other than the traditional SROs . . . The goal was to provide a more supportive and stable environment.

 —Director and care coordinator, housing services center, New York, NY

The purpose of this chapter is to establish an historical context for case management. It will describe four perspectives on case management that have evolved in the past 40 years: case management as a process, client involvement, the role of the helper, and utilization review and cost–benefit analysis. A brief history of case management in the United States follows, including its evolution to broader service coordination responsibilities within the last decade, as well as an introduction to managed care.

 By the end of each section of the chapter, you should be able to accomplish the performance objectives listed.

PERSPECTIVES ON CASE MANAGEMENT
- Identify four perspectives on case management.
- Trace the evolution of case management.
- Describe the impact of managed care organizations on case management and service delivery.

THE HISTORY OF CASE MANAGEMENT
- Assess the contributions of the pioneers in the areas of advocacy, data gathering, recordkeeping, and cooperation.
- Using the Red Cross as an example, describe casework during World Wars I and II.
- Name the key pieces of federal legislation that spurred the development of case management.

THE IMPACT OF MANAGED CARE
- List the goals of managed care.
- Summarize the impact of managed care on human service delivery.
- Differentiate between these types of managed care organizations: HMOs, PPOs, and POS.

EXPANDING THE RESPONSIBILITIES OF CASE MANAGEMENT
- Trace the shift in emphasis in case management.
- Explain the strengths and weaknesses of managed care.

Case management has long been used to assist human service clients. Today, professionals are discovering new and more effective ways to deliver services, and there is no longer a standard definition. Modern-day case management does resemble the practice of the past, but many dramatic changes have occurred. Among them are the changing needs of individuals served; financial constraints on the human service delivery system; the increasing number of people needing services; and the growing emphasis on client empowerment, evaluation of quality, and service coordination.

One consistent theme that pervades the study of human service delivery is diversity. The three helping professionals quoted at the beginning of this chapter describe the services of their agencies. The caseworker from a family services agency describes the work of her agency as providing financial assistance and advocacy for families for whom there is no other support. The counseling center, on the other hand, gives patients just discharged from the hospital only the aftercare services they need. This often includes support to meet psychological, social, medical, financial, and daily living needs. The housing services center began by providing temporary housing with accompanying support services.

Much of the foundation of case management service delivery developed to serve those people with mental illness who were deinstitutionalized in the 1970s. Illustrating our discussion here is the story of Sam, who was diagnosed as mentally ill and promptly institutionalized. Sam has received many services since that first diagnosis, and his history with the human service delivery system reflects the evolution of service delivery from the traditional form of case management to the new paradigm that is applied today.

Perspectives on Case Management

This section explores four different perspectives on case management, which together illustrate the development of case management since the 1970s.

Case Management as a Process

In the 1970s, the mental health community was involved in the process of **deinstitutionalization**: the movement of large numbers of people from self-contained institutions to community-based settings, such as halfway houses, family homes, group homes, and single residential dwellings. A member of the American Psychiatric Association's Ad Hoc Committee on the Chronic Mental Patient offered the following definition of case management.

> My view is that [case management] is a vital, perhaps the most primary, device in management for any individual with a disability where the requirements demand differential access to and use of various resources. Far from a new concept, it has long been the central device in every organized arrangement that heals, rehabilitates, cares for, or seeks change for persons with social, physical or mental deficits. . . . Case management is a key element in any approach to service integration. . . . A counselor manages assessment, diagnosis, and prescription . . . synthesizes information, emerges with a . . . treatment plan; and then purchases one or more interventions. (Lourie, 1978, p. 159)

Many clients need assistance in gaining access to human services. Often they have multiple needs, limited knowledge of the system, and few skills to help them arrange services. Sam's case reflects the experiences of many clients who were institutionalized in the 1950s and 1960s and were later deemed appropriate for discharge during deinstitutionalization. The case management process illustrated next is an elementary one: limited assessment followed by placement.

---

*After several ear infections, Sam became severely hearing impaired when he was three. He was the youngest of four children and lived in a small town with his mother, two sisters, and a brother. His mother took care of him, and he became dependent on her. They learned to communicate with each other using a sign language they devised themselves. None of his siblings learned to sign. Sam was often unruly and found that tantrums would get him what he wanted. The older he got, the harder to handle he became. When Sam was fifteen, his mother died. None of his siblings would assume responsibility for him, so they decided to have him admitted to the state mental hospital in the capital. This occurred in the early 1950s; the exact date is unknown because a fire at the institution in the 1960s destroyed the records of those who were admitted previously. Sam was in the mental institution for many years before the deinstitutionalization movement started. At that point, Sam's long odyssey began. (**See Table 2.1.**)*

TABLE 2.1   SAM'S JOURNEY

Sam's age	Provider	Activity
3	Mother	Care at home
15	Institution	Residential care
26	First case manager	Limited assessment
	Sister	Lived with sister for one weekend
	Group home	Lived in group home for six days
	Institution again	Residential care
27	Second case manager	Assessment
	Third case manager	Learning American Sign Language
		Contact with deaf community
		Planning, involving Sam in process
28		Independent living skills
29	Day care program	Socialization skills
30		Fluent in American Sign Language
		Developed friendships
30–40	Lois Abernathy (care coordinator)	Regular visits and assessment
	Interdisciplinary team	3 halfway houses; 4 group homes;
		School for the day
		Living with siblings
		Living in apartment
		Vocational training
40	Rehabilitation counselor and	Vocational training
	Care coordinator	Living with friend in apartment
41	Rehabilitation counselor	Full job responsibility
		Termination of services

Sam's first case manager was an employee of the institution, and his job was to identify patients who could function in a community setting. Limited assessments of Sam's mental and emotional state indicated that he was not mentally ill but simply hearing impaired. Unfortunately, his time in the institution had compounded his problems: He did not know American Sign Language (ASL), and he had begun to behave like other patients who did have mental illness. The case manager decided that Sam should be moved from the institution to another setting. The case manager located Sam's oldest sister, but he stayed with her for only one weekend. She returned Sam to the institution on Monday morning, saying that she couldn't handle him and his presence was too disruptive to her family. None of his other siblings was willing to help, so Sam remained in the institution while his case manager searched for a group home that had an opening. Eventually, Sam did move into a group home, but he lived there for just six days before returning to the institution. According to the home's director, no one could communicate with Sam, his behavior was inappropriate, and he needed constant supervision.

The responsibility of Sam's case manager was to find Sam an environment that could foster his growth and development. Unfortunately, the search for such an environment was quite difficult, given Sam's institutional behavior and the limited assessment of his abilities. Sam is one of those clients who need access to multiple services before they can make the transition from an institution to a community setting.

One way to think about the case management process is to examine the key elements for success: responsibility, continuity, and accountability (Ozarin, 1978). *Responsibility* means that one person or team assesses the client's problem and then plans accordingly. Linkages "must be established to form a network of service agencies which can provide specific resources when called upon without assuming total responsibility for the client, unless responsibility for carrying out the total plan is also transferred and accepted" (Ozarin, 1978, p. 167). In other words, there must be a clear line of responsibility for the case and the client.

Continuity is another significant element of good case management. Planning is the key that ensures continuity. It is important not only during the intensive treatment phase but also in aftercare. To foster *accountability*, methods "must be in place to assure the patient is not lost …. The case management process must help the client increase the ability to function independently and to assume self-responsibility. The client should be involved in all aspects of decision making" (Ozarin, 1978, p. 168). Guided by these goals, organizations and professionals at every level work hard to develop systems that participants understand, working together to serve and involve the clients.

Case management can be seen as a "set of logical steps and a process of interaction within a service network which assure that a client receives needed services in a supportive, effective, efficient, and cost-effective manner" (Weil & Karls, 1985a, p. 2). In a more recent definition, a social work best practice white paper (n.d.) described case management as "a method of providing services whereby a professional Social Worker collaboratively assesses the needs of the client and the client's family, when appropriate, and arranges, coordinates, monitors, evaluates, and advocates for a package of multiple services to meet the specific client's complex needs" (Social Work Best Practice: Health Care Management, n.d.).

Case management is an important and necessary component of the human service delivery system, for it provides a focus and oversees the delivery of services in an orderly fashion. As you read about Sam's case, you will see case management evolve into a more logical and complex process that focuses on client participation, integration of services, and cost effectiveness.

Client Involvement

In the 1980s, client involvement came to be emphasized more strongly. A model of case management was proposed based on the concept of enabling clients "to solve problems, meet needs, or achieve aspirations by promoting acquisition of competencies that support and strengthen functioning in a way that permits a greater sense of individual or group control over its developmental course" (Dunst & Trivette, 1989, p. 93). Sam's experience in the human service delivery system reflects the beginning of changes in service provision.

Sam remained institutionalized for the next four months because there was a shift in the case management process. The institution decided to contract with a local mental health agency for case management services. Case management was a new role and responsibility for this agency; it assigned two individuals fifty cases each, with few guidelines for performing this new function. Sam's new case manager, his second, spent some time assessing Sam's needs, getting to know him, and talking with the mental health professionals within the institution. In concert, they determined that Sam needed a very structured environment in the community if his deinstitutionalization was to succeed. He also needed to learn sign language and to begin to communicate with others using this medium. Because he was hearing impaired but did not have mental illness, he needed to be in contact with the deaf community, where he could find support and role models for independent living. At the age of twenty-seven, he had little ability to care for himself.

Unfortunately, his case manager left her position before she had the opportunity to implement the plan. A third case manager assumed responsibility for Sam's case, with much determination. Her own brother had been hearing impaired since birth, and she recognized Sam's potential. She could communicate with him using ASL. She was also committed to planning, documenting her work, involving Sam in the case management process, and following through on referrals and the involvement of other professionals.

The goals of the plan included having Sam learn ASL, teaching socialization skills and independent living skills, and introducing him to members of the deaf community. His case manager was able to find a day care program where Sam could learn independent living skills. Three times a week, he went to the local school for the deaf for ASL lessons. Once a week, Sam and his case manager joined other hearing impaired adults for a special community program and social hour. Sam still lived at the institution. By the end of the second year, the case manager was able to include Sam in the process of setting priorities and planning for his treatment.

After two years, Sam had made considerable progress. He was able to use ASL to communicate his needs, and he had developed several friendships with people he met at the school for the deaf and at the community programs. On his thirtieth birthday, he celebrated with his friends from the school. Tantrums continued to occur, but less frequently.

After a rocky beginning, Sam benefited from the service delivery process as it evolved. His third case manager assumed responsibility for his case, provided the continuity needed for him to make progress, and was accountable for his care. She used a process of logical steps to establish goals and set priorities. She also established a partnership with Sam by involving him in problem identification, plan development, and service provision. By learning daily living skills, Sam reinforced his ability to care for himself, and his increasing mastery of ASL gave him a new medium for self-expression and communication.

The Role of the Case Manager

Chapter 1 lists an array of job titles that have emerged to reflect the new goals of service delivery. Traditionally, terms such as *caseworker* and *case manager* described the efforts of helpers. Today, job titles include *service coordinator, liaison worker, counselor, case coordinator, healthcare case manager,* and *care coordinator*. These new job titles represent not only the diversity of service delivery today but also the broader range of responsibilities and the different ways case managers perceive their roles. The change in job titles reflects the evolution of case management and, in a larger context, of service delivery. The emphasis shifted from what was previously understood to be case management, when it meant the skills of managing someone, to terminology reflecting a more equitable relationship, such as *coordination* and *liaison*. A change in philosophy had occurred regarding the role of the case manager, emphasizing working with other professionals, coordinating care and other services, and empowering individuals to use the system to help themselves. The focus became the client's ability to develop the skills needed to work within the human service network.

The next ten years were a struggle for Sam and for those who worked with him. His case manager of two years left her job for a promotion in a nearby city, and his case was transferred to Lois Abernathy, a care coordinator at a different agency. Because of increasing pressure to deinstitutionalize, it was decided to move Sam to a halfway house before helping him establish residence in the local community. Over the course of the decade, Sam lived in three halfway houses, in four group homes, at the school for the deaf, with his siblings, and in an apartment with a roommate. Ms. Abernathy was the link between Sam and each of these placements. Her responsibilities included meeting with Sam regularly to review his needs, problems, and successes and arranging any additional services for him. Often, she and Sam would meet with other professionals who were involved with his case. Ms. Abernathy was committed to giving Sam choices about his future. When he expressed the desire to work at a job, she helped him determine exactly what he would like to do. After exploring the options available to him, Sam decided that he would like to work with a local vending machine business program that Ms. Abernathy knew about. After Sam received some education and training, his responsibilities with this program came to include stocking machines, collecting money, and minor repair work.

The assignment of a *care coordinator* to Sam's case signaled a shift in the role of the case manager—from management to coordination. The client's participation in the process also became significant; a partnership emerged to identify, locate, link, and monitor needed services. In this way, case management built on Sam's strengths and empowered him to help himself.

Utilization Review and Cost–Benefit Analysis

One result of the spiraling cost of medical and mental health services and the push for health care reform has been the growth of the managed care industry. The purpose of managed care is to authorize the type of service and the length of time care is provided and to monitor the quality of care. In the managed care environment, case managers function very differently from those described earlier.

What makes case management in managed care distinctive is the emphasis on efficient use of resources. Case managers are involved in utilization review and have the responsibility to authorize or deny services. Also, they must know how to interact with insurance providers and how to process claims through the insurance system.

This new case manager is also responsible for cost–benefit analysis. Such an analysis does not include the traditional reporting that is found in a case history in the form of notes or recommendations and follow-up. It is focused on the financial matters of the case, specifically the cost and efficiency of services.

As we leave Sam at the age of forty-two, he is two months away from assuming responsibility for all the vending machines in a nearby neighborhood. It has taken two successive six-month training sessions to teach Sam the necessary job skills and repair techniques. Sam is living with a new friend in a small apartment near his vending area. This friend is also a client at the rehab center and is helping Sam train for his new job. Sam's rehabilitation counselor and his care coordinator from the mental health center have met and developed a coordinated plan, with input from Sam. Because of his recent success in rehabilitation, mental health services are no longer authorized for Sam.

Sam's experience with a care coordinator has led him in a new direction. The emphasis on tapping Sam's potential and coordinating care has given him a major voice in decision making. This requires coordination between two systems: rehabilitation and mental health. Sam does make progress, and the managed care case manager decides to discontinue mental health services. As we leave Sam, his rehabilitation counselor supports him with job training and housing. If he should again need professional mental health support, it is hoped that the rehabilitation counselor can arrange these services for him.

 # The History of Case Management

As important as various current perspectives on case management are, the historical roots are equally informative. The pages that follow trace the history of case management from its origins in institutional settings through the work of early pioneers, the impact of the American Red Cross, and the influence of federal legislation. The chapter concludes with a discussion of case management as it is practiced today.

Documenting the history of case management, Weil and Karls (1985a) stated, "The process of service coordination and accountability has a century-long history in the United States" (p. 1). As first used in institutional settings, case management included the responsibilities of intake, assessment of needs, and assignment of living space. These institutions provided residential services to people incarcerated for crimes, orphans, people with mental illness, people with disabilities, and elderly people. Which professionals performed the case management function depended on the particular institution; among them were doctors, nurses, psychiatrists, psychologists, counselors, and teachers (Weil & Karls, 1985b).

A Pioneering Institution

One example of an institution with an early commitment to case management was the Massachusetts School for Idiotic and Feebleminded Youth, established in 1848. This school promoted the belief that people categorized as "idiotic" or "feebleminded" could improve if they were given appropriate clinical, social, and vocational services and support (Weil & Karls, 1985b). In 1839, a child who had mental retardation as well as vision impairment had come to the Massachusetts institution for the blind. It was clear that the child had needs beyond the expertise of the institution, and the director, Samuel Howe, was determined to help this child and others with similar needs. He convinced the state that he could improve these children in three areas: bodily habits, mental capacities, and spirituality (Winsor, 1881). The institution he founded was the Massachusetts School for Idiotic and Feebleminded Youth.

CONTRIBUTIONS TO CASE MANAGEMENT

This Massachusetts school provided services in case management, such as tracking student progress, providing follow-up services, and managing information. Early services at the school included observation and diagnosis of physical and mental behavior. The helping professionals tracked clients' progress, and they soon began differentiating and individualizing treatment: "...[W]e cannot properly care for a young and helpless idiot in the houses devoted to the brighter moron children" (Trustees, 1920, p. 17).

When demand for the services increased, the school established outpatient clinics in several cities (Trustees, 1919). These clinics supported families who cared for children at home. Aftercare was also an important service, provided by trained "visitors" who helped plan the transition from the institution to the home or other setting. The visitor would gather information to determine whether the child should be sent home on trial or be released for vacation with family or friends. They would also follow up after transition to determine whether the release and placement were appropriate. This emphasis on aftercare was the forerunner of modern-day continuity of care, as well as to today's commitment to provide services as the client makes a transition from the treatment setting to a less restrictive one.

In the early 1900s, the school made two improvements in the management of information, which is an important component of case management. In 1916, the institution began an evaluation of its services. This included a study of clients who had been discharged and those in aftercare. The information gathered included where patients lived, with whom they were living, whether they supported themselves, and, if so, how. The information was gathered first by survey, and then patients, families, and friends were interviewed in their homes. In 1919, new legislation established a Registry for the Feebleminded in an effort to catalog the state's population of people with retardation (Trustees, 1920). These advances in recordkeeping and information management contributed to case management as we know it today.

The next section discusses the work of some early pioneers in human services who developed case management further, especially regarding coordination of services and interagency cooperation.

Early Pioneers

Early case management took one of two forms: a multiservice center approach or a coordinated effort of service delivery. Jane Addams, Lillian Wald, and Mary Richmond were three early pioneers who contributed to the development of the emerging case management process.

HULL HOUSE

Jane Addams and Ellen Starr, classmates at the Rockford Female Seminary, founded Hull House in Chicago in 1889. In their twenties, while traveling in England, they visited Toynbee Hall, a university club that had established a recreation club for the poor. Addams and Starr were committed to increased communication between social classes, and the activities of Toynbee Hall inspired them. They returned to America and moved into a poor section of Chicago, hoping to improve that environment (Addams, 1910).

They bought Hull House, an older home on Chicago's West Side. Committed to sharing their home and their love of learning, they opened the house to the neighborhood. They acquired collections of furniture, art, and literature; soon they became involved with music and crafts. The purposes of Hull House were threefold: to provide a center for civic and social life, to improve conditions in the neighborhood, and to provide support for reform movements (Addams, 1910).

CONTRIBUTIONS TO CASE MANAGEMENT

As the number of services expanded, Hull House's need for effective administration and recordkeeping increased. Many participants would have been suspicious if formal files were kept, so an informal card system was housed in the administrative office, and its contents were shown only to people who could establish a need to see it (Woods & Kennedy, 1911). Information about demographics, participation, and attendance was gathered. In addition to recordkeeping, advocacy was a case management function that was integrated into the work of Hull House. Jane Addams and her colleagues were involved in many efforts to improve the living and working conditions of the neighborhood and its inhabitants. Promoting better housing and improving sanitation services were two areas of focus (Polacheck, 1989). **(See Box 2.1.)**

BOX 2.1   A Hull-House Girl
● ● ● ● ● ● ● ● ● ● ●

Hilda Satt Polacheck was born in 1882 in Wolclawek, Poland. She was the eighth of twelve children born to a well-to-do Jewish family. Because of oppression by the Russian government, her family immigrated to America in 1892. During their early years, Hilda and her sister attended the Jewish Training School on Chicago's West Side. After the death of their father, however, they joined the ranks of the working poor. Hilda began working in a knitting factory when she turned 13. She first came to Hull House for a Christmas party in 1896, and it soon became the center of her social life. She attended classes and club meetings, read literature, exercised, and performed in plays there. Later she worked there as a receptionist and a guide. Hilda grew up with Hull House and spent much of her time there from 1895 until 1912. In her autobiography, she provides the following description of Jane Addams and her advocacy work.

Bad housing of the thousands of immigrants who lived near Hull-House was the concern of Jane Addams. Where there were alleys in the back of the houses, these alleys were filled with large wooden boxes where garbage and horse manure were dumped. . . . When Jane Addams called to the attention of the health department the unsanitary conditions, she was told that the city had contracted to have the garbage collected, there was nothing it could do. . . . She was appointed garbage inspector for the ward. I have a vision of Jane Addams . . . following garbage trucks in her long skirt and immaculate white blouse. . . .

Hull-House was in the Nineteenth Ward of Chicago. The people of the Hull-House were astonished to find that while the ward had 1/36 of the population of the city, it registered 1/6 of the deaths from typhoid fever. Miss Addams and Dr. Alice Hamilton launched an investigation that has become history in the health conditions of Chicago. . . . Whatever the causes of the epidemic, that investigation, emanating from Hull-House, brought about the knowledge of the sanitary conditions of the Nineteenth Ward and brought about the changes we enjoy today.

SOURCE: From I *Came a Stranger: The Story of a Hull-House Girl*, by H. Polacheck, pp. 71–72. Copyright © 1989 University of Illinois Press.

HENRY STREET SETTLEMENT HOUSE

Lillian Wald and Mary Brewster, who were nurses, established the Henry Street Settlement House organization in 1895 in New York City. Early in their careers, they decided to provide services to the city's Lower East Side, home to a large number of immigrants. Wald and Brewster lived in the neighborhood and provided health care services.

They established a system for nursing the sick in their own homes, promoting the dignity and independence of the patient. According to Wald (1915), "the nurse should be as ready to respond to calls from the people themselves as to calls from physicians … she should accept calls from all physicians, and with no more red tape or formality than if she were to remain with one patient continuously" (p. 27). There was an explicit focus on accessibility.

CONTRIBUTIONS TO CASE MANAGEMENT

The work at Henry Street led to two significant innovations: the designation of the visiting nurse and the development of the Red Cross. Both of these services were important in promoting public health. One important function of the visiting nurse was to establish an organized system of care and instruction for people with tuberculosis and their families. Early in its history, the Red Cross facilitated the use of the public schools as recreation centers, taught housekeeping skills to women, and provided penny lunches for children (Wald, 1915).

MARY RICHMOND

Mary Richmond, a social reformer at the turn of the century, had a similar commitment to bettering the lives of individuals and families living in poverty. She, too, made significant contributions to the development of case management. Richmond promoted the idea that each person was a unique individual whose personality, family, and environment should be respected. Working with immigrants, she emphasized the need for social workers to resist the tendency to stereotype or overgeneralize (Lieberman, 1990). Richmond wrote, "[T]he social adjustment cannot succeed without sympathetic understanding of the old world backgrounds from which the client came" (1917, p. 117). She also believed that professionals should work with clients rather than doing things to them.

Want More Information?

• • • • • • • • • • • •

The Internet provides in-depth resources related to the study of case management. Search the following terms to read more about individuals and agencies who provided the foundation for case management practice.

- Hull House
- Henry Street Settlement House
- Mary Richmond

CONTRIBUTIONS TO CASE MANAGEMENT

One method Richmond developed to focus on the individual was **social diagnosis,** a systematic way for helping professionals to gather information and study client problems. She established a series of methods for gathering information about individuals, assessing their needs, and determining treatment. This process is often referred to as social casework, and it became a part of the case management process. Richmond contributed a case record form designed to focus on individuals and their unique problems (Pittman-Munke, 1985; Trattner, 1999).

She also recognized that gathering data is a complex process and urged the use of different methods for different individuals. According to Richmond (1917), some clients should be interviewed in an office; for others, the home is the preferred location. She believed in multiple sources of information and warned that data gathering was a complex and often incomplete process.

During the early part of the 20th century, the Red Cross emerged to meet the multiple needs of individuals. The subsection that follows describes its involvement in providing assistance, primarily to servicemen and their families, during World War I and World War II.

The Impact of World Wars I and II and the American Red Cross

During the First World War, there was an increased interest in casework as developed by Mary Richmond. The American Red Cross, whose roots are found in the work of Clara Barton and the Civil War, used casework to address individuals' problems and their psychological needs. The use of a casework approach to assist individuals began during the Mexican civil war (1911–1917), when the American Red Cross provided a variety of services to support the daily life of civilians and troops along the Mexican border (Dulles, 1950). Subsequently, the services were extended to dependents of military personnel in army installations throughout the country. These services, performed by the Home Corps of the Red Cross, helped address the needs of the families of military personnel. In World War I, the Home Service Corps (later known as the Social Welfare Aide Service) provided help to families experiencing problems such as illness and marital difficulties.

The Home Service Corps addressed a wide variety of problems. Dulles (1950, pp. 391–392) presents the following example of messages received and sent by the Home Service Corps.

Incoming: MILITARY AUTHORIZE INFORM FAMILY SERVICEMAN WELL ON ACTIVE DUTY NOT REPEAT NOT THE MAN THEY SAW IN NEWSREEL

Outgoing: MESSAGE DELIVERED FAMILY MUCH RELIEVED MOTHER IMPROVING

Incoming: SERVICEMAN REQUESTS MATERNITY REPORT WIFE EXPECTING CONFINEMENT EARLY JULY

Outgoing: SON BORN JULY SEVEN BOTH WELL

Incoming: SERVICEMAN INFORMED BY FRIEND MOTHER DIED TWO MONTHS AGO STILL RECEIVING LETTER FROM HER REGULARLY INVESTIGATE

Outgoing: MOTHER DIED CANCER BREAST APRIL TWENTY SEVENTH SISTER FEARED SHOCK TO SERVICEMAN HAS BEEN WRITING IN MOTHERS NAME WILL WRITE IMMEDIATELY

Incoming: SERVICEMAN BEGS WIFE DISREGARD HIS LAST LETTER MAILED JULY SEVEN RECEIVED ONE HUNDRED FIFTEEN LETTERS FROM HER YESTERDAY

Outgoing: MESSAGE RECEIVED WIFE WILL WRITE

Incoming: SERVICEMAN REQUESTS HEALTH CONFINEMENT REPORT WIFE

Outgoing: WIFE DIED CHILDBIRTH JULY TWENTY SEVEN SON WELL WITH SERVICEMAN'S MOTHER-IN-LAW GETTING GOOD CARE MOTHER-IN-LAW WILL KEEP CHILD UNTIL SERVICEMAN'S RETURN

CONTRIBUTIONS TO CASE MANAGEMENT

The Home Service Corps workers made two contributions to the development of service delivery. First, they offered extended help to individuals and their families. The intervention was problem focused but not time bound. A Home Service Corps volunteer helped identify the problem and worked with the family until it was resolved. Second, the volunteer did not just solve problems, but also became a broker of services. He or she would often coordinate communications and requests for services between the family and the agencies that could provide the help and support. The work often involved helping families communicate with the military (Dulles, 1950; Hurd, 1959).

After World War II, it became increasingly evident that many people needed assistance to improve their quality of life. In the early 1960s, the federal government became increasingly involved in helping people in need.

The Impact of Federal Legislation

Several pieces of federal legislation were passed between the mid-1960s and the late 1980s and modified in recent years. These legislative efforts recognized the need for a case management process to provide social services to people in need. Serving older adults, children with disabilities, and the families and children who live in poverty requires the services that case management represents.

CONTRIBUTIONS TO CASE MANAGEMENT: THE OLDER AMERICANS ACT OF 1965

The Older Americans Act of 1965 (Public Law 89-73) focused on providing services for older individuals in order to improve their quality of life. Among its contributions to the development of case management was an emphasis on the multiplicity of human needs. This act advanced case management by recognizing the need to coordinate care. Section 101 of the act describes its goals and the services to be provided. **(See Box 2.2.)** The services were designed to meet a variety of needs: financial, medical, emotional, housing, vocational, cultural, and recreational.

Today the need for support and care of the elderly has increased, because the number of individuals in the United States over the age of 65 is projected to more than double by 2050 (Clubok, 2001). Plagued by chronic illness, depression, insufficient financial resources, and many other difficulties, this population is already placing demands on many families and communities, and on the human service delivery system. Even though federal programs such as Social Security, Medicaid, and Medicare support many individuals, these resources need to be used effectively and efficiently. **(See Box 2.3.)** Case management

---

BOX 2.2   The Older Americans Act of 1965

Title I—Declaration of Objectives: Definition

Sec. 101. The Congress hereby finds and declares that, in keeping with the traditional American concept of the inherent dignity of the individual in our democratic society, the older people of our Nation are entitled to, and it is the joint and several duty and responsibility of the governments of the United States and of the several States and their political subdivisions to assist our older people to secure equal opportunity to the full and free enjoyment of the following objectives:

1. An adequate income in retirement in accordance with the American standard of living.
2. The best possible physical and mental health which science can make available and without regard to economic status.
3. Suitable housing, independently selected, designed, and located with reference to special needs and available at costs which older citizens can afford.
4. Full restorative services for those who require institutional care.
5. Opportunity for employment with no discriminatory personnel practices because of age.
6. Retirement in health, honor, dignity—after years of contribution to the economy.
7. Pursuit of meaningful activity within the widest range of civic, cultural, and recreational opportunities.
8. Efficient community services which provide social assistance in a coordinated manner and which are readily available when needed.
9. Immediate benefit from proven research knowledge which can sustain and improve health and happiness.
10. Freedom, independence, and the free exercise of individual initiative in planning and managing their own lives.

SOURCE: Older Americans Act of 1965, Public Law 89-73.

BOX 2.3   Federal Programs

SSDI Program

Title II of the Social Security Act establishes the Social Security Disability Insurance Program (SSDI). SSDI is a program of federal disability insurance benefits for workers who have contributed to the Social Security trust funds and become disabled or blind before retirement age. Spouses with disabilities and dependent children of fully insured workers (often referred to as the primary beneficiary) also are eligible for disability benefits on the retirement, disability, or death of the primary beneficiary. Section 202(d) of the Social Security Act also establishes the adult disabled child program, which authorizes disability insurance payments to surviving children of retired or deceased workers or workers with disabilities who were eligible to receive Social Security benefits, if the child has a permanent disability originating before age 22.

Hereinafter in this policy brief, the term SSDI refers to all benefit payments made to individuals on the basis of disability under Title II of the Social Security Act.

SSDI provides monthly cash benefits paid directly to eligible persons with disabilities and their eligible dependents throughout the period of eligibility.

SSI Program

Title XVI of the Social Security Act establishes the Supplemental Security Income (SSI) Program. The SSI program is a means-tested program providing monthly cash income to low-income persons with limited resources on the basis of age and on the basis of blindness and disability for children and adults. The SSI program is funded out of the general revenues of the Treasury.

Eligibility for SSI is determined by certain federally established income and resource standards. Individuals are eligible for SSI if their "countable" income falls below the federal benefit rate ($512 for an individual and $769 for couples in 2000). States may supplement the federal benefit rate. Not all income is counted for SSI purposes. Excluded from income are the first $20 of any monthly income (i.e., either unearned, such as Social Security and other pension benefits, or earned) and the first $65 of monthly earned income plus one-half of the remaining earnings. The federal limit on resources is $2,000 for an individual and $3,000 for couples. Certain resources are not counted, including, for example, an individual's home and the first $4,500 of the current market value of an automobile.

The Ticket to Work and Work Incentives Improvement Act of 1999

On December 17, 1999, President Clinton signed into law the Ticket to Work and Work Incentives Improvement Act of 1999 (Public Law 106-170). Hereinafter in this policy brief, Public Law 106-170 will be referred to as *the Act*. The Act has four purposes [Section 2(b) of the Act]:

- To provide health care and employment preparation and placement services to individuals with disabilities that will enable those individuals to reduce their dependency on cash benefit programs.
- To encourage states to adopt the option of allowing individuals with disabilities to purchase Medicaid coverage that is necessary to enable such individuals to maintain employment.
- To provide individuals with disabilities the option of maintaining Medicare coverage while working.
- To establish a return-to-work ticket program that will allow individuals with disabilities to seek the services necessary to obtain and retain employment and reduce their dependency on cash benefit programs.

This Act improves work incentives under the SSDI and the SSI and expands health care services under Medicare and Medicaid programs for persons with disabilities who are working or who want to work but fear losing their health care.

SOURCE: Center on State Systems and Employment (RRTC). (2000). Policy Brief: Improvements to the SSDI and SSI Work Incentives and Expanded Availability of Health Care Services to Workers with Disabilities under the Ticket to Work and Work Incentives Improvement Act of 1999. Policy Brief, Vol. 2, No. 1. [Online]. Available: http://www.communityinclusion.org/publications/pdf/pb2.pdf

Medicare

Medicare is a health insurance program for:

- People 65 years of age and older.
- Some people with disabilities, under 65 years of age.
- People with end-stage renal disease (permanent kidney failure requiring dialysis or a transplant).

Medicare provides hospital insurance. Most people do not have to pay for this insurance. Medicare also provides medical insurance and most people pay monthly for this insurance. There are several ways that individuals can access their Medicare benefits.

- The Original Medicare Plan—This plan is available everywhere in the United States. It is the way most people get their hospitalization and medical insurance benefits. They can go to any doctor, specialist, or hospital that accepts Medicare. Medicare pays its share and the recipient pays a share. Some services are not covered, such as prescription drugs.
- Medicare Managed Care Plans—These are health care choices (like HMOs) in some areas of the country. In most plans, individuals can only go to doctors, specialists, or hospitals that are part of the plan. Plans must cover all Medicare hospitalization and medical insurance benefits. Some plans cover extras, like prescription drugs. Individual out-of-pocket costs may be lower than in the Original Medicare Plan.

(continues)

(continued)

- Private Fee-for-Service Plans—This is a new Medicare health care choice in some areas of the country. Individuals may go to any doctor, specialist, or hospital. Plans must cover all hospitalization and medical insurance benefits. Some plans cover extras like extra days in the hospital. The plan, not Medicare, decides how much you pay.

SOURCE: Medicare: The Official U.S. Government Site for Medicare Information. (2000). Available at http: www.medicare.gov

Medicaid

Title XIX of the Social Security Act is a program that provides medical assistance for certain individuals and families with low incomes and resources. The program, known as Medicaid, became law in 1965 as a jointly funded cooperative venture between the Federal and state governments to assist states in the provision of adequate medical care to eligible needy persons. Medicaid is the largest program providing medical and health-related services to America's poorest people. Within broad national guidelines that the federal government provides, each of the states:

- Establishes its own eligibility standards.
- Determines the type, amount, duration, and scope of services.
- Sets the rate of payment for services.
- Administers its own program.

Thus, the Medicaid program varies considerably from state to state, as well as within each state over time.

SOURCE: Health Care Financing Administration. (2000). Available at http://www.cms .hhs.gov/medicaid/default.asp?

will continue to be used in elder care to serve these individuals with complex short- or long-term needs, especially as the numbers of seniors needing support increases.

CONTRIBUTIONS TO CASE MANAGEMENT: REHABILITATION ACT OF 1973

Client involvement, a basic principle of case management today, was also an important theme in the Rehabilitation Act of 1973 and its subsequent amendments through 1986. These pieces of legislation promoted consumer involvement while serving individuals with severe disabilities (Rubin & Roessler, 2001), particularly in eligibility determination and plan development. Client satisfaction and the adequacy of services also received attention. Client involvement was further strengthened when the Rehabilitation Act Amendments of 1992 and 1998, emphasizing consumer choice and control in setting goals and objectives, were passed.

CONTRIBUTIONS TO CASE MANAGEMENT: CHILDREN WITH DISABILITIES, EDUCATION FOR ALL HANDICAPPED CHILDREN ACT OF 1975

Public Law 94-142, the Children with Disabilities, Education for All Handicapped Children Act of 1975, included an explicit case management process to treat the client as a customer. The client was to be involved in identifying the problem, given complete information about the results of the assessment of needs, and empowered to help determine the type of services delivered. The client also participated in the evaluation of the helping process and in any decision to terminate or redirect activities (Jackson, Finkler, & Robinson, 1992).

The passage and subsequent implementation of this act serves as an excellent example of how federal legislation applies the case management process (Weil & Karls, 1985a). One tool to assist with planning, implementation, and evaluation was the *individualized educational program* (IEP). The IEP articulates a plan of intervention for each child based on goals and recommended intervention strategies. Although educational agencies have implemented the IEP in numerous ways, it has always been critical that the case manager assume the leadership role on the team of participants who develop the IEP.

On June 24, 1997, President Bill Clinton signed into law the IDEA amendments. These amendments reflect changes in the values and process of case management and the planning and implementation of the IEP, focusing on four important areas: strengthening parental participation, creating accountability for student participation in general education, addressing remediation and behavior problems within the educational environment, and preparing students for independent living. Client empowerment was strengthened and there was an emphasis on client strengths. These are two values and goals of the case management process.

CONTRIBUTIONS TO CASE MANAGEMENT: THE FAMILY SUPPORT ACT OF 1988 AND THE PERSONAL RESPONSIBILITY AND WORK OPPORTUNITY ACT

In 1988, the Family Support Act was passed. As described here, the act mandated that case management be applied to the process of serving those who were deemed eligible. This marked a new status for the case manager. The act was passed with the express goal of increasing the economic self-sufficiency of families who receive Aid to Families with Dependent Children (AFDC). In 1996, AFDC was replaced by the Personal Responsibility and Work Opportunity Act. This new welfare legislation required that young mothers receive financial support for two years while they get vocational education and training to join the work force. Case managers became a key component in these welfare-to-work programs, as they helped develop and coordinate the plans that move young mothers toward self-sufficiency. For example, in New Jersey, Atlantic and Cape May counties developed a collaborative case management model that included the New Jersey departments of labor, human services, and health and senior services. The collaboration included representatives from state and local government agencies as well as more than 42 community-based organizations. They developed "One Ease E Link," an electronic linkage that serves as the basis for an elaborate collaborative case management and referral system (Welfare & Workforce Development Partnerships, 2000). Congress reauthorized the Personal Responsibility and Work Opportunity Act in the Deficit Reduction Act of 2005.

Just as social legislation was a major factor in the development of case management in the 1960s, 1970s, and 1980s, the advent of managed care during the 1980s expanded the range of case management in the 1990s. The next section describes managed care and explores its impact on service delivery.

 # The Impact of Managed Care

The emergence of managed care as a model of health care delivery has increased the demand for case management services and provided new models and definitions of service delivery. To understand its impact, one must first grasp what managed care is.

History of Managed Care

Until the 1930s, most medical care in this country was provided on a **fee-for-service** basis. This means that a patient would be assessed a fee for each health or mental health service provided by a professional. When a client went for her annual checkup, she might receive, for example, a bill for the doctor's consultation time plus additional services and tests such as a tetanus injection or an EKG.

In the early 1930s, physicians implemented prepaid group plans, which were managed plans for medical services. This was an alternative way of organizing medical care. The basic concept of a prepaid plan was to guarantee a defined set of services for a negotiated fee. On such a plan, the client would pay a yearly fee that covered a set of services such as those provided at her annual checkup.

The growth in prepaid group plans was relatively slow until the 1970s. Then the Health Maintenance Organization Act of 1973 (Public Law 93-222) allowed managed medical plans to increase in number and expand the numbers of patients being served (MacLeod, 1993). The prevalence of managed care is now commonly regarded as connected to the rising cost and decreasing quality of health care and mental health care. The escalation of costs reflects many trends: improved technology, shifting of costs from nonpaying patients to paying patients, an older population, higher expectations for a long and healthy life, increased administrative costs, and varying standards of efficiency and quality care.

Defining Managed Care

There are several ways to define **managed care**. First, the term may simply refer to an organizational structure that uses prepayment rather than fee-for-service payment. Second, it can designate an array of different payment plans, such as prepayment and negotiated discounts. It may also imply the inclusion of quality assurance practices, such as agreements for prior authorization and audits of performance (Mullahy, 2010). Third, *managed care* may refer to the policy of restricting clients' access to providers such as physicians and other health professionals. Instead, the providers or professionals are paid a flat fee to provide service to a certain group of patients or clients. Most simply stated, managed care is an agreement that health providers will guarantee services to clients within specified limits. The restrictions are intended to improve efficiency of services (Mullahy, 2010). The goals of managed care are as follows:

- To encourage decision makers (providers, consumers, and payers) to evaluate efficiency and priority of various services, procedures, and treatment.
- To use the concept of limited resources in making decisions about services, procedures, and treatment.
- To focus on the value received from the resources as well as the lower cost.

Models of Managed Care

Three types of managed care models have evolved to meet the goals stated in the previous paragraph: health maintenance organizations (HMOs), preferred provider organizations (PPOs), and point-of-service (POS). Each has a particular strategy for maintaining cost and ensuring quality.

HEALTH MAINTENANCE ORGANIZATIONS

The HMO managed care model, which is the most structured and controlled, emphasizes positive health promotion. **HMO** is a generic term covering a wide range of organizational structures; unlike traditional fee-for-service health care systems, it combines delivery and financing into one system. Most HMOs have three characteristics (John Hopkins AIDS Service, 2004):

- An organized system for providing health care or otherwise assuring health care delivery in a geographic area.
- An agreed-upon set of basic and supplemental health maintenance and treatment services.
- A voluntarily enrolled group of people. The HMO assumes the financial risk for providing the contracted services.

There are advantages and disadvantages to the HMO model. One immediate benefit is that the HMO constantly monitors both the services available and the cost of providing them. Physicians and other health professionals must establish a rationale for recommending services, procedures, and treatment, and their rate is monitored. Client spending is also monitored, since enrolled members can receive services only from the professionals participating in the plan. In some instances, the clients must obtain preauthorization from someone outside the plan. Through its leverage at the site of services, the HMO can control utilization and improve the efficiency of service delivery (Trends in Health Care Costs and Spending, 2009).

One special version of the HMO is the independent practice association (IPA). The IPA hires physicians to provide services for HMO members. Physicians may contract with several HMOs. In most instances, the fee for service is negotiated between the HMO and the physicians. In many areas of the country the IPA advocates for quality services for clients and uses collaborations to improve client care.

From the client's perspective, the site-of-services restrictions also represent the greatest disadvantage of an HMO. Clients do not like limits on their use of providers; they wish for more freedom to choose. In response to members' demands for freedom to choose their own providers, two other managed care systems have emerged: PPOs and POS.

PREFERRED PROVIDER ORGANIZATIONS

The term **preferred provider organization (PPO)** does not describe any single type of managed care arrangement. Rather, this plan falls between the traditional HMO and the standard indemnity health insurance plan. The following characteristics apply to PPOs (Joint Interim Committee on Managed Care, 2000):

- Contracts are established with providers of medical care.
- These providers are referred to as preferred providers.
- The benefit contract provides significantly better benefits for services received from preferred providers.
- Covered persons are allowed benefits for nonparticipating providers' services.

PPOs point with pride to their prompt payment of claims. The providers accept a negotiated discount, which represents the PPO fee, and they do not bill patients an additional amount. Based on the negotiated fee, both the clients and the PPO can anticipate their costs, and providers can anticipate their income. From the providers' perspective, they are assuming a business risk in terms of the fees that they agree to accept. On the other hand, they expect to increase the number of patients under their care. Many providers also maintain independent medical practices.

POINT-OF-SERVICE

The third option of managed care offered today is the POS. It is often adopted by traditional HMO members who want more flexibility than the HMO or the PPO provide. The following features characterize a **point-of-service** plan:

- Customers are allowed to use out-of-plan providers, but if they do, they receive reduced coverage.
- To participate in a POS plan, clients pay higher premiums, higher deductibles, and a higher percentage of the medical fees.
- Clients are encouraged to use the providers in the managed care system, but they receive partial benefits if they choose medical care outside the system.

Voices from the Field: Research and Practice

● ● ● ● ● ● ● ● ● ● ● ●

Case Management Society of America: Our History

The **Case Management Society of America** is an international, non-profit organization founded in 1990 dedicated to the support and development of the profession of case management through educational forums, networking opportunities and legislative involvement. Unique in its composition as an international organization, over 70 affiliated and prospective chapters in a tiered democratic structure, CMSA's success and strength is its structure as a member-driven society.

CMSA's emergence as a prominent national organization is in large part due to its unique and involved membership . . . To enhance this process, the society provides ongoing leadership training seminars geared for the local leaders.

In nationally recognized innovation, CMSA developed the nationally recognized Standards of Practice for Case Management. This publication was officially released in early 1995, and is a forerunner in establishing formal, written standards of practice from a variety of disciplines. During that same year, the National Board approved a peer-reviewed Ethics Statement on Case Management Practice, a base foundation from which to apply ethical principles to the practice of case management. The Standards of Practice and Ethics Statement are both available from CMSA National.

In response to payer and purchaser expectations for demonstrating value in the marketplace, CMSA created the Council for Case Management Accountability. This new division of CMSA will establish evidence-based standards of practice and help its members achieve those standards through the measurement, evaluation and reporting of outcomes.

Education, research, and networking continue to be top priorities sought by CMSA's members. Proactive measures from grassroots lobbying to national briefings have been at the forefront of CMSA legislative activity. Government Affairs committees are currently active in most of the local and state chapters across the country.

Through the support of a certification program, CMSA continues to enhance the level of case manager professionalism - furthering the development of a new, higher level of industry expertise. For more information, please contact the Commission for Case Manager Certification at (856) 380-6836. **CCMC is a separate entity and is independent from CMSA.**

Ultimately, the quality and productivity of CMSA's services rely upon the commitment of its membership. Because case managers are effective communicators, problem solvers, and visionaries, CMSA offers an opportunity for members to utilize their skill sets and maximize their talents. The enormous success of the organization lies in one simple concept—professional leadership.

SOURCE: Case Management Society of America: Our history. Retrieved from http://www.cmsa.org /Home/CMSA/OurHistory/tabid/225/Default.aspx

Managed care has emerged as a response to the fact that employers, governments, payers, clients, and providers are all seeking ways of containing health care costs. All three types of plans emphasize management of medical cases, review and control of utilization, and incentives for or restrictions on providers and clients to reduce costs and maintain quality. Managed care systems are achieving many of these goals, but some clear advantages and disadvantages have emerged. Proponents of the system point to the following advantages of managed care:

- Providers, clients, and payers involved in health care are beginning to prioritize among services, procedures, and treatments provided.
- Providers are more thoughtful about any plan of action prescribed for the client, since they must provide justification for each component of the plan.
- Efficiency of service delivery improves because services are to be provided in the shortest time needed to meet the goals established.
- Evidence-based practice standards emerge that articulate what defines quality care.
- From the client's perspective, there is a single, coordinated point of entry into the system.
- Resources are saved.
- Resources are spent according to priorities.

Those who question the virtue of the managed care system focus their concerns on two areas: the quality of services delivered and the efficiency of service. They see the following disadvantages of delivering services through managed care.

- Professionals do not believe that they are offering the best services available, because they are constrained by resource limitations.
- Managed care staff are making judgments about the suitability of proposed treatment without adequate training or professional knowledge.
- Services are not delivered in a timely manner because of the extra layer of bureaucracy.
- Managed care organizations require paperwork that limits the amount of time the professional can give the client.
- Clients worry about the quality of the care they are receiving.
- Clients do not have access to all the services they believe they should have.
- Clients cannot choose their service providers.

Institutions and organizations dedicated to the professionalization of case management emerged during the late 1980s and the 1990s. One example is the Case Management Society of America (Case Management Society of America, n.d.). This organization commits to the development of professional case management, especially as it relates to managed care (see Voices from the Field).

In response to professional and patient or client frustrations with managed care, several advocacy efforts have evolved. One example is a patient bill of rights, developed by the American Association of Marriage and Family Therapy, American Counseling Association, American Family Therapy Academy, American Nurses Association, American Psychological Association, American Psychiatric Association, American Psychiatric Nurses Association, National Association of Social Workers, and National Federation of Societies for Clinical Social Work, which articulates recommendations for clients and patients.

- Individuals have a right to know and understand the benefits of the managed care plan in which they are enrolled. Managed care organizations must provide this information and explain the information when asked.
- Individuals must have full access to the names of the professionals eligible to provide treatments. They must also have access to the professionals' qualifications, experience, and areas of expertise.
- Individuals must know if the medical professional has an agreement with the managed care organization to limit treatment options, and if the professional receives monetary incentives for restricting treatments.
- Individuals must know the methods to appeal or to grieve a decision.
- Individuals must know the policies on confidentiality of decisions and records.
- Individuals must know the parameters of choice of professionals.
- Individuals must know all of the individuals involved in making treatment decisions.
- Individuals must know if substance abuse and mental health treatments are part of the plan.
- Individuals must know if the professionals are liable for their actions.
- The concept of a bill of patient rights has evolved. (Patient Bill of Rights, n.d.).

In spite of the criticisms raised, managed care is no longer just one alternative in the health care delivery system; it is part of the structure of service delivery. It has expanded beyond the traditional medical arena. This model of oversight regulates employment assistance programs, long-term services for people with mental retardation, child and adult rehabilitation, behavioral health (mental health), and child welfare. The educational arena and the criminal justice system interact with managed care when services for their clients cross into these areas.

The use of managed care in human services will continue to influence the delivery of services. Throughout this textbook you will see how managed care has influenced the practice of case management.

The Professionalization of Case Management and Expanding Responsibilities

The professionalization of case management within human service delivery includes a national certification offered by several professional organizations and states. For example, the Human Services Board-Certified Practitioner (HS-BCP) certification includes demonstrated competence in case management, professional practice, and ethics as one of the four knowledge and skills assessment components (Center for Credentialing and Education, 2011). Eligibility includes a degree "earned at a regionally accredited college or university, or a state-approved community or junior college at the Technical Certificate level or above. Applicants must also have completed the required Postdegree Experience" (Center for Credentialing and Education, 2011) ranging from one to five years and contingent upon educational level. A few competencies related to the case management function relevant to the HS-BCP follow:

- Collaborate with professionals from other disciplines.
- Identify community resources.
- Utilize a social services directory.
- Coordinate delivery of services.
- Participate as a member of a multidisciplinary team.
- Determine local access to services.
- Maintain a social services directory.
- Participate in case conferences.
- Serve as a liaison to other agencies.
- Coordinate service plan with other service providers. (Center for Credentialing and Education, 2011)

In addition, the National Association of Social Workers offers BSW social worker case managers the Certified Social Work Case Manager (C-SWMC) credential (National Association of Social Workers, 2011). The National Association of Social Workers (NASW) developed areas of specialization in 2000 that included the C-SWMC. The purpose of this specialty area is to provide visibility to the function and role of the case manager and to develop recognition of social workers in the role. Eligibility requirements include the following:

- A baccalaureate degree in social work from an accredited university
- Documentation of at least three (3) years and 4,500 hours of paid, supervised, post-BSW professional experience in an organization or agency that provides case management services
- Current state BSW-level license or an ASWB BSW-level exam passing score
- Adherence to the NASW *Code of Ethics* and the NASW *Standards for Continuing Professional Education*. (National Association of Social Work, 2011)

In the area of substance abuse, the Substance Abuse and Mental Health Services Administration (SAMHSA), specifically the Center for Substance

Abuse Treatment (CSAT), provides a comprehensive guide for the case manager and case management function (Siegal, 1998). This set of standards and competencies for case managers include an understanding of managed care, cultural competence, service coordination, planning, practice guidelines, and treatment. Many states developed their own case management certification on the roles, responsibilities, and competencies and skills outlined in the SAMHSA Treatment Improvement Protocols for addiction-related and other human service professionals (Kansas Association of Addiction Professionals, 2011; Oklahoma: Behavioral Health Case Management Certification, 2012). For example, Kansas Association of Addiction Professionals first offered the Person-centered Case Manager (PCCM) credential. Focused primarily on case management for health care and mental health care related to Medicaid, the case manager competencies include but are not limited to screening, patient education/self-management, medication, psychotherapy, coordinated care, clinical monitoring, medical adherence, standardized follow-up, formal stepped care, and supervision (Kansas Association of Addiction Professionals, 2011).

Given the various perspectives on case management, its historical development, and the impact of managed care, and the professionalization of case management, it will be necessary to revise the case management process for the future. Shifts are evident in client involvement, the roles of the helper, and the emphasis on cost containment. Historically, the case management process has emphasized coordination of services, interagency cooperation, and advocacy, but the process of service delivery is expanding. Other trends have emerged from federal legislation, including coordination of care, integration of services, and the client as a customer. More recently, professionals providing services have been encouraged to empower clients, contain costs, and ensure quality services. Accountability reflected in the new certifications for case managers is also an important trend. These shifts in emphasis are reflected in the roles and responsibilities of the people delivering service today.

Deepening Your Knowledge: Case Study

Nancy, a 42-year-old white female, began her career in case management eleven years ago in a Cincinnati, Ohio, welfare office. She earned her bachelor's degree in sociology and married soon after college. She and her husband decided that they wanted to start a family soon after, so they agreed that he would work and she would stay at home to help with the children, who are now seventeen and fifteen. When her youngest child was a year away from kindergarten and had settled into preschool, Nancy decided that she was ready to enter the workforce. She had thought a lot about the helping profession while her children were growing up and knew from her involvement in church and community service that she felt called to social work. With her education in sociology, she found a position working with the many clients who needed guidance and direction navigating the welfare system.

After spending three years working with this population, she thought that she would be happier with more flexibility and diversity in her work. She enjoyed discussing approaches to working with clients with her coworkers and thought on many occasions that her direct supervisors did not do enough to help the case managers or the clients they served. Nancy began to realize that she desired to help clients and other social workers more fully by assuming a managerial role in the office.

At this time Nancy decided to enroll in a local master's of social work program and took classes at night over the next three years. During this time, she continued her work for two years at the welfare office until she became aware, through one of her professors, of an opening at a local center for adults with pervasive developmental disorders. She investigated this opportunity and realized that it would let her work with fewer clients in both group and individual settings; accordingly she could address their needs in greater depth with their needs and have more impact on their lives. Nancy interviewed and accepted the position with one year of coursework remaining on her master's degree.

She loved the work and, upon completing her degree, made plans to pursue licensure with the National Association of Social Workers as a Certified Social Work Case Manager (C-SWMC). Her supervisor was supportive of this decision and helped her with the supervised hours requirement. By working with her previous supervisor at the welfare office as well as taking and passing the ASWB-BSW exam, Nancy was able to receive her certification one year after pursuing the credential. The license, coupled with her master's degree, allowed Nancy to ascend to the level of program coordinator at the center in her ninth year as a social worker.

Morgan, C. (2012). Unpublished manuscript, Knoxville, Tennessee. Used with permission.

Discussion Questions

1. Identify the key steps that Nancy took towards enhancing her professional identity as a social worker. What further steps could Nancy take in order to develop her role and career?
2. If Nancy had decided to explore work in substance abuse after the welfare position, how might her path toward enhancing her professional identity have differed?
3. Think about your current stage in your professional development. What steps would you like to consider and pursue over the next ten years? Who might you consult to help you achieve these steps?

Chapter Summary

This chapter establishes a historical context for case management by describing four perspectives on case management that have evolved in the past thirty years. A brief history of case management, the impact of managed care, and expanding case management responsibilities complete the historical perspective.

The four perspectives on case management illustrate the recent development of the process. Case management as a process was recognized during deinstitutionalization, when limited assessment was followed by placement. Client involvement, the second perspective, evolved during the 1980s, when caseworkers encouraged clients to become partners in the process. A shifting emphasis from management to coordination characterized the third perspective, which focused on the role of the case manager. The final perspective concerns utilization review and cost-benefit analysis, which resulted from spiraling costs and health care reform.

The history of case management traces the use of advocacy, data gathering, recordkeeping, and cooperation. Federal legislation, such as the Older Americans Act of 1965, the Individuals with Disabilities Education Act, the Family Support Act of 1988, and the Personal Responsibility and Work Opportunity Act of 1996, further spurred the development of case management.

By the 1970s, managed care was seen as the answer to the rising cost and decreasing quality of health care. It is now part of the structure of service delivery, and it affects both social and mental health services. Case management responsibilities continue to evolve. Empowering clients, containing costs, and ensuring quality services are reflected in case management responsibilities today. The professionalization of case management, evidenced by new certification at the national and state level, enhance the distinctiveness of the role and the emphasis on quality and accountability. An increasing emphasis on delivering services in a manner that respects client culture represents an important shift in case management.

 ## Chapter Review

Key Terms . . . .

Deinstitutionalization   Health maintenance organization (HMO)
Social diagnosis   Preferred provider organization (PPO)
Fee-for-service   Point of service (POS)
Managed care

Reviewing the Chapter . . . .

1. Trace the development of the case management function since the 1970s.
2. Describe the shift in the meaning of the term *case management* that occurred in the 1980s.
3. Why have *service coordination* and *care coordination* become common terms for case management?
4. How has the managed care movement influenced case management?
5. Describe the influence on case management of the early organization of the Massachusetts School for Idiotic and Feebleminded Youth.
6. What emerging case management functions do Hull House and the Henry Street Settlement House illustrate?

7. How did immigrant populations benefit from the form of case management practiced in the settlement houses?
8. Describe the case management functions of the American Red Cross.
9. How did federal legislation contribute to the evolution of case management?
10. Trace the development of managed care in the 20th century.
11. Explain the various ways the term *managed care* is defined.
12. Identify and describe the three models of managed care.
13. Define the purpose and components of a patient bill of rights for those involved in the managed care system.
14. Describe the professionalization of the case management function.

● *Questions for Discussion* . . . .

1. What part of the history of case management most influences the process today? Cite the reasons for your answer.
2. Why do you think case management developed in the United States over the past century?
3. If you could choose a managed care organization with which to work, what criteria would you use to make your decision?
4. What do you predict will be the future of case management in the United States?

● *References* . . . .

Addams, J. (1910). *Twenty years at Hull House*. New York: New American Library.

American Psychological Association (n.d.). Mental health patient's bill of rights. Retrieved from http://healthyminds.org/Main-Topic/Patient-Bill-of-Rights.aspx

Case Management Society of America (n.d.). Our history. Retrieved from http://www .cmsa.org/Home/CMSA/OurHistory/tabid/225/Default.aspx

Center for Credentialing and Certification (2011). Human services board-practitioner application packet. Retrieved from http://www.cce-global.org/credentials-offered /hsbcp

Clubok, M. (2001). The aging of America. In T. McClam & M. Woodside (Eds.), *Human service challenges in the 21st century* (pp. 352–358). Birmingham, AL: EBSCO.

Dulles, F. (1950). *The American Red Cross*. New York: Harper.

Dunst, C., & Trivette, C. (1989). An enablement and empowerment perspective of case management. *Topics in Early Childhood Special Education, 8*(4), 87–102.

Hurd, C. (1959). *The compact history of the American Red Cross*. New York: Hawthorne.

Jackson, B., Finkler, D., & Robinson, C. (1992). A case management system for infants with chronic illnesses and developmental disabilities. *Children's Health Care, 21*(4), 224–232.

John Hopkins AIDS Service. (2004). Managed care. Retrieved from http://www .hopkins-aids.edu/manage/glossary_e.html.

Kansas Association of Addiction Professionals (2011). *KAAP Certification Commission: Person-centered case manager*. Retrieved from http://www.ksaap.org/casemanager .shtml

Lieberman, F. (1990). The immigrants and Mary Richmond. *Child and Adolescent Social Work, 7*(2), 81–84.

Lourie, N. (1978). Case management. In J. Talbott (Ed.), *The chronic mental patient* (pp. 159–164). Washington, DC: American Psychiatric Association.

MacLeod, G. (1993). An overview of managed healthcare. In P. Kongstvedt (Ed.), *The managed health care handbook* (pp. 1–3). Gaithersburg, MD: Aspen.

Mullahy, C. M. (2010). The case manager's handbook (4th ed.). Sudbury, MD: Jones and Bartlett.

National Association of Social Workers (2011). The certified social work case manager. Retrieved from http://www.socialworkers.org/credentials/specialty/c-swcm.asp

Ozarin, L. (1978). The pros and cons of case management. In J. Talbott (Ed.), *The chronic mental patient* (pp. 165–170). Washington, DC: American Psychiatric Association.

Pittman-Munke, P. (1985). Mary E. Richmond: The Philadelphia years. *Social Case Work*, 66(3), 160–166.

Polacheck, H. (1989). *I came a stranger: The story of a Hull-House girl*. Chicago: University of Illinois Press.

Richmond, M. (1917). *Social diagnosis*. New York: Russell Sage.

Rubin, S. E., & Roessler, R. T. (2001). *Vocational rehabilitation process*. Austin, TX: PRO-ED.

Siegal, H. A. (1998). *TIP 27: Comprehensive case management for substance abuse*. Retrieved from http://www.ncbi.nlm.nih.gov/books/NBK14516/

Social Work Best Practice (n.d.). *Healthcare Case Management Standards*. Retrieved from http://www.sswlhc.org/docs/swbest-practices.pdf

Trattner, W. I. (1999). *From poor law to welfare state: A history of social welfare in America*. New York: Free Press.

The Kaiser Family Foundation (2009). Trends in health care costs and spending. Retrieved from http://www.kff.org/insurance/upload/7692_02.pdf

Trustees of the Massachusetts School for the Feebleminded. (1919). *Seventy-second annual report*. Boston: Wright and Potter.

Trustees of the Massachusetts School for the Feebleminded. (1920). *Seventy-third annual report*. Boston: Wright and Potter.

Wald, L. (1915). *The house on Henry Street*. New York: Dover.

Weil, M., & Karls, J. (1985a). *Case management in human service practice*. San Francisco: Jossey-Bass.

Weil, M., & Karls, J. (1985b). Key components in providing efficient and effective services. In M. Weil & J. Karls (Eds.), *Case management in human service practice* (pp. 29–71). San Francisco: Jossey-Bass.

Welfare & Workforce Development Partnerships. (2000). Atlantic and Cape May Counties, New Jersey. Retrieved December 16, 2004, from http://wtw.doleta.gov /wwpartnerships

Winsor, J. (Ed.). (1881). *The memorial history of Boston 1630–1880* (Vol. IV). Boston: Ticknor.

Woods, R., & Kennedy, A. (1911). *Handbook of settlements*. New York: Russell Sage.

CHAPTER THREE

Models of Case Management

Thriving and Surviving as A Case Manager

Katie

I work for a nonprofit mental health agency. It provides a lot of different services, such as social services and mental health services. I work in the children's services in one of our school-based programs providing intermediate-level care. The goal is to get the children's behaviors under control and get them into a condition where they can be successful at school.

We call the agency's lowest level "integrated services team." It's a team approach and the case manager spends time with the kid maybe a couple of times a month and does one home visit. That might be three to five hours a month. This means just checking in, saying, "How are things going?" Our services at the next level are called intensive case management. My specific program is school based because we are bridging the gap between home and school. I see my kids at school once a week and communicate with their teachers or principal or school counselor or whoever is involved in their school life. Also I meet with their parents at least twice a month, so there's more communication, more support. Our next level of case management is continuous treatment. Case managers spend six to eight hours a month or maybe more than that. They conduct one individual visit and one family visit per week, sometimes more than that; they are in the home and with the child individually multiple times in one week so there's more support for working on behavior. Most of the kids I work with have emotional and behavior problems but they also are related to the environment. For most of the kids that get referred up to continuous treatment there is more of a mental health component to their treatment. There is still another level up from that, which is called intensive case management. Case managers are with the child every day. They see each child every day during the week. This is the most support that we can offer outside of residential treatment; the next step after intensive case management is residential treatment.

—Permission granted from Katie Ferrell, 2012, text from unpublished interview.

I enjoyed my work as a rehab case manager. Sometimes it was really challenging. I remember one of my clients had had a heart attack and she had physical limitations after that. Her three adult children lived out of town and she did not have anyone to help her. So I worked with friends and other agencies to provide her services after she left the rehabilitation center. She was a good client. We got home health to come to the house to continue her rehab. And some members from her church took her to doctor's appointments and church services and events. She just could not live independently without support. The office on aging provided meals and friendly visitors for her. I think that the transportation piece was the most difficult.

—Rehabilitation case manager, rehabilitation services agency, Kennesaw, GA

Our primary focus is providing housing for the HIV and AIDS clients. We pride ourselves on offering a safe, comfortable place for these clients to live. Short- and long-term housing is available. Our services have changed over the years. The changes reflect the changing populations we serve and the medical treatments that allow clients a longer life. We continue to grow and now we offer crisis intervention and prevention services. Our case managers are the lifeblood of our service delivery.

—Director, housing services nonprofit, New York, NY

Parole is an important part of the criminal justice system. We work hard to help parolees before they are released so they can have a smooth transition. Part of the information I use to support the release process is on the release plan. This is a form that inmates fill out; it helps me know about where they might want to live and work when they leave prison. Once I get this information, then I begin to explore living accommodations and possible employment opportunities. It is my job to make sure that rooms can be rented and the landlords will accept the ex-inmate. I also talk with the employer. It is important I find out if the work is legitimate. If I cannot verify housing or employment, then I contact the institutional parole officer and he or she helps the inmate submit another plan. We try to make sure that the parolee has an important responsibility during the transition process.

—Parole officer, adult corrections parole, Knoxville, TN

This chapter introduces the models of case management, reflecting the creative ways in which services are delivered. The roles that helpers assume within the context of case management follow.

The preceding quotations relate to both the roles that case managers perform in service delivery and the models used to deliver services. The director of housing services in New York, who works with individuals and families affected by HIV/AIDS, supports clients by providing a variety of services including housing, recreation, and vocational rehabilitation. The case managers at her agency serve as coordinators, planners, and problem solvers. The rehabilitation agency provides services both in the hospital and after hospital discharge. The program makes a long-term commitment to clients, regardless of their abilities or status. This requires case managers to be flexible in their roles to accommodate the needs of the client. Working from a very different model, the parole officer's roles are defined by the state and shaped by her large caseload. Because the parolees assume most of the responsibility for themselves, the parole officer is primarily a recordkeeper, monitor, and problem solver.

For each section of this chapter, you should be able to accomplish the following objectives.

MODELS OF CASE MANAGEMENT
- List reasons why it is important to understand the different models of case management.
- Name the three models of case management.
- Illustrate each model with an example.

ROLES IN CASE MANAGEMENT
- Identify the roles in the case management process.

MULTICULTURAL PERSPECTIVES
- Provide two reasons why a multicultural perspective is important for case management.
- Explain multiple levels of identity as they relate to the multicultural perspective.
- Articulate three ways power influences the case manager's professional experience.

 # Models of Case Management

This section introduces three modern models of case management. It is important to understand the different models for several reasons. First, the existence of three separate models demonstrates that service delivery can occur in a variety of

TABLE 3.1   MODELS OF SERVICE DELIVERY

	Models		
	Role-based	*Organization-based*	*Responsibility-based*
Examples	Generalist	Comprehensive service center	Family
	Broker	Interdisciplinary team	Supportive care
	Primary therapist	Psychosocial rehabilitative center	Volunteer
	Cost containment		Client as case manager

© Cengage Learning 2006

ways; it is a flexible process. Second, the different goals that characterize each model provide a perspective on the case manager's responsibilities, roles, and length of involvement with the client. Third, each model has particular strengths and weaknesses. Knowing these models helps determine the setting in which each model is most relevant.

The three models presented have their bases in roles, organizations, and responsibilities (**see Table 3.1**). The **role-based case management** model focuses on the roles the case manager is expected to perform. One case manager may act primarily as a broker of services, whereas another is a therapist who has occasional brokering and coordinating responsibilities. Another case manager may concentrate on cost containment and cost efficiency. The rehabilitation case manager's chapter-opening quote exemplifies the broker model: Her major responsibility is linking the client to needed services.

The **organization-based case management** model focuses on providing a comprehensive set of services and meeting the needs of clients with multiple problems. This model can be applied to work in a variety of situations, such as a comprehensive service center, a psychosocial rehabilitative center, or an interdisciplinary team. A quotation at the beginning of the chapter described an agency that serves individuals with HIV/AIDS; the organization-based model applies in this case. The agency provides a range of services, all available in one location. Some clients live at the center.

In the **responsibility-based case management** model, family members, a supportive care network, volunteers, or the client may perform the case management function. In the third chapter-opening quotation, the parole officer described how the parolee assumes responsibility for a release plan. At her agency, the client has many responsibilities and is empowered by the case manager to find a job and housing.

In the material that follows, each model is described according to the following characteristics: the goals of the process, the responsibilities and roles of the case manager, the length of the process, and the strengths and weaknesses of the model. Illustrations of each model follow the description of its characteristics.

Role-Based Case Management

The role assumed by the designated case manager characterizes this model of case management. Roles may vary according to the function and the services provided (Mullahy, 2010). Following the characteristics of this model, we present four illustrations: the case manager as a generalist, a broker, a therapist, and a cost container.

This model is also called the collaborative or integrated care model.

Goal   The case manager attempts to meet all the needs of the client through a single point of access. This may require the case manager to serve as the link to a variety of needed services, to be the provider of therapeutic care, or to monitor the efficiency and quality of services.

Responsibilities   The case manager assumes a broad set of responsibilities, including intake interviewing, data gathering, planning, linking to services, coordinating or delivering services (or both), referral, and evaluation.

Primary Roles   The roles can include broker, coordinator, counselor, planner, problem solver, and recordkeeper.

Length of Involvement   The duration of case management varies according to the complexity of the client's needs, the limitations managed care places on service delivery, and the need for short-term or long-term care.

Strengths   In many cases, there is a single point of access for the client. The case manager assumes the various roles as needed. Together they identify problems and develop a plan of services. The case manager remains closely involved as both provider and coordinator of services. This involvement promotes a strong relationship with the client, who has regular access to services. The case manager also provides assistance with financial aspects of the medical and mental health systems.

Weaknesses   Some agencies limit the services they provide; case managers may have large caseloads, no backup, and limited time for community involvement. Limited services may also mean more referrals, incomplete assessment, and a narrowing of the focus to only one perspective on service delivery. Where the case manager's role is oriented toward cost containment, there may be a focus on managed care concerns or a standard of care that does not correspond to the client's needs.

One organization, the National Association of Geriatric Care Managers (NAPGCM), outlines a mission and vision for their organization. Its stated vision is "Defining excellence in geriatric care management," and its mission is "[t]o advance professional geriatric care management through education, collaboration, and leadership" (National Association of Geriatric Care Managers, 2012). Within this framework, NAPGCM describes a role-based approach to the work of the care manager.

Standard 6—Definition of Role to Other Professionals

Standard

The GCM should clearly define his/her role and scope of practice to clients and others involved with the client system.

Rationale

GCMs are professionals with diverse educational backgrounds and skill sets. Therefore each GCM should define his/her scope of practice and the particular roles he/she will accept in assisting clients and those involved in the client's care.

Guidelines

- The GCM should provide a clear, comprehensive explanation of his/her role and responsibilities to clients and the client system.
- The GCM should accept only those roles and responsibilities for which he/she has the skills, knowledge and training. He/she should recommend consultations with other experts as needed. (National Association of Geriatric Care Managers, 2012)

Let's look at several cases that illustrate how role-based case management is delivered.

ILLUSTRATIONS

The first illustration of role-based case management involves the case manager as a generalist. This role, widely used in human service delivery, focuses primarily on providing the services that can be delivered by a helper with knowledge and skills applicable to a range of clients in various settings. Brokering or linking the client to other services occurs infrequently.

Azzurra Given's life revolved around Florida State University, where she taught French and Italian literature for more than 50 years. When she retired three years ago, she lost her focus, became depressed, and ended up in a hospital and then a rehabilitation facility. Her daughter, Iona, who is forty-six and an only child, worried about the next step. Thanks to Home Instead, a national franchise that provides nonmedical caregivers by the hour, the next step was for Azzurra to return to her own apartment. Initially, a caregiver visited every day to help with meal preparation, monitor her medications, drive her to the grocery store, and accompany her to doctor's appointments. Once Azzurra was feeling better, the visits were cut back to three times a week.

Iona is pleased that she has found a way to keep her mother in the garden apartment she loves, close to the university and the local senior center, where she teaches Italian. Iona relies on the home-care agency to act as her eyes and ears. She also has a maid service clean the apartment twice a month, pays $15 a month for an emergency-response bracelet that her mother can use to summon help, and phones every day—even if her mom doesn't always remember the calls. (Franklin, 2000, p. 88)

When the case manager's work focuses on the role of *broker*, the emphasis is on linking the client to other services. Interaction with the client is done primarily to assess the person and the environment, link him or her to services, and monitor the services delivered by others. The case manager in this role provides fewer direct services than does the generalist. For example, individuals who are responsible for elderly family members or those who are disabled or ill often find it difficult to fulfill their responsibilities when they live in different towns, states, or parts of the country. Case-managers-for-hire is a new trend that helps meet these needs.

Professional geriatric-care managers, who typically charge from $100 to $250 an hour, can be particularly helpful when you're trying to provide care across the miles. They can assess your parent's condition, acquaint you with local services and residential facilities, and recommend solutions in keeping with your parent's—or your—financial resources. You can get a list of managers at the National Association of Professional Geriatric Care Managers' Web site, http://www.caremanager.org/, or by writing the group at 1604 N. Country Club Rd., Tucson, AZ 85716. (Franklin, 2000, p. 90)

In some cases, the case manager is a counselor or a therapist, personally providing this service. Often in the mental health field, the client seeks the help of the counselor for a presenting problem. The counselor presents credentials and areas of expertise and works with the client to determine whether there is a match with the client's problem. Referral elsewhere occurs when there is not a good match, or if the client later experiences a crisis during the helping process.

Sena and Renaldo Peterson work with a case manager to coordinate the care that their son, Carter, needs. At the age of seven he was diagnosed as bipolar. For three years prior to the diagnosis and for six years after the diagnosis, they struggled to provide him with the services that he needed. They fought with the educational system and the juvenile justice system, both of which pegged him as a troublemaker. He was expelled from school three times, and placed in "special" programs designed to keep him out of the regular classroom. He broke into the school twice and was caught stealing from a local liquor store. He has been hospitalized twice. Finally, Sena and Renaldo found a therapist willing to provide counseling first and coordination of care second.

There are no services in their town that provide multiagency intervention. The therapist, willing to work with various agencies, is interested in developing a plan. In the town there are services for persistently mentally ill adults, but none for children.

In many situations, clients need services because they become physically or mentally ill, sometimes catastrophically so. There is often a case manager responsible for cost containment. The case manager must make recommendations after considering outcomes, levels and quality of care, and expense. The case manager gets the client, family, and professionals involved in decisions. Negotiating services to find the best quality for the lowest price is one responsibility of the case manager. He or she also documents the negotiated arrangements, prepares a cost analysis statement, and tracks all interactions between the client and the providers. **Box 3.1** lists the information included in a standard cost-containment analysis; **Box 3.2** gives an example.

Organization-Based Case Management

In this model, the nature of case management is determined by the organizational structure of the agency or organization. In other words, the way in which services are arranged determines how services are delivered. The three examples of organization-based case management presented here illustrate the collaborative atmosphere that exists among many helping professionals. Each person has a specific assignment and responsibility; the services are organized so that relationships among professionals are integrated to serve the clients' needs better.

Goal   Case managers meet multiple needs through a single point of access, with one location for service delivery. This comprehensive service delivery sometimes resembles that provided by the traditional extended family.

Responsibilities   The organization-based model provides comprehensive case management: Each client receives an individual assessment and plan, which may

---

BOX 3.1	**Summary of Cost-Containment Analysis Variables**

1. Identifiers such as name and date client enters the system
2. Description of the case management plan
3. Description of the goals, services, and outcomes
4. List of the costs of the services
5. List of how case management process saved resources
6. How much was spent on the client
7. Is the case still in process or has the case been terminated?

BOX 3.2   Case Summary
●●●●●●●●●●●●

The case number: 00057

Company: Black Brokers

Case Opened: xx/xx/xxxx

Diagnosis: Pneumonia and depression

Status: Case has been ongoing for 4 weeks

The client is no longer hospitalized. The medical staff is providing care for the client in the home of the client's sister. The medical supplies have already exceeded the $50,000 limit. This includes both physical and mental health care.

The client has been referred to two services, one to address the pneumonia and the other to address the mental health issues. Both of these companies will prorate the cost of services: the insurance company will pay 90% of the services and the family will pay 10% of the services.

When analyzing the potential costs if the patient goes to the hospital, managing the case and serving the client at home saves over 50% of the potential hospitalization costs.

The family has asked for increased services but this request has been denied.

Avoidance of potential charges:

Potential hospitalization 30 days @ $1,000	$30,000.00
Cost of medical and mental health services	3,000.00
Total potential charges	$33,000.00

Actual charges:

Physical health and mental health care	$13,800.00
Supplies	1,000.00
	$14,800.00
Case management fees	$ 1,000.00
Total actual charges:	$15,800.00
Net savings (potential charges minus actual charges):	$17,200.00

include social support, housing, recreation, work, and time to integrate into the community. The case manager's responsibilities range from coordinating services (supervision of intake, assessment, planning, brokering, monitoring, and termination) to leading a team of professionals who provide services to the client. Within the second scenario, the case manager role varies: Sometimes there is a professional whose primary responsibility is management of the case, and at other times the one who initiates services also assumes the management role.

Primary Roles   Advocate, broker, coordinator, planner, problem solver, and recordkeeper.

Length of Involvement   The duration varies. If the case is complicated and several specialists are needed, services are provided for a longer time. In other cases,
short-term service is adequate.

Strengths   Services are provided on an inpatient, outpatient, or residential basis,
but all are provided in one location. Client assessment is multifaceted, with a
holistic approach. The plan is individualized and easily monitored. Staff members function as a team, with a common goal, regular meetings, and a common
reporting scheme.

Weaknesses   Resource availability may be a problem if the client needs services
not available in the center. Service integration depends on clear organizational
structure and lines of authority; the staff must agree on the problem, the plan,
and the implementation. Resource availability can be a problem. The family of
the client may be less engaged in the helping process than in other models. In
addition, the client may become accustomed to the environment and never grow
beyond it.

One example of organization-based services, where comprehensive services
are provided, occurs in the state of Wisconsin. The Wisconsin Works (W-2) program provides services to parents with children. To be eligible for the program
these parents must earn a salary that is 115% below the federal poverty level.
Case management is a key component of the services available to individuals who
qualify for the program. Read further to understand the case management functions related to comprehensive case management for Wisconsin's W-2 program.

ILLUSTRATIONS

Many case managers see the multiservice center as the ideal. They believe it reduces the risk of people getting lost in the system, clients' feelings of frustration,
and duplication of information, forms, and services. Clients can have a single
intake interview and go through just one assessment. They have one case manager who has access to all the professionals involved in the delivery of services.
An example of the use of comprehensive service centers is a new approach to
public welfare that has transformed the eligibility worker into a case manager.
The ultimate goal is to move families toward self-sufficiency. Case managers collect information to determine eligibility and assess the client's use of other social
support services, such as child care, health services, housing, employment, and
medical benefits. All these services are available in the neighborhood or a short
bus ride away.

The case management team members are all employees: managers, supervisors, case managers, specialists, and clerks. Everyone from the same office is
trained together in the operations of case management, problem solving, and
teamwork, and all are committed to helping clients become self-sufficient.

Comprehensive Case Management: Wisconsin's Model Approach for W-2 Participants

The comprehensive case management model is a process, not a program or a type of service. It represents a fundamental change in the way services are designed and delivered. In a comprehensive case management model, families in W-2 with substance abuse issues are to receive individualized wraparound services. It is value-based and has an unconditional commitment to customize services on a "one-family-at-a-time" basis to support normalized and inclusive options for families with complex needs. At its core, the comprehensive case management model is based on interventions, which are collaborative, community-based, emphasize the strengths of families, and include the delivery of highly coordinated, individualized services for families. This process addresses the unique needs of families with a focus on achieving positive and effective partnerships with families, the community, and agencies that provide children's and family services.

For successful treatment programs that focus on recovery to work, they should not only offer a continuum of services but also integrate these services within the larger community. Because many factors affect a woman's substance abuse problem, the purpose of a comprehensive case management approach is to address a woman's substance abuse in the context of her health and her relationship with her children and other family members, the community, and society. This type of case management can generally be described as a coordinated approach to the delivery of health, substance abuse, mental health, vocational, and social services, linking participants with appropriate services to address specific needs and stated goals. All services and supports must be culturally competent and tailored to the unique values and cultural needs of the family, and of the culture that the family identifies with. When implemented to its fullest, comprehensive case management will enhance the scope of substance abuse treatment and the recovery continuum, and will stress the following goals:

- Provide the participant with a single point of contact for multiple health and social services systems
- Advocate for the participant
- Be flexible, community-based, and family-focused
- Assist the participant with needs generally thought to be outside the realm of employment and training and substance abuse treatment
- Develop a universal service plan that integrates activities from all service providers and is outcome based. (Wisconsin Department of Children and Families, 2012)

Another example of the organization-based model is the interdisciplinary team, which is particularly effective for clients with complex problems that require the involvement of several professionals. This approach brings specialists together with the mutual responsibility to help the client. Difficulties may arise when they do not agree, when the case manager or the specialists have too heavy a caseload, or when resources are insufficient for the services needed.

The following example provides a description of how comprehensive care was provided to Audrey, a senior living in Victoria, Australia (Aged and Community Services: Australia and Case Management Society of Australia, 2006).

Audrey is a seventy-eight-year-old pensioner who lives alone in rural Victoria. Audrey has two daughters—one lives in another state and the other one lives overseas. Audrey has osteoarthritis, rheumatoid arthritis, cardiac problems, and glaucoma. Audrey suffers from insomnia, anxiety, and depression. Six months ago she fractured her hip and she experiences chronic and ongoing pain as a result.

Audrey has many medical appointments. Audrey has been missing appointments because she is unable to access public transport. Her telephone has recently been disconnected because she has not been able to pay the bills.

Audrey had been completely independent but since fracturing her hip she has lost her confidence and become overwhelmed and anxious in managing her day-to-day affairs.

Audrey's GP believes that she should be assessed for low residential care because he is concerned that she can no longer manage on her own. The GP contacted the local Community Options (COPS) package service, and after meeting Audrey they agreed to provide a package in the interim. As a result of this Audrey was assigned a case manager. The case manager spent time getting to know Audrey and identifying the various issues that were making Audrey feel that she could no longer cope at home.

Audrey did not really want to go into residential care but felt she had no alternative. The Case Manager believed that with some limited services and increased confidence Audrey would be able to continue to live independently. Armed with Audrey's preferences and her own knowledge of resources available, the Case Manager did the following:

- *Arranged for a home cleaning service to visit fortnightly*
- *Introduced the idea of supported/assisted living*
- *Assisted Audrey in finding an independent living unit that catered to people on low incomes in the closest regional city*
- *Assisted Audrey in obtaining the correct advice to complete the relevant paperwork for the independent living unit*
- *Sourced an affordable relocation service, using the Salvation Army*
- *Arranged for a financial counselor to work with Audrey to develop a budget*
- *Advocated on Audrey's behalf with the phone company to address unpaid bills and reconnection costs*
- *Organized an ongoing transport service through the Red Cross to ensure Audrey was able to keep medical appointments. (Aged and Community Services: Australia and Case Management Society of Australia, 2006.)*

Many professionals regard the approach to care Audrey received as an excellent treatment approach. As with other organization-based services, the care is comprehensive, and the client receives social support. Among the criticisms of these organizations are that they may resemble institutions and can be very expensive.

The following provides another illustration of a case that is best served by the comprehensive care model.

A car hit John H., a senior at a college in New England, when he was attending Mardi Gras in New Orleans. After hospital discharge, John entered a rehabilitation center for the assessment of job skills and personal and social adjustment. The brain damage he suffered is permanent, and he must cope with a number of limitations. The center provides vocational assessment, counseling, independent living skills, and education. Staff also dispense medication and make referrals when necessary.

At times organizational case management services focus on issues that begin with a crisis, but later demand longer-term care. Using this model, these services identify immediate needs and plan to support clients on a short- and long-term basis. In an effort to service victims of domestic violence, many agencies use the organizational model.

Safe Horizon is an organization that addresses violence in our society and serves adults and children in New York City. One of its programs, which addresses domestic violence, is a comprehensive program that supports women living in abusive situations. The agency is structured so that women and their children receive full support once they decide to leave their relationships. Services, delivered within the case management framework include short and long-term housing, child care, vocational support, legal aid, transportation, and other services. This center serves women by using the "one-stop" model. The case management also encompasses counseling and support groups. Services range from crisis support to long-term care. (Safe Horizon, 2012)

Responsibility-Based Case Management

The focus of responsibility-based case management is the transition of care from human service professionals to nonprofessionals. Often clients continue to need assistance long after the professional case managers have terminated their work with clients and families. To meet the ongoing or recurring need for case management services, teams of family, friends, or community volunteers are trained to provide continuing case management. Professional case managers support the caretaker and are available to help during emergencies, crises, and other stressful times.

Goal   The responsibility-based model emphasizes both short- and long-term involvement of the case manager, the coordination of services, the help of volunteers, and the empowerment of clients.

Responsibilities   The individual or group responsible for case management provides coordination, finds assessment services, and networks with others in the human service delivery system to provide access to needed specialists and services. Problem identification, plan development, and implementation are other responsibilities. The case manager also provides support and assistance in making and maintaining other linkages.

Primary Roles   Broker, coordinator, planner, problem solver, and recordkeeper.

Length of Involvement   The involvement may be short-term, during a crisis or developmental problem, or long-term, as with a physical or mental illness, a disability, or geriatric problems.

Strengths   The responsibility-based model allows various individuals or groups to assume case management responsibilities, including family, neighbors, volunteers, and the client. In many cases, the designated case manager may already have an established relationship with the client. Involving family, neighbors, and volunteers at times improves client access to services. Under this model, service delivery is cost effective, the community is involved, and independence is encouraged.

Weaknesses   In some cases, the person designated as case manager may not have the client's best interests at heart, may lack the necessary knowledge and skills, or may be ineffective in monitoring service provision. Training and supervision may be costly. Accepting a family member or volunteer as case manager may be difficult for the client. Case managers who are not part of the human service delivery system may have trouble coordinating and gaining access to services.

ILLUSTRATIONS

It is a current trend in human services to ask families to act as case managers and then provide them with the support to do so. With costs escalating and institutional care being replaced by community care, it is cost effective for families to perform this central role. However, for such a system to be effective, the family must receive continuing education about the human service delivery system and must have professional help available when crises arise.

Elaine Mayer from Outland, Kansas, assumed the role of long-distance caregiver about five years ago, when her mother received a diagnosis of colon cancer. Ten years prior, her mother became the caregiver of Elaine's stepfather when he was diagnosed with early-onset Alzheimer's. Elaine drives to her mother and stepfather's home to check on them every four weeks. She

leaves after work on Friday, drives 200 miles, and returns late Sunday eve-
ning. With the help of the pastor at her parents' local church, Elaine formed a
group of people she calls the "sandlot crew," a name she adopted from her
early childhood memories of summer softball. She got the idea for the team
from the American Medical Association's caregiver health Web site
(http://www.ama-assn.org/ama/pub/physician-resources/public-health
/promoting-healthy-lifestyles/geriatric-health/caregiver-health/family
-caregiving-topics.page). The sandlot crew helps her support her mother and
her stepfather. She was worried about her mother long before the diagnosis of
cancer. She asked her mother to fill out the caregiver assessment (AMA,
2012) she downloaded from the web. The results of the assessment confirmed
to her that she needed to support her parents more actively even though she
lives quite a distance from them.

Members of the team include the local pastor, her mother's doctor, her moth-
er's hairdresser, the couple's attorney, and a case manager Elaine hired to
check in with her parents three days a week, make sure there are groceries in
the house, take them to the doctor, and call team members if needs or a crisis
arises. An accountant manages their day-to-day finances. Each month this
team gets together for lunch after church on Sunday to talk about care of
Elaine's parents. Elaine maintains a diary at her mother's home and each
member of the team visiting makes notes about needs and duties. Elaine be-
lieves that what has worked for her and is the single most important facet of
parent care—especially from a distance—is communication.

The responsibility-based case management model also applies to supportive
care, which is used most often when clients cannot travel to receive services.
Most affected are people located in rural areas and those who live alone—often
persons who are disabled or elderly. Sometimes the services provided are minimal, but they are helpful to people who otherwise would not be served at all.

An example of a supportive care program is one that uses *gatekeepers*—
individuals who have contact with physically disabled clients in their day-to-day
work.

Clients who live alone and use wheelchairs are one type of client served by a
gatekeeper program. Ms. H. is forty-five years old and has been widowed for
six years. Four weeks ago, she developed a severe respiratory problem and has
not left her apartment since. Her apartment manager, realizing that he had
not seen her for a while, visited her one afternoon. During the visit, it became
clear to him that she was having difficulty caring for herself. She had lost
weight and was unkempt; her apartment was dirty and messy. After the visit,
he called the Gatekeeper Program to get assistance for her.

Because of the rising cost of service delivery, many agencies and communities have developed strong volunteer programs. The agencies provide excellent training, ongoing education, and good supervision, thus allowing volunteers to assume case management responsibilities. In this way, the volunteers are able to contribute to the welfare of their local community.

Many social service organizations use volunteers to help immigrants when they first enter the country. Because immigrants and their families have multiple needs, a team consisting of a nurse, social worker, environmental assessment specialist, volunteer coordinator, and program administrator works to provide a positive entry into this country. Because the case managers for these social service agencies have caseloads greater than 200 clients, this program supplements the care provided by volunteers.

The day-to-day contact with the immigrants is maintained by volunteers, who are supervised and assisted by community workers—paraprofessionals who themselves live in the neighborhoods. The services the volunteers provide might include grocery shopping, cooking, cleaning, companionship, and making others aware of any special needs or emergencies. One volunteer, a Cuban immigrant, was once a client in the program. Now, she uses her language skills to help clients who speak only Spanish.

The responsibility-based case management model is applicable when the client is the case manager. This approach develops the client's maximum potential, which is emphasized by strengths-based case management (Rapp & Goscha, 2006). This type of case management stresses building on the strengths and resources of individuals, distinguishing it from more traditional approaches that focus on deficits and needs. Self-determination—the right to establish one's own goals and to have an active role in problem resolution—is of primary importance here. The belief is that a client who has learned to act as manager can provide long-term care for himself or herself and for others.

With the support of professional case managers, breast cancer survivors may serve as care managers for themselves or for newly diagnosed patients. They complete an in-service training program at the hospital's breast cancer center. Although they do not provide medical services, they do counsel, arrange for prosthetic consultations, refer for group discussion and counseling, and talk with family members (Power & Hergarty, 2010). The focus of this type of peer support, offered both on site and online, is to reduce social isolation, increase patients' ability to handle stress, and provide information.

Voices from the Field: Research and Practice

The Reach to Recovery International Network's mission is to "improve the quality of life for women with breast cancer and their families" (Bloom, 2011). One online service recently developed in Australia provides a case management function, that of linking with services, to all who visit the website, along with peer counseling and support. An announcement of this new case-management related service follows:

> Not only does the Network offer members an online peer-support network where they can find and connect with others affected by breast cancer, it also actively seeks out recommendations from members about support services in their local communities that they have found beneficial.
> The idea for the Local Services Directory, as it is called, came from BCNA's members. It has been designed to help people find services such as

- A hairdresser who is understanding and will provide assistance and a private space for a woman who wants her hair shaved
- A specialist lymphedema massage therapist
- Counseling services that extend to families as well as those who have been diagnosed with breast cancer
- Wig and prosthesis suppliers
- Breast care nurses and other health professionals

> Local knowledge is key to the success of the Directory, so members are encouraged to submit and maintain their own entries, and service providers can also submit entries for the directory.
> Users can search the directory by

- Keyword e.g. wigs, hairdresser
- Categories e.g., Health services, Emotional wellbeing, Physical wellbeing, Practical support, or Products
- Location e.g., distance of 3, 5, 10, 30 and 100 Kms from a specific postcode

> The online network is another key initiative that helps BCNA members connect and stay in touch. Members can use the network to:

- Connect with others of similar experience, regardless of location
- Set up a profile and personal blog to keep a record of their breast cancer journey
- Create and join online interest and support groups
- Use the privacy settings to control who can access their information and the content they create

> BCNA has provided these projects as part of the Supporting Women in Rural Areas Diagnosed with Breast Cancer' Program, funded by the Australian Federal Government.

From Online Support Connects Survivors to Services, from Bloom, November, 2011. Retrieved from http://www.reachtorecoveryinternational.org/bloom/issue_10_2011/files/bloom%20-%20 november%202011.pdf

Roles in Case Management

Each of the models of case management represents goals and responsibilities assumed by the helper. The responsibilities, described as roles, constitute the tasks—the actual work—that case managers do when they provide services to their clients. The next section describes examples of case managers who are performing those roles.

Advocate

An **advocate** speaks on behalf of clients when they are unable to do so, or when they speak but no one listens. The case management process presents many opportunities for advocacy. Working at various levels, the case manager represents the interests of the client, helping him or her to gain access to services or improve their quality. At the organizational level, the case manager serves as a community organizer who influences the policies that control eligibility and access to services. The case manager also works with the clients within the legal system, helping them make or defend their case. The case manager helps agencies work together to assess the needs of the community and plan how the local human service delivery system will meet those needs. At the legislative level, case managers can work to influence government policies and programs that serve the needs of their clients, which include addressing issues of inequality and discrimination. Case managers also help clients to become advocates for themselves and their families. This is one way to empower clients.

Jim was an advocate for Bryan, a nineteen-year-old recently admitted to a Veterans Affairs inpatient program for treating individuals with a dual diagnosis of substance abuse and depression. Bryan, a veteran of the conflict in Afghanistan, had participated twice before in a short-term inpatient program, and each time he had returned after six months. This time he was admitted after attempting suicide. As Jim, his care coordinator, evaluated Bryan's past history, it became obvious that the previous treatment had not been effective. Jim petitioned the managed care team and the alcohol treatment team to develop an individualized program for Bryan. Jim also decided that the VA standard treatment for individuals with dual diagnosis was not effective. He met with four case managers from three counties. Together they wrote a paper that outlined more current best practices for the dual diagnosis of depression and substance abuse and presented it to the regional director.

Broker

As a **broker**, the case manager links the client with needed services. Once the client's needs are clear, the broker helps the client choose the most appropriate service and negotiates the terms of service delivery. In this brokering role, the case manager is concerned with the quality of the service available and any difficulties the client may have in accessing it.

When Jo Sinclair assumed the brokering responsibilities for her son, Jasper, who had just left an inpatient mental health treatment facility, she found she needed help. Jo had decided to take a leave of absence from her job so she could provide major support for Jasper. His father had left the home the previous month, and Jasper had had several anxiety attacks. As the broker, Jo arranged appointments for Jasper to see a psychologist, a physician, and a lawyer. When Jo needs professional help, she calls the outpatient facility located next to the inpatient facility. The broker role is a difficult one for Jo because she is unaware of which services are available and their terms of eligibility. Jo would like more help from the inpatient team that worked with Jasper, but team members are restricted because of the fee structure and how services are delivered.

Coordinator

Many clients have multiple problems and need more than one service to meet their needs. In the role of **coordinator**, the case manager works with other professionals and agency staff to ensure that services are integrated and to expedite service. The case manager must know the current status of the client, the services being delivered, and the progress being made. Monitoring the client's progress and interfacing with professionals are important roles for the case manager. In this way the case manager can help the client with problems such as ineligibility, seemingly closed doors, poor service quality, and irrelevant services. Case managers also collaborate with other professionals during team staffing and program planning.

Jamie Wolfenbarger assumes the coordinator role for the local hospital's long-term care clients. In this role, she plans the aftercare for patients who will require long-term care. She coordinates previously unrelated services performed by professionals from different agencies. For Rose Woodson, a patient soon to be released from the cardiac observation unit, Ms. Wolfenbarger has arranged home health care visits once a day, a housekeeper to clean twice a week, meal deliveries at noon each day, and special ambulatory equipment. Ms. Wolfenbarger will contact these professionals each week for the next month for feedback on Ms. Woodson's progress.

Consultant

Often an outside professional can help solve case management problems. An organization may need assistance with such matters as cost analysis, quality control, and organizational structure. A **consultant** may have the expertise to identify the problem, study it, and make recommendations. Consultants can also assist with the case management of individual clients when special information or expertise is needed. This is especially true in small agencies that employ only generalist case managers.

Ann Marsella is a well-known expert on the treatment of young children with developmental disabilities. She is often called in for consultation on particularly challenging cases. Her expertise is in the legal and ethical aspects of serving these children and their families; she is respected for her ability to clarify a situation's ethical issues and the logical consequences of the proposed alternatives.

Counselor

The case manager who is a **counselor** or therapist maintains a primary relationship with the client and his or her family. Having a thorough understanding of the client's mental health and medical history, this professional can tell what aspects of his or her current situation support or discourage progress.

David Tanaka maintains therapeutic relationships with fifteen sexual assault victims for whom he also serves as a caseworker. He sees each client once a week for an hour and talks weekly with other professionals involved in each case. These clients were considered high risk prior to the sexual assault. They are enrolled in a special program designed to provide more than crisis care. He expects to retain these fifteen clients for the next year, without increasing his caseload.

Planner

One of the primary responsibilities of the case manager as **planner** is preparing for the service or treatment that the client is to receive. Planning is directly connected to the findings of the assessment phase of case management. The planner evaluates the client to determine his or her functioning and to assess service provision. Then the planner compiles data in medical, psychological, financial, social, and vocational areas that inform the implementation phase of case management. This phase includes setting goals, determining outcomes, and implementing the plan with input from the client, family members, other professionals, and other agencies. The case manager's planning role begins in the early stages of the helping process and continues until services are terminated. Planning may include a transition period that lasts until the client is able to manage his or her own case.

Tony Nix is in her first year as a case coordinator for the state parole board. She works with juvenile offenders after they have been paroled. Her first interaction with her clients occurs before they are released. At this time she evaluates the status of the client. The evaluation includes a mental status exam, a physical exam, a social history, and an assessment of their family and environmental setting. Then she develops a plan for their integration into their home environment. One of her greatest challenges is to plan for the first weeks after release, since these young people often believe that being released means they can do anything they wish. To solidify their commitment to the stated goals and outcomes, she includes them in the planning process.

Problem Solver

The goal of the **problem solver** is to make clients self-sufficient by helping them determine their strengths, find alternatives to their current situations, and learn to solve their own problems. One area of problem solving is clarifying the roles of the client, the family, the caregiver, and the case manager. Disagreements about services, the direction of case management, or the plan often lead to conflicts. The case manager is continually involved in problem solving; many problems arise unexpectedly, and time must be allotted each day for them.

Sonja McCreless has always admired her direct supervisor, Jim Fitzpatrick, because he is an expert problem solver. She remembers his work with a very difficult client, Sue D'Ambrosio, a scattered and unfocused mother who lost custody of her children. Jim Fitzpatrick was able to work with Sue to determine her own strengths, which included accomplishing very short-term tasks that would demonstrate that she could provide her children with a stable environment. So he modified the case management process into small tasks while clearly spelling out both of their responsibilities. Together they decided what the outcomes were to be and what behavior was acceptable, guided by the restrictions of the program. Components of the work were learning to manage money, making better decisions about her sexual behavior, and controlling her alcohol consumption. Sue responded positively to structured problem solving and eventually learned how to use the process without the guidance of her case manager.

Recordkeeper

Throughout service delivery, it is necessary to document assessment, planning, service provision, and evaluation. As a recordkeeper, the case manager maintains detailed information relating to all contracts and services. This is important for providing long-term care, communicating with other professionals and agencies, and monitoring and billing for services. Good documentation constitutes the linking element in the case management process. Many electronic systems of information management can help record, track, plan, monitor, and evaluate client progress, but the key is the quality of the data entered. This type of record-keeping is essential for program evaluation. Agencies and individual case

Want More Information?

• • • • • • • • • • • •

The Internet provides in-depth resources related to the study of case management. Search the following terms to read more about how roles of case management are defined.

- Case management roles
- Case management responsibilities
- Case management jobs

managers can determine if goals are being reached and if quality services are being provided. These data help determine if changes to organization, staffing, or service delivery need to be made.

Eli Brawley works with families and children having severe medical problems. He makes detailed computer records of his activities. Each client has a file that contains a record of every interaction and action taken for or with that client. This serves as the official record of service delivery, as well as the basis for accountability and quality assurance evaluations. The cases of these children are very complicated. Many of them have received human services since their birth. One of the challenges is to keep the record current. Clients work with a variety of professionals (beyond the case management system), and family members do not always inform the case manager about the other services they are receiving.

Wanted: Case Managers

There are an increasing number of agencies that are using case managers to provide services for their clients. The vignettes in this chapter illustrate case managers working with the aging, those with developmental disabilities, youth and adults in the criminal justice system, children and families in foster care, families on welfare, and in many other settings. Described below are two job announcements for case managers. Let's look at these to see what models are used and what roles the case managers are being asked to assume.

Job Announcement #1: Case Manager

Summary

The Innovative Drug Team is part of the local government's program to serve those in the criminal justice system who have a dual diagnosis of substance abuse and mental health issues. The case manager will work with a team to prepare treatment plans with multiple-problem clients who are facing adjudication.

Duties and Responsibilities

Provide case management and substance abuse/mental health counseling to clients in the Innovative Drug Team

Work with team to create treatment plans

Monitor and document implementation

Write reports

Conduct assessments and present recommendations to the court

Provide counseling to inpatient clients

Perform drug testing

Be involved in professional development

Minimum Job Requirements

Work experience with diverse client populations. At least two years working in community-based setting. At least one year in residential or inpatient treatment. Be familiar with case management; be familiar with substance abuse; be familiar with mental health issues. Bachelor's degree in psychology, social work, human services, public health, or related field, or certification as a drug and alcohol counselor.

This case manager position reflects services based upon the organizational model. The services are provided on an inpatient basis. Services are available while the client is in residence. Once treatment is determined, the client may receive services from other models of case management presented in this chapter. The responsibilities of this case manager include developing an assessment plan and working with a team to make recommendations to the court. This particular case manager also provides short-term counseling services. The roles for the Innovative Drug Team include coordinator, counselor, and planner. It is a short-term role before the client receives a court sentence (Monchick, R., Scheyett, A., & Pfeifer, J., 2006).

Job Announcement #2: Case Manager/Administrator

Summary

Oversees and administers all aspects of a specialized, community-focused case management program designed to serve the needs of a selected target population, to include supervision, training, and support of the operational and administrative duties of case managers. Provides case management services to individuals referred from domestic violence shelters. Compiles and prepares educational materials as necessary.

Duties and Responsibilities

Supervises and trains case managers and associated support staff

Administers the day-to-day activities of the program

Oversees quality of case management program

Carries a caseload of clients as a case manager

Reviews case management records to evaluate quality

Maintains confidentiality of records

Develops educational materials

Supervises staff

Performs miscellaneous job-related duties as assigned

Minimum Job Requirements

Bachelor's degree in human services, social work, psychology, nursing, or directly related behavioral health field; at least three years' experience directly related to the duties and responsibilities specified.

Knowledge, Skills, and Abilities Required

Administrative skills

Communication skills

Ability to understand and implement confidentiality policies

Ability to supervise case managers for organizational and time management skills

Knowledge of community resources and how to use them

Knowledge of case management skills and abilities

Ability to gather data, assess information, and write reports

Ability to supervise staff

Ability to prepare reports and other written documents

Ability to foster cooperation

The second job description is for an individual to work as an administrator of a case management service and as a provider of case management services for women and their children who are preparing to leave short-term shelters. This service illustrates the role-based case management model, since this case manager is the center point for these services. The case manager will work with the individual clients to determine needs. In this situation the case manager serves as the link to a variety of needed services, provides therapeutic care, and monitors the efficiency and quality of services. What roles will the case manager be performing? This case manager has multiple responsibilities to plan and problem solve, write and keep records, coordinate care, and broker additional services. Of course, the job is complicated by these additional administrative responsibilities.

 # Multicultural Perspectives

The importance of multicultural perspectives for all helping professionals cannot be overstated. For several reasons this is especially true for case managers. First, the basic values that guide the case management process support advocacy.

Social justice facilitates clients' self-determination and empowerment. Case managers are committed to knowing their clients from a holistic perspective, as well as understanding the individual issues and context that they bring to the case management process. This means it is essential to attend to the ethnic and cultural dimensions of who the client is, the experiences the client brings, and the direction the client wishes the process to take. Multiculturalism is a relatively new force in the helping professions. As cultural diversity continues to increase globally and in the United States, its importance expands.

Cultural competence encompasses three aspects of professional development: values, knowledge, and skills. This includes being aware of one's own assumptions, values, and biases; understanding the worldview of culturally diverse clients; and providing culturally sensitive interventions during the helping process (Sue & Sue, 2007).

For those who practice case management, there is an ethical commitment to delivering ethnic and multicultural services. For example, the certification requirements for the Human Services Board-Certified Practitioner (HS-BCP) include as standards: 1) competence in case management, as well as 2) knowledge and skills needed to work with cultural diverse populations (Center for Credentialing and Education, 2011). In addition, the National Association of Social Workers (NASW) provides a separate set of standards for multicultural competence (National Association of Social Workers, 2001). The standards follow on the next page.

Aside from the very practical considerations of how to deliver culturally sensitive case management services that are integrated into each chapter of this text, it is important to note three concepts that influence taking a multicultural perspective: multidimensionality of culture, identity, and worldviews; advocacy; and power. Understanding these concepts will help case managers to provide culturally competent services to clients and their families.

 # Multidimensionality of Identity

Embedded in the case management process is the commitment to help a client; a first step is to begin to understand the client and develop an empathetic response to the client's situation. Understanding the client from the contextual standpoint provides the case manager an opportunity to know the multiple dimensions of that client and begin to understand the unique identity the client

Standards for Cultural Competence in Social Work Practice

Standard 1. Ethics and Values Social workers shall function in accordance with the values, ethics, and standards of the profession, recognizing how personal and professional values may conflict with or accommodate the needs of diverse clients.

Standard 2. Self-Awareness Social workers shall seek to develop an understanding of their own personal, cultural values and beliefs as one way of appreciating the importance of multicultural identities in the lives of people.

Standard 3. Cross-Cultural Knowledge Social workers shall have and continue to develop specialized knowledge and understanding about the history, traditions, values, family systems, and artistic expressions of major client groups that they serve.

Standard 4. Cross-Cultural Skills Social workers shall use appropriate methodological approaches, skills, and techniques that reflect the workers' understanding of the role of culture in the helping process.

Standard 5. Service Delivery Social workers shall be knowledgeable about and skillful in the use of services available in the community and broader society and be able to make appropriate referrals for their diverse clients.

Standard 6. Empowerment and Advocacy Social workers shall be aware of the effect of social policies and programs on diverse client populations, advocating for and with clients whenever appropriate.

Standard 7. Diverse Workforce Social workers shall support and advocate for recruitment, admissions and hiring, and retention efforts in social work programs and agencies that ensure diversity within the profession.

Standard 8. Professional Education Social workers shall advocate for and participate in educational and training programs that help advance cultural competence within the profession.

Standard 9. Language Diversity Social workers shall seek to provide or advocate for the provision of information, referrals, and services in the language appropriate to the client, which may include use of interpreters.

Standard 10. Cross-Cultural Leadership Social workers shall be able to communicate information about diverse client groups to other professionals.

Prepared by the NASW National Committee on Racial and Ethnic Diversity. Adopted by the NASW Board of Directors June 23, 2001. Retrieved from http://www.socialworkers.org/practice/standards/NAswculturalstandards.pdf

has. This commitment presupposes that every helping encounter is a multicultural encounter; each client represents a perspective different from any other.

Let's look at the many dimensions that comprise such a complex identity. Sue and Sue (2007) presented three levels of dimensionality: universal level; group level (similarities and differences); and individual level (uniqueness). All of three of these characteristics form an individual's identity. Using the three levels to understand clients help case managers understand the similarities and differences between themselves and their clients.

Universal Level   At this level all of us belong to the same biological genus and species, *Homo sapiens*. What distinguishes us from other animals are our abilities to use symbols to communicate with each other and ourselves and to self-reflect. We share common experiences of birth and death and other biological similarities.

Group level   Each individual has beliefs, values, rules, and social practices (Goffman, 1959) presenting a cultural milieu or cultural context in which he or she lives. Gender, socioeconomic status, age, geographic location, ethnicity, disability/ability/race, sexual orientation, marital status, religious preference, and culture all shape similarities and differences at a group level. Based upon these characteristics, a sense of belongingness and membership in groups becomes available. Group membership also presents another set of similarities and differences, for one can be "outside" the group. Stereotyping, bias, and oppression are linked to treatment of those unlike ourselves.

Individual level   At a genetic basis, human beings resemble each other, and in fact, we are more alike than different from animals outside our species (Guffey, 2012). This common genetic, combined with what Sue and Sue (2007) called "nonshared experiences," represent the individual level of identity. No individual shares with another exactly the same experiences. We become most unique at this level of individual differences.

The import of identity within the case management process is profound. First, all human have much in common, especially at the level of genus and species. It is being human that allows us to enter a helping relationship, form bonds, communicate, and think critically about how to solve problems and challenges. Second, if we acknowledge the impact of the variables associated with group-level identity (e.g., gender, race, ethnicity) on how we think, behave, and value, helping others becomes more complicated. As case managers we find ourselves working with others very different from ourselves. In this context of delivering case management services, we may believe we understand helping. But, clients from differing perspectives may perceive helping and case management differently. Third, understanding clients and experiences for which we have little knowledge continues to be a critical component in the case management process. We describe in detail in Chapter 6 how to talk with clients to better understand their perspectives.

 # Power

Power is both implicit and explicit in human culture. According to sociologists, power and hierarchy result in the "classification of persons into groups based on shared socio-economic conditions . . . a relational set of inequalities with economic, social, political and ideological dimensions." (Barker, 2003, p. 436.) The power aspects of identity are formed within the variables of the group level described (Sue & Sue, 2007). Although a part of our personal ideas about power and hierarchy stems from individual experience, power is directly related to how society views groups, not individuals. Social stratification carries from one generation to the next. For example, if a culture values the contributions of its elderly members, this honoring may continue over time; likewise, if infant girls have less value than infant boys, the practice of female infanticide may be common. In addition, within all cultures, there exist power differentials, dominance, and oppression, but salient characteristics that inform the social stratification may vary. In Western culture, male-dominated power structures pervade, while in several societies in Asia, Africa, the Americas, and Oceania are more matriarchal in nature (Goettner-Abendroth, 2005). Finally, reliance on power dimensions to define who we are, where we belong, who others are, and where they belong manifests itself in both behaviors and beliefs. What this means for us as case managers is that our thoughts and values, as well as actions, translate into acts of power and control. Implicit in our professional responsibilities to help others lies a subtle statement about what we "have" and what they "have not." In all eleven chapters of the text, we provide you with ways to become more sensitive to the culture of others and open the way for empowerment.

Advocacy

Advocacy and the importance of advocacy is a theme that runs through this case management text. It is critical work from the multicultural perspective. Simply put, advocacy from the case management viewpoint, as stated in Chapter 1, means that case managers insure that everyone in need of assistance has the same opportunity to approach, apply for, and use case management services without regard to ethnicity, race, religion, sexual orientation, socioeconomic class, or disability. In addition, case managers enter into the service delivery system at various levels (individual, community, policy, legislative) to promote these client rights.

Advocacy related to cultural identity and power allows us to examine the multiple ways we promote cultural advocacy. These ideas, inspired by Sue and Sue (2007), frame individual, professional, organization, and societal actions. They are in keeping with the values of case managers, the goals of case management, and each of the models of case management presented earlier.

Individual level   At this level, the professional case manager understands his or her cultural attitudes, beliefs, emotions, and behaviors as they relate to others. Case managers continually reflect on the biases, prejudices, and misinformation they hold and promote. As professionals, they understand that this self-reflection is a lifelong process.

Professional level   According to Sue and Sue (2007), each profession is culturally bound. Note that the case management process introduced in Chapter 1 and its history, described in Chapter 2, reflects values, goals, and beliefs about human nature, client rights, and social responsibility. In addition, helping processes are based upon psychological principles, guided by ethical codes of conduct and standards of practice. It is important for the case manager to understand how the individual and professional levels of culture influence the case management process.

Now you will read about Ellen, who encountered a situation that challenged her advocacy at an individual level. She ended with more questions than answers, many of which were shaped by her own culture, agency culture, and client culture.

Ellen

It was windy. I was downtown. I asked the workers at the shelter, "What's your best suggestion on how I can get this medication to her?" And they said, "She usually hangs out here." So I took off on foot, looking for my client and poking into alleyways and looking in empty door wells and trying to find this young lady so I could give her some medication, and I didn't find her. I remember walking down the street with a bag of drugs in my hand and just stopping. It was one of those moments where I thought, "I didn't ask for this. This wasn't in my job description, so what do I do?" Because, my internal script is, "She is in my charge. She is one of my clients and my primary responsibility is her safety." And I'm stuck. "What do I do?" I remember that I stood in the middle of the sidewalk and started crying.

—Permission granted from Ellen Carruth, 2012, text from unpublished interview

---

Organizational level   As case managers our work is embedded within the context of the agency in which we work. In some sense, regardless of the agency, there exists a specific, defined, agency culture. It is represented in its policies, programs, structure of services, and treatment and practice. It is important for case managers to recognize the culture of the agency (see Chapter 10 for more information) and understand how the culture serves some clients and fails to serve others. Chapters 5, 6, 7, 8, and 9 provide information about developing a culturally sensitive approach to case management.

Societal level   Within each society there exists an ethnocentric, monocultural ideal defined by the majority culture. It is dictated by social, economic, political, and religious sectors and excludes the rights of those who do not meet its membership standards. Within this culture the majority or powerful social unit maintains its power to define reality, what is good and what is just. In the West, the social culture retains a Euro-American historical bias. The case manager has a responsibility to advocate for those within the society who are not part of the majority and whose needs are rarely reflected in the culture of human services and the broader social, economic, political, and religious cultures.

Read through this story told by Jessica. See if you can identify any organizational and societal cultural dimensions present in this case.

Jessica

About a year and a half ago, Clyde came into custody. His mother is an addict and his father is also an addict, in and out of jail. Clyde was born prematurely and had a lot of health issues and so he was in the hospital for a long time with heart problems and respiratory issues, and when he was really young he wasn't properly taken care of. For some reason, he was born with drugs in his system and they didn't take custody away from his mom. I don't know why. But he would come back to the hospital throughout his infancy for colds, sinus infections, and asthma. He was about three and a half when he came into custody, and we paired him up with this wonderful woman named Sarah [name changed]. She was of a different race and different ethnicity than he was. Although she is one of my favorite foster parents and she and Clyde immediately bonded, the agency does not want to set up cross-cultural foster parents. Clyde's mother died of an overdose. His maternal aunt came forward and said, "I would like to take care of Clyde. I would like to take custody of him." And that was really hard on Sarah, because Clyde called her Mommy. But she said, "Okay, if the biological family wants a part, we are going to let that happen." So we let the aunt take custody. When they let the family take custody they staffed Clyde down to a lower level and that worried me because that meant that the worker rather than a case manager would have to follow up with his appointments. He was in speech, occupational, and physical therapy and he also has a lot of ENT issues. So while he was with the aunt none of that was taken care of, of course, as I had feared. And then we found out— well, Sarah told me that Clyde was being dropped off at her house randomly. It was on Christmas Day where she called the aunt to ask to wish Clyde a merry Christmas and the aunt said, "Oh, he's not here. I dropped him off at an old babysitter's house."

> —Permission granted from Jessica Brothers-Brock, 2012, text from unpublished interview

 # Deepening Your Knowledge: Case Study

This chapter focuses on models of case management. The case study in this chapter reflects one aspect of human service: providing effective services for individuals involved in drug-related crimes. These efforts result from the Anti-Drug Abuse Act of 1988, an act that established the White House Office of National Drug Control Policy (ONDCP). The ONDCP's mission is "to establish policies, priorities, and objectives for the Nation's drug control program. The goals of the program are to reduce illicit drug use, manufacturing, and trafficking, drug-related crime and violence, and drug related health consequences" (U.S. Department of Justice, National Drug Court Institute, 2011, first paragraph). Goals for the drug court programs include decreasing "illicit drug use, manufacturing, and

trafficking, drug-related crime and violence, and drug related health consequences" (U.S. Department of Justice, National Drug Court Institute, 2011, first
paragraph).

Case management for clients involved with the drug court is the centerpiece
of intervention services. According to Monchick, Scheyett, & Pfeifer (2006) in a
report published by the Department of Justice, "Case management is essential to
carrying out the mandate of the key components of drug court. Without case
management, the integration of AOD treatment services and justice system processing would be limited" (p. ix). Read through the Key Components of Case
Management described in *Drug Court Case Management: Role, Function, and Utility*
(p. 1–3) with the models of case management described earlier. Then answer the
questions that follow.

Key Components of Case Management

**#1: Drug courts integrate alcohol and other drug (AOD) treatment services
with justice system processing.** *This component highlights the necessity of a multifaceted, collaborative "team" approach for integrating the delivery of services into the
administration of justice and enhancing the justice and treatment systems' joint mission
of promoting abstinence and law-abiding behavior. It underscores the need for collaborative goal setting and program monitoring through ongoing communication and continuous processing of timely and accurate information about each participant's
performance in the program. It is the case manager who coordinates the flow of drug
court information across and within the treatment and justice systems.*

**#2: Using a nonadversarial approach, prosecution and defense counsel promote
public safety while protecting participants' due process rights.** *The case manager
assists in keeping these traditionally adversarial parties focused on the primary purpose
of the program: the participant's movement toward fulfilling his or her recovery plan. As
an advocate for the participant's recovery, the case manager supports due process, ethical, and strengths-based treatment, and confidentiality while simultaneously promoting
individual accountability and community safety. It is in this sense that the case manager
helps bridge the traditional gap between the coercive traditions of justice, the protection
of the public, the privacy mandates of treatment, and respect for individual rights.*

**#3: Eligible participants are identified early and promptly placed in the drug
court program.** *The case manager helps ensure the coordination of this process by
"tracking" and facilitating the prompt sharing among the team of all relevant information arising from the initial referral, eligibility screening, and assessment process.*

**#4: Drug courts provide access to a continuum of alcohol, drug, and other re-
lated treatment and rehabilitation services.** *The case manager identifies and monitors each participant's unique needs for support and rehabilitation services, coordinates
participant access to these services, and ensures linkage and coordination among the
drug court service providers. The case manager works closely with the clinical treatment provider(s) and community supervision officers to provide ongoing assessment*

and communication of the participant's progress and to coordinate referrals to appropriate ancillary service providers.

#5: Abstinence is monitored by frequent alcohol and other drug testing. *The case manager ensures that drug test results, whether obtained by probation, treatment, law enforcement, or other court partners, are promptly and accurately recorded and disseminated to the drug court team.*

#6: A coordinated strategy governs drug court responses to participants' compliance. *As the central person responsible for coordinating team information flow, the case manager tracks and monitors the court's allocation of sanctions and incentives to each participant to help ensure that subsequent sanctions, incentives, and interventions are graduated, treatment-relevant, strengths-based, and otherwise consistent with the program's philosophy.*

#7: Ongoing judicial interaction with each participant is essential. *As the primary link between the treatment and justice systems, the case manager serves as the bearer of much participant information and, in this role, can give critical insight and input to the drug court judge.*

#8: Monitoring and evaluation measure the achievement of program goals and gauge effectiveness. *The case manager ensures that all relevant information is accurately, promptly, and systematically documented so that ongoing monitoring of the participants and evaluation of the program can occur.*

#9: Continuing interdisciplinary education promotes effective drug court planning, implementation, and operations. *Because the case manager deals daily with clinical and ancillary service providers as well as justice system partners, he or she is well situated to facilitate interdisciplinary education within the drug court team. In some jurisdictions, case managers integrate interdisciplinary training into drug court meetings by periodically enlisting an ancillary service provider or justice system professional to address the team and, if appropriate, participate in the staffing process.*

#10: Forging partnerships among drug courts, public agencies, and community-based organizations increases the availability of treatment services, enhances drug court effectiveness, and generates local support. *While all drug court team members contribute to the formation and maintenance of these critical partnerships, it is the case manager that sustains ongoing contact with key line staff of the partnering agencies and organizations. This consistent and direct contact with other community-based service delivery professionals puts the case manager in a position to learn about the policies, procedures, capacities, strengths, and limitations of existing support service organizations. With this knowledge base, the case manager is well positioned to identify service gaps and community needs, and offer strategies to facilitate collaboration between the court and the community.*

Case Study Questions

1. Read each of the Key Components. Explain how each component relates to one or more models described in Chapter 3.
2. List roles needed to perform each Key Component.

Chapter Summary

Models of case management contribute to an understanding of case management. They demonstrate that case management is a flexible process; illustrate the different goals that determine roles, responsibilities, and length of client involvement; and identify strengths and weaknesses. The three models are role-based, organization-based, and responsibility-based case management. Role-based case management focuses on case manager roles in service delivery. The configuration of services to meet multiple client needs defines organization-based case management. Family members, a supportive care network, volunteers, and the client are examples of case managers in the responsibility-based model.

Case managers perform many roles that form the basis of their professional responsibilities. Eight roles that are integral to case management are discussed in this chapter. A case manager may function primarily in one role and engage in other roles sporadically.

The multicultural perspective is critical for the case manager. Understanding dimensions of identity (of both oneself and the client), how power relates to the helping process, and the importance of advocacy help the case manager better relate to the client and provide services that meet client goals, not just agency and societal goals.

Chapter Review

Key Terms ....

Role-based case management   *Consultant*
Organization-based case management *Counselor*
Responsibility-based case management *Planner*
Advocate   *Problem solver*
Broker   *Recordkeeper*
Coordinator

Reviewing the Chapter ....

1. Why is it important to understand the different models of case management?
2. How are the three models of case management different?
3. Describe a situation in which the role-based model of case management is applicable.
4. How does the helping professional function in the organization-based model?
5. Describe the characteristics of the responsibility-based model.
6. Compare and contrast the roles of case managers and clients in each of the three case management models.
7. Describe situations in an agency setting in which you might function in the following roles: planner, advocate, broker, problem solver, and recordkeeper.

8. Explain how you will develop a multicultural perspective and a sense of advocacy for clients.

Questions for Discussion ....

1. Now that you have completed your reading about the models of case management, what can you conclude about the definition of case management?
2. Suppose you wanted to use case management to deliver services to elderly people in your community. How would you determine which model would apply?
3. If you were working as a case manager in children's protective services, how would you determine which case management roles would be needed?
4. Why do you think so many roles are involved in case management? Cite an example to illustrate your answer.
5. Why is taking a multicultural perspective important for a case manger?
6. How would you apply the concepts of identity, power, and advocacy when working with clients of a different gender, race, ethnicity, social class, or sexual orientation?

References ....

Aged and Community Services: Australia and Case Management Society of Australia, (2006). Case management and community care: A discussion paper. Retrieved from http://www.cmsa.org.au/FinalcaseManagementDiscussionPaper.pdf

American Medical Association (2012). Caregiver health: Caregiver self-assessment. Retrieved from http://www.ama-assn.org/ama/pub/physician-resources/public-health/promoting-healthy-lifestyles/geriatric-health/caregiver-health/family-caregiving-topics.page

Barker, C. (2003). *Cultural studies: Theory and practice.* London: Sage.

Goffman, I. (1959). *The presentation of self in everyday life.* New York: Anchor.

Goettner-Abendroth, H. (2005). Modern matriarchal studies: Definition, scope, and topicality. Austin, TX: Societies of Peace: 2nd World Congress. Retrieved from http://second-congress-matriarchal-studies.com/goettnerabendroth.html

Guffey, S. (2012). Personal communication on June 5, 2012. Knoxville, Tennessee.

Hemming, M., & Yellowlees, P. (1997). An evaluation study of clinical case management using clinical case management standards. *Journal of Mental Health, 6*(6), 589–598.

Monchick, R., Scheyett, A., & Pfeifer, J. (2006). *Drug court case management: Role, function, and utility.* U.S. Department of Justice, Monograph Series 7. Retrieved from http://www.ojp.usdoj.gov/BJA/pdf/Drug_Court_Case_Management.pdf

Mullahy, C. M. (2010). *The case manager's handbook* (4th ed.). Sudbury, MD: Jones and Bartlett.

National Association of Geriatric Care Managers (2012). About care management. Retrieved from http://www.caremanager.org/about/mission-vision/

National Association of Social Workers. (2001). NASW standards for cultural competence in social work practice. Retrieved from http://www.socialworkers.org/practice/standards/NAswculturalstandards.pdf

Power, S., & Hergarty, J. (2010). Facilitated peer support in breast cancer: A pre- and post-program evaluation of women's expectations and experiences of a facilitated

peer support program. *Cancer Nursing, 3*(2), 9–16. Retrieved from http://www.ncbi
.nlm.nih.gov/pubmed/20142745

Rappy, C. A., & Goscha, R. J. (2006). *The strengths model: Case management with people
with psychiatric disabilities.* New York: Oxford.

Safe Horizon. (2012). Home page. Retrieved from http://www.safehorizon.org/index
/get-help-8/for-domestic-violence-35.html

Sue, D. W., & Sue, D. (2007). *Counseling the culturally diverse* (4th ed.). Boston: Wiley.

U.S. Department of Justice, National Drug Court Institute (2011). *A national report on drug
courts and other problem-solving court programs in the United States.* Retrieved from
http://www.ndcrc.org/sites/default/files/pcp_report_final_and_official_july_2011.pdf

Wisconsin Department of Children and Families (2012). Wisconsin Works W2 and
Other Programs. Retrieved from http://dcf.wisconsin.gov/w2/default.htm

. . . .

CHAPTER FOUR

Ethical and Legal Perspectives

Thriving and Surviving as a Case Manager

Sara

One really challenging ethical issue for me, and maybe more so now that I am a therapist doing case management, is confidentiality. The line gets blurred when I am seeing the child in therapy but I'm also supposed to facilitate their child and family team meeting. I have two goals. I need to protect that kid, but the Department of Human Services says that it's her legal guardian. I've learned how to be very vague while still providing the department with information that it needs. And something that I've also learned to do is to try to empower the kids. This means letting them know we are going to prepare for their discharge meeting. I'll tell them, "I've sat through enough team meetings to know that they are going to ask what you have been doing. So what would you tell them?" I really try to call upon the child to speak about what they he or she has been doing. That way I'm not breaking any confidences. We can let the team know what is going on because that is important. Confidentiality within the treatment team is difficult too. We are all working with the same kids but others don't necessarily need to know everything I know. I guess that's more the case now than when I was strictly a case manager. As a therapist doing case management, I realize that there are things that people should know. But I don't always know that they are going to keep these things to themselves or that they are not going to confront the child about something that I say.

—Permission granted from Sara Bergeron, 2012, text from unpublished interview

*M*ost of our clients do not want us to share their information with anyone else. For kids, adults, especially adolescents, it is the first question they ask. "You won't tell my parents, will you?" There are some things we can keep confidential, but there is lots of information we need to share or report.

—Case manager and counselor, family services, Bronx, NY

*W*hat is most difficult for me is giving the client room to make mistakes or to refuse service. I see all of the ways a client's life could improve "if only." But if the client doesn't want to help herself, then we are powerless. Every once in a while I want to push the client harder than I should.

—Case manager, intensive case management, Los Angeles, CA

*T*here are some things that my clients need—services, help—that I am not supposed to do or provide. One day I was visiting a young boy and his mother. He was absent from school. His mother had been beating him before I arrived and she opened the door with a strap in her hand. I told his mom I needed to take him to school. Legally I am not supposed to transport clients.

—Case manager, school-based intervention services, New York, NY

*F*amilies have lots of influence with our clients, even our adult clients. We use groups to help expand client points of view beyond the family stance. Sometimes clients begin to see another perspective when they interact with their peers. They see family as not always having the final say.

—Case manager and counselor, family services and addiction treatment, Knoxville, TN

Effective human service delivery often requires a delicate balance of consideration for the client, the agency for which the case manager works, laws and regulations, court rulings, and professional codes of ethics. These conflicting interests can create crises that require the case manager to make difficult choices. The chapter-opening quotations reflect some of the tensions that case managers face. The case manager at family services describes the conflicts that arise related to confidentiality. Sometimes client confidentiality can be maintained. But there are other times that the case manager must break confidentiality, especially when suspicion of harm to self or others is involved.

The case manager from intensive case management speaks of a different type of dilemma, involving the mandate to grant clients autonomy whenever possible. Case managers may see clients choosing alternatives that are not in their best interests. Sometimes, it is very difficult to let clients make these choices. The case manager from school-based intervention services talks about the difficulties involved in providing services according to legal or ethical guidelines. Clients often violate rules just so they can maintain their stability and have their needs met. Sometimes professionals violate or think about violating policies. Professionals always have to assess what they think and how they will behave in these situations.

Working with families is often integrated into the case management process. Sometimes families can enrich and support the helping process; sometimes they can cause difficulties. The case manager from family services and addiction treatment describes one of the times when family pressure threatened implementation of the case management plan and how the agency uses client participation in groups to help clients expand beyond the family's influence.

In situations such as those just described above, as well as many others, finding the appropriate resolution is difficult and challenging. Case managers must constantly ask themselves certain questions: What is in the client's best interest? What is the right choice ethically? Am I operating within the guidelines of the agency that employs me? Case managers look to codes of ethics, the law, and agency policies and procedures to guide their practice. Mediating disagreements among family members and clients, working with potentially violent individuals, honoring client preferences, maintaining confidentiality, and upholding complicated rules and regulations are among the thorny issues with which case managers grapple. Several pressures increase the challenges case managers face. Clients are becoming more aware of their right to make decisions about their own care, and families are becoming more involved in their relatives' care. New technological, psychological, and economic interventions are continually being developed. Dealing with finite resources, case managers must control costs and allocate resources equitably.

For each section of the chapter, you should be able to accomplish the objectives listed.

FAMILY DISAGREEMENTS
- Describe what happens when family members disagree about the care of a family member.
- List guidelines to follow to encourage positive participation by families.

WORKING WITH POTENTIALLY VIOLENT CLIENTS
- Describe why violence is becoming more prevalent in modern society.
- Apply the steps in addressing issues of violence in the workplace to a specific case management situation.

CONFIDENTIALITY
- List reasons why the issue of confidentiality is so difficult.
- Define ways in which managed care and technology have affected confidentiality.
- Describe guidelines for discussing confidentiality with clients.
- Describe the ways that technology affects client confidentiality.

WORKING IN THE MANAGED CARE ENVIRONMENT
- Identify two difficult dilemmas case managers encounter working with managed care organizations.
- List three ways that case managers might respond to these situations.

DUTY TO WARN
- Define the duty to warn.
- Demonstrate how the case manager works with a team on issues involving the duty to warn.

AUTONOMY
- Describe the difficulties that arise with regard to granting client preferences.
- Explain how guidelines can help a case manager who faces issues of autonomy.
- Describe how case managers can support autonomous end-of-life decisions.

BREAKING THE RULES
- List the sources of rules and regulations.
- Determine when to advocate rather than break a rule.

Family Disagreements

Working with clients and their families is an important part of case management. Often, families can provide support to clients that will help them meet their goals. The American Counseling Association supports the idea that counselors should work with families of the clients they serve (American Counseling Association, 2005). End-of-life care is one of the most difficult ethical areas for the case manager. One such issue occurs when there are family disagreements over the care of an incompetent client. The following example illustrates the difficulties faced by the family and the professionals involved in the case.

The client, Mrs. X, was a woman fifty years of age with a history of severe depression. She had been cared for by her twenty-year-old daughter and placed in a hospital from time to time when the daughter could no longer care for her. The mother lost the ability to live independently, was unable to work, and had lost a majority of her social support. When the mother attempted suicide, she suffered brain damage. The daughter recommended that her mother be given comfort care or medical attention so that she was comfortable and out of pain. There was no written directive from the mother so the request went to the hospital ethics board. The board concurred with the request. The daughter was sure that this measure coincided with her mother's wishes.

Complications arose when the mother's son (the daughter's brother), who had not seen his mother in seven years, arrived in town. The son demanded that his mother receive aggressive medical intervention. He was furious with his sister and told the medical staff that he knew what was best for his mother. He told them that he knew she could be saved, and then could return home. In the next few treatment team meetings, he did not allow his sister to speak, and he refused to listen to the other professionals. He then left town abruptly.

The treatment team then again asked for a recommendation from the daughter. She again asked that her mother be taken off medical intervention and be made comfortable and pain free. The mother died two weeks later, and the son did not return.

This case is one example of the complex decision making that must occur when the wishes of the client, the family, and helping professionals are not in agreement. The situation becomes more complicated when the goals of the organization or the ethical standards of the professionals are also being challenged (Pecchioni & Sparks, 2007). Here is a suggested hierarchy of decision making to help professionals proceed when the patient is incompetent and the family cannot agree on the treatment (Lang & Quill, 2004). Other sources for how to make the decision might be state law or agency/organizational policy.

1. The patient's wishes are followed if there is an advance directive.
2. When patient does not have a directive, then the family makes the decision. If there is disagreement in the family, then the spouse is consulted first.
3. At times the medical staff must ask the court to intervene if there are irreconcilable family differences.
4. When the patient cannot speak for him- or herself and there is no one else to consult, the medical staff makes the decision.

Interdisciplinary teams need strategies for working with difficult families, especially when the client is incompetent. Here are some guidelines suggested by Sparks (n.d.):

- Ask all family members to attend a meeting with the case management team.
- At the meeting, provide a summary of the client's status. Use language all members of the family can understand.
- Identify the issue(s).
- Outline the decisions that need to be made and the options available.
- Be clear about the challenge or issue. Ask members of the family to talk over the issue and choose an option. Work with the family to achieve consensus.

- The family can appoint a member to attend treatment team meetings.
- If the family cannot reach consensus, then ask only one member of the family to choose an option. The spouse is the first person to ask.
- Continue to re-evaluate the decision. Situations are likely to be dynamic.

There are other situations in which case managers are working with families, especially when the client's skills are limited, but the client is not incompetent. Examples are clients who are young, have limited mental capacity, are homeless, and/or are mentally ill. In working with each population, the case manager must maintain a delicate balance between advocating for the client, engaging the support of the family, and assessing what is in the client's interest. Many times families have invested years of effort and care in assisting the client. They may feel they deserve to have a primary role in the decision making.

The following case illustrates some of the difficulties involved.

Reid has been homeless for fifteen years. Now thirty-five years of age, he continues to struggle to meet basic needs each day. In the past, his parents and younger sister begged him to live at home. He used to come home for a day or two, create chaos, and then leave. Today Reid and his case manager have broached with the family the possibility of his living with them. The family's reaction was one of shock and anger. They refused even to discuss the issue.

The case manager can use some guiding principles to think through situations in which families and clients do not agree on treatment plans. The case manager has several issues to consider when working toward a decision or a choice for the client. The mission and goals of the organization guide the case manager; ethical guidelines, ethical codes, state statutes or law also provide a foundation for decisions. It is also important to allow each member of the family to express his or her opinion. In Reid's situation, the case manager works to establish clear communication in which the family can outline current conflicts and establish a plan supported by all parties. More about how to work with families as a team is presented in Chapter 9, focusing on service coordination.

Multicultural issues further complicate work with families. Attitudes about health, illness, treatment, lifespan, and death and dying are basic cultural orientations that the case manager must consider. Situations become even more complex when one considers the relationship between the culture of the client and the culture in which the service occurs. Within the Western model, the family becomes a focal point of decision making when the client is not competent to make judgments; when the client is competent, the individual retains his or her rights. The Eastern model, represented by a triad of client, professional, and family, provides an alternative perspective for family involvement (Aslam, Aftab, & Janjua, 2005). For example, in Pakistan, the Muslim family provides structure for social and economic interaction. Decision making related to medical care lies

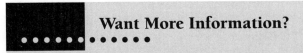

Want More Information?

Understanding the issues surrounding end of life care is important. Search the Internet using the terms presented below to better understand these issues:

- End of life care
- Dying with dignity
- Dilemmas in end of life
- Five wishes
- Hospice
- Medical power of attorney
- Living will

with the family, not with the individual. In addition, the role of the doctor is paramount in making the decisions. Together, the doctor and the family control information, decisions about treatment, and finally end-of-life care (Aslam, Aftab, & Janjua). This is but one example of how culture influences ethical decision making.

Working with Potentially Violent Clients

Cases of violence are increasingly common in the human service delivery system, especially since treatment occurs within community settings including neighborhoods and homes. **Violence**, in this context, can mean actual physical assault, verbal assault, or the destruction of property. Oftentimes, helping professionals sense the threat of violence, even though they are not actually physically assaulted. Individuals who have the potential for violence are often human service clients. Case managers are often the ones who work with difficult clients and those with complicated cases. And case managers often meet clients in the client's environment. Here is an example of one case manager's experience with a violent client:

Susan is a case manager for the persistently mentally ill. She has clients on a long-term basis. At a minimum she visits her clients at least once a week in homes or places of work. Last week Susan experienced a scare that she believes she will never forget. She arrived at George's home at three p.m. on Thursday afternoon. She had tried to find George at work but was told that he did not show up. George had a habit of taking his meds regularly for several months and then discontinuing the use of them. Susan suspected that this was one of those times. Early in her relationship with George, he was violent and had to be restrained. For the first two years she did not visit with him alone; another member of the case management team accompanied her on the visits. But in the past three years, Susan had seen no signs of violence in George.

Susan approached the door, knocked, and heard a faint sound inside. George quickly opened the door and beckoned her in. He had a gun; he grabbed her. She was surprised and very frightened.

Case managers have a responsibility—to themselves and everyone else involved—to look for violent tendencies in their clients. Because it is the case manager who gathers much of the information about the client and monitors client treatment and progress, he or she is in a position to know of any history of violence or warning signs during treatment. The case manager then warns the interdisciplinary team of the possibility. If any other members of the team report danger signals, the case manager passes that information to all involved. This may also require changes in the treatment plan or the addition of other professionals to the team.

Our society is more violent today than ever before. Stresses are exacerbated by personal pressures, including divorce, illness, feelings of hopelessness, and lack of skills for coping with difficulties. In addition, the policies of deinstitutionalization and treatment within the least restrictive environment have released into the community many individuals who do not have resources to care for themselves. The problem is compounded by a lack of social support for many individuals, changing values, easy access to weapons, and a tendency in our culture to condone violence.

Some human service settings are more susceptible to violence than others. The most vulnerable seem to be psychiatric settings, nursing homes, emergency departments, and outpatient settings. In these settings, there is often insufficient support for staff who work with potentially violent clients.

Many agencies do not recognize the potential for violence in their clients. Service providers are not educated about this potential, and protocols for working with violent clients are seldom established. Many agencies just expect their clients to be violent. It is "part of their culture," and staff members are simply expected to deal with it. Some agencies reward staff for keeping things quiet, putting the emphasis on controlling clients rather than working with them in a therapeutic way. In such an environment, professionals may hesitate to report any signs of violence, fearing that they will be accused of not doing their jobs. Administrative factors like understaffing and insufficient supervision may also contribute to poor management of violent clients: (OSHA, 2004).

The case manager can support the work of the team by acknowledging the potential for violence and preparing for manifestations of it. What exactly can the case manager do to address the issue of violence? The following steps can be used as guidelines.

1. Members of the team need to identify potentially violent clients and vulnerable situations. To do this, they need to understand and be able to conduct a violence risk assessment on each client.
2. Members of the team need to understand several triggers of violence, such as lack of control.

3. Members of the team should receive training to de escalate a potentially violent situation.
4. The team needs to know how to write a report on violence in the workplace.
5. The agency needs policies and procedures that protect clients, employees, and family members from violent acts.

Voices from the Field: Research and Practice: Steps to Reduce Violence in the Workplace

Recognizing that the inherent potential danger of violence exists when working with some clients in the social services sector, the Department of Labor, the Occupational Safety and Health Administration (OSHA) outlined *Guidelines for Preventing Workplace Violence for Health Care and Social Service Workers* (2004). Two areas of particular relevance for case managers are the risk factors related to the work and suggested administrative policies and procedures that reduce that risk.

The Risk Factors

Health care and social service workers face an increased risk of work-related assaults stemming from several factors. These include

- The prevalence of handguns and other weapons among patients, their families or friends
- The increasing use of hospitals by police and the criminal justice system for criminal holds and the care of acutely disturbed, violent individuals
- The increasing number of acute and chronic mentally ill patients being released from hospitals without follow-up care (these patients have the right to refuse medicine and can no longer be hospitalized involuntarily unless they pose an immediate threat to themselves or others)
- The availability of drugs or money at hospitals, clinics, and pharmacies, making them likely robbery targets
- Factors such as the unrestricted movement of the public in clinics and hospitals, and long waits in emergency or clinic areas that lead to client frustration over an inability to obtain needed services promptly
- The increasing presence of gang members, drug or alcohol abusers, trauma patients or distraught family members
- Low staffing levels during times of increased activity such as mealtimes, visiting times and when staff are transporting patients
- Isolated work with clients during examinations or treatment
- Solo work, often in remote locations with no backup or way to get assistance, such as communication devices or alarm systems (this is particularly true in high-crime settings)

- Lack of staff training in recognizing and managing escalating hostile and assaultive behavior
- Poorly lit parking areas

Administrative and Work Practice Controls to Minimize Risk

Administrative and work practice controls affect the way staff perform jobs or tasks. Changes in work practices and administrative procedures can help prevent violent incidents. Some options for employers are the following

- State clearly to patients, clients, and employees that violence is not permitted or tolerated.
- Establish liaison with local police and state prosecutors. Report all incidents of violence. Give police physical layouts of facilities to expedite investigations.
- Require employees to report all assaults or threats to a supervisor or manager (for example, through a confidential interview). Keep log books and reports of such incidents to help determine any necessary actions to prevent recurrences.
- Advise employees of company procedures for requesting police assistance or filing charges when assaulted and help them do so, if necessary.
- Provide management support during emergencies. Respond promptly to all complaints.
- Set up a trained response team to respond to emergencies.
- Use properly trained security officers to deal with aggressive behavior. Follow written security procedures.
- Ensure that adequate and properly trained staff are available to restrain patients or clients, if necessary.
- Provide sensitive and timely information to people waiting in line or in waiting rooms. Adopt measures to decrease waiting time.
- Ensure that adequate and qualified staff are available at all times. The times of greatest risk are during patient transfers, emergency responses, and mealtimes and at night. Areas with the greatest risk include admission units and crisis or acute care units.
- Institute a sign-in procedure with passes for visitors, especially in a newborn nursery or pediatric department. Enforce visitor hours and procedures.
- Establish a list of "restricted visitors" for patients with a history of violence or gang activity. Make copies available at security checkpoints, nurses' stations, and visitor sign-in areas.
- Review and revise visitor check systems, when necessary. Limit information given to outsiders about hospitalized victims of violence.
- Supervise the movement of psychiatric clients and patients throughout the facility.
- Control access to facilities other than waiting rooms, particularly drug storage or pharmacy areas.
- Prohibit employees from working alone in emergency areas or walk-in clinics, particularly at night or when assistance is unavailable. Do not allow employees to enter seclusion rooms alone.
- Establish policies and procedures for secured areas and emergency evacuations.
- Determine the behavioral history of new and transferred patients to learn about any past violent or assaultive behaviors.

(continues)

(continued)

- Establish a system—such as chart tags, log books, or verbal census reports—to identify patients and clients with assaultive behavior problems. Keep in mind patient confidentiality and worker safety issues. Update as needed.
- Treat and interview aggressive or agitated clients in relatively open areas that still maintain privacy and confidentiality (such as rooms with removable partitions).
- Use case management conferences with coworkers and supervisors to discuss ways to effectively treat potentially violent patients.
- Prepare contingency plans to treat clients who are "acting out" or making verbal or physical attacks or threats. Consider using certified employee assistance professionals or in-house social service or occupational health service staff to help diffuse patient or client anger.
- Transfer assaultive clients to acute care units, criminal units, or other more restrictive settings.
- Ensure that nurses and physicians are not alone when performing intimate physical examinations of patients.
- Discourage employees from wearing necklaces or chains to help prevent possible strangulation in confrontational situations. Urge community workers to carry only required identification and money.
- Survey the facility periodically to remove tools or possessions left by visitors or maintenance staff that could be used inappropriately by patients.
- Provide staff with identification badges, preferably without last names, to readily verify employment.
- Discourage employees from carrying keys, pens, or other items that could be used as weapons.
- Provide staff members with security escorts to parking areas in evening or late hours. Ensure that parking areas are highly visible, well lit, and safely accessible to the building.
- Use the "buddy system," especially when personal safety may be threatened. Encourage home health care providers, social service workers, and others to avoid threatening situations.
- Advise staff to exercise extra care in elevators, stairwells and unfamiliar residences; leave the premises immediately if there is a hazardous situation; or request police escort if needed.
- Develop policies and procedures covering home health care providers, such as contracts on how visits will be conducted, the presence of others in the home during the visits and the refusal to provide services in a clearly hazardous situation.
- Establish a daily work plan for field staff to keep a designated contact person informed about their whereabouts throughout the workday. Have the contact person follow up if an employee does not report in as expected.

Confidentiality

In the helping professions, the obligation of **confidentiality** is fundamental to developing a relationship between the helper and the client. When the client is sure that information disclosed during the helping process will be kept in confidence, he or she feels freer to share concerns and issues. The fuller the disclosure, the greater the opportunity for the case manager to gather valuable information about the client and his or her situation. This facilitates assessment and treatment planning. Trust between the helper and the client is a prerequisite to the success of their relationship. Case managers are in a unique position with respect to confidentiality, for they work with the family and friends of the client as well as with professional colleagues.

One of the first points of discussion between case manager and client must be confidentiality and its meaning within the case management process. Five standards for confidentiality must be stated.

1. The case manager keeps client information confidential except when the client intends to harm self or others.
2. When the client needs to share information with colleagues, the case manager will inform the client of three factors: (1) who will be told; (2) the reason for the disclosure; and (3) what information will be disclosed.
3. If the client consents, some information will be disclosed to family and friends.
4. The case manager must testify in court regarding information about the client, unless the case manager is protected by the state. This legal privilege of communication is usually reserved for patient-therapist communication.
5. The case manager must ask for the client's permission to release information.

There are exceptions to a case manager's maintaining confidentiality, such as when the case manager is entering data into a electronic system, talking with colleagues either during a staffing or consultation, or working under supervision. If case managers intend to share client information in any of these settings or in the course of performing job responsibilities, they should let clients know.

Even with these guidelines articulated, dilemmas concerning confidentiality often arise. Consider the following situations.

An eighteen-year-old has just been diagnosed as pregnant. There is no legal obligation to inform the parents of the young woman or the father of the child, but the care coordinator always encourages clients to disclose this information. This young woman refuses. The discussion is closed and the matter remains confidential, even though the young woman's mother requests the information.

An elderly man is furious with his social service coordinator because she told his daughter that he was dying of pancreatic cancer. The service provider knew that the daughter's husband had just asked for a divorce. The daughter was devastated because she had been counting on the father's support.

A counselor has been asked by his minister for information about a client who is a member of their congregation.

Another consideration for the case manager is collecting information under the guidelines of the Health Insurance Portability and Accountability Act of

1996 (HIPPA) (Department of Health and Human Services, 2003; Remley & Herlihy, 2010). The purpose of this legislation is to protect the confidentiality of client health records by allowing patients access to their medical records, requiring health professionals to explain how they will use patient information on health records, limiting the personal information a health professional may share with others, and allowing patients to request confidential communications. The act went into effect in April 2003. Compliance with this act may be integrated into the standard practice of case management. For example, if the case manager is seeing a client for the first time, the case manager may explain the HIPPA regulations to the client. Each client must sign a statement indicating he or she understands how HIPPA regulations are followed in that agency or organization.

The Family Educational Rights and Privacy Act (FERPA) protects the educational records of children (U.S. Department of Education, n.d.). This act allows parents and children to view their educational records, to request copies, and to amend these records. It also requires the school to gain permission from parents to release the records. Schools may provide records without parental consent in connection with specific circumstances such as school audits, accreditation, health and safety issues, and, if allowed by the state, use by the juvenile justice system.

Having sensitive information about clients often causes dilemmas for case managers. To avoid problems, some decide to limit what information they seek to gather about a client, because the less information case managers have, the fewer confidentiality problems they will encounter. Certainly, case managers must carefully choose what information to seek, but strict limits on information gathering are not always recommended. A complete history of the client enables the case manager to develop a treatment plan that will meet the needs of that client. *Relevant* information must be gathered, which helps the case manager understand how to work with family members and friends. Since the case manager is also coordinating care and monitoring progress, abundant information helps him or her give better guidance to other professionals on the team.

The case manager should address certain matters before collecting any information, to anticipate any confidentiality problems.

1. What does the case manager need to know to do the job, and why?
2. What should become part of the permanent record?
3. What understanding should exist between case manager and client about why the information is being sought, how it will be used, and the client's right to refuse to answer?
4. Under what circumstances should information be shared with third parties?

At the beginning of the process, the case manager determines the information needed to determine eligibility, conducts a comprehensive assessment, sets priorities, and develops a treatment plan. During the course of any case, information will emerge that the case manager needs to know but would not have thought to ask about at the outset. A good outline of the information to be gathered keeps the case manager focused on appropriate and relevant areas to probe. The client often asks the case manager why certain information is necessary; at other times, he or she discloses much more than is needed for the process to proceed.

Sometimes the case manager needs information about the client from other agencies or institutions. It is important for the case manager to obtain written consent from the client to receive this information. The client may also have to provide a written consent to release the information to the agency or institution holding the information. To obtain client consent, the case manager discusses the need for the information with the client and helps the client with the appropriate paperwork. At this time, it is appropriate for the case manager to talk with the client about security of records, including those acquired from another agency.

Another dilemma is how much information to put in the permanent record. The case manager must record all information that documents the work with the client and describes his or her history. Not everything the client reveals needs to be recorded, however, especially if it is not relevant to the presenting problem. Unfortunately, the confidentiality of written information is not always secure, regardless of whether the record is on paper or computerized.

Once the case manager has decided what information to gather and what to record, he or she explains why each piece of information is important. Then he or she decides what parts of the information will be shared with family, friends, and other professionals and discusses this with the client. However, professionals and family members are bound to ask questions beyond what the client and case manager have agreed on. The client and the case manager negotiate about each such piece of information, and the client must give permission for disclosure. He or she also has a right to refuse to answer any question; if the information is necessary for determining eligibility, the case manager explains this. He or she can ask the client whether there is another way to obtain the necessary information.

Even following these guidelines, the case manager is bound to face issues that warrant further consideration. Three examples are the short cases presented earlier. Let's consider how these situations might be resolved. In the first case, the care coordinator gathers confidential information about the 18-year-old's pregnancy. The information remains confidential unless the agency has a policy mandating disclosure to parents or to the father of the child. If this is so, the coordinator should have informed the young woman at the time of intake. The care coordinator's obligation to inform supersedes the confidentiality guarantee. Such a policy is less likely to apply here since the young woman is of legal age. The situation becomes complicated if the young woman's mental competence is in question or if her health or that of the fetus is threatened in any way. If there is no legal or policy mandate, the coordinator must not inform the mother of the pregnant young woman, even though she is the potential grandmother.

In the second situation, the social service coordinator is torn between her allegiance to the dying father and her responsibility to his daughter. The difficulty here hinges on the coordinator's definition of who the client is. She has chosen to behave counter to the wishes of the father by breaking confidentiality with regard to his physical condition. Before doing so, she should ask herself the following questions: Is the father competent to request that his daughter not be told? Did he give the coordinator other information indicating that the daughter should be told, in spite of his reaction after the fact? Does she believe that it is in the father's best interest for the daughter to have this information? Does the co-ordinator see the daughter as the primary client? If so, why? The coordinator should not violate the father's request for confidentiality unless the answers to these questions provide sufficient justification.

In the third case, the counselor is asked for information by a professional who is not involved with the client's case, at least within the established service delivery system. The counselor is under no obligation to give the information unless there is an established need to know and the counselor gains the client's consent to share the information. It would be a different matter if the counselor had reason to believe that the client might harm himself or others. There would then arise a *duty to warn,* changing the counselor's obligation from confidentiality to a duty to share information. Before discussing the duty to warn in more detail, let's look at the client confidentiality issues that have developed in the last decade relative to technology.

Confidentiality and Technology

As case managers increase their use of technology to communicate, store, and retrieve information and care for clients, considerable concern is developing about client confidentiality. Remley (2011) outlined the following issues related to technology:

- Electronic storage of records
- Rendering of counseling or supervision services via electronic correspondence (i.e., telephone, Skype, two-way video, or e-mail)
- Helping professionals communicating with clients or other professionals via e-mail, text, or other electronic means
- Client use of social networking
- Storage and transmission of recordings of client interviews
- Electronic signatures on documents
- Technology related to federal and state statutes and regulations; licensure and certification board rules; and agency operating procedures (p. 2).

All but one of these aspects of delivering social services involve confidentiality. For example, many case managers today are using e-mail to communicate with other professional colleagues, to contact case management team members, and to explore referral services. By using the management databases established by human service agencies and organizations at the national, state, and local levels, case managers are able to enter and retrieve information from online databases. This means that while working online, case managers have access to comprehensive, up-to-date information about their clients. Case managers are also using organizational and community computer networks to facilitate the sharing of information. As use of the technology expands, as indicated by Remley (2011), there are three areas of security and confidentiality to address: security of the data; use of e-mail and social media; and securing the communication site.

The computer has already become an integral part of the case manager's day-to-day work. As the use of technology continues to expand, case managers use guidelines to address the concern for and protection of client confidentiality. Several measures address security issues. Organizations can use systems with special encryption programs that scramble data and protect the data en route. They can install firewall systems for data stored on the Internet to prevent casual users and hackers from retrieving this information. To ensure the legitimacy of remote users, they can use special devices to authenticate the individuals who are using the system. With the advent of off-site storage (the cloud), it is easier to share data with other agencies and other professionals. Complicating this ease of access is the "guarantee" of privacy and confidentiality.

There are two other considerations about confidentiality for case managers to consider. The first focuses on the use of e-mail and social media (e.g., Facebook, Twitter) in service delivery. Because e-mail and social media have become such popular media of communication, many patients are beginning to communicate with their case managers online and through social media sites. Sometimes clients are following up on a referral or on their treatment, just checking in, or requesting their own records. Indeed, in many instances, access to e-mail and social media has increased the efficiency of communication between the case manager and the client. For clients who live in isolated, remote rural areas, live alone, or are homebound, e-mail and social media allow a new type of access to services. However, if a case manager decides to use e-mail or social media as a method of communication, some decisions must be made that address issues of confidentiality.

Many agencies have developed or are developing a policy related to agency use of e-mail or social media. The issue is complicated because of the benefits and drawbacks of using these tools as a means of advocacy, outreach, and communication. An agency's social media agency policy may include features such as the ones developed by the Mid-Ohio Regional Planning Commission (2012). Its policy illustrates many of the issues to be considered.

Specifically, case managers need to determine at the outset what types of case management services will be provided via e-mail or social media. They need to obtain informed consent from the patient to communicate via e-mail or social media, including acknowledgement by the patient that at times communications, especially e-mail, may be sent to an incorrect address. Case managers can state that once information leaves the case manager's office, its security cannot be guaranteed. Professionals might want to inform the client when a message is received and when the client might expect a reply. Finally, it is important to be professional in correspondence.

A second consideration also relates to technology, the definition of a "work site," and its influence on patient confidentiality. Technology enables case managers to work at home or at a remote site. Using fax machines, telephone answering machines, smartphones and tablets, and computers with Internet access, case managers can write reports, make referrals, monitor client progress, and talk with clients while offsite. Using the home or a nonoffice site requires sensitivity to client confidentiality, and being especially

MORPC Social Media Guidelines

MORPC does not seek to control, through this policy or otherwise, the purely personal online content posted by MORPC staff members, when that content is posted during non-working time, is posted using the staff member's own equipment, is unrelated to and does not identify MORPC or . . . staff member's position with MORPC, and is not . . . disruptive to MORPC's mission. However, the following rules apply to any online post by a MORPC staff member that (a) occurs during working hours; (b) is posted using MORPC equipment; or (c) identifies MORPC, links to information about MORPC or identifies the staff member's position. . . .

RULES

1. Online activity during working hours or using MORPC equipment must be primarily business related. . . .
2. Respect audience.
3. Don't use religious, racial, or ethnic slurs, personal insults, or obscenity. . . .
4. Show proper consideration for others' privacy and for topics . . . considered objectionable or inflammatory. . . .
5. Do not participate in any political activity or . . . communication during work hours . . . .
6. MORPC staff members are personally responsible for the digital content they publish . . . .
7. Protect your privacy . . . determine what personal information you are comfortable sharing. . . .
8. Identify yourself. Anonymous postings are rarely beneficial to anyone . . . .
   a. If you are speaking for yourself. . . . "The views in this post are my own and don't necessarily represent my employer's positions, strategies or opinions."
   b. If you identify yourself as a MORPC staff member, ensure your profile and related content are consistent with how you wish to present yourself . . . .
   c. The lines between public and private, personal and professional are blurred in online and mobile interactions. . . .
9. Respect copyright, fair use and public records laws. . . .
10. Protect confidential and proprietary information. Social media blurs many of the traditional boundaries between internal and external communications.
11. Prior to MOPRC events, acquire
    a. Written agreements from speakers that all event materials be available to the public. . . .
12. Don't pick fights. Be the first to correct your own mistakes. . . .
13. Consider the purpose of the social media use. Online communications must not negatively impact achievement of . . . assigned tasks. . . .
14. The availability of personal information on social media sites (both for the professional and the client) provide public self-disclosure unprecedented. . . .

SOURCE: The Mid-Ohio Regional Planning Commission.

watchful for violations of confidentiality that are inadvertent and unintentional (Laidlow, 2009).

Whether at home or at a remote location, what makes confidentiality more difficult is that the work space is usually shared. For example, when a case manager uses computers or Internet accounts that are also used by other family members or professionals, there is a risk to client confidentiality. Handwritten notes, reports, and files may also be left in this shared space. Care must be taken by the case manager to store these papers, either by securing a file or drawer or by taking the records to a secure place. Several safeguards for a shared work space include compiling a list of all technology you use, evaluating the risks of each to patient confidentiality, making changes in the procedures that contribute to the risk, and discussing with those with whom space is shared. Specific, yet simple, suggestions from Laidlow (2009) help those who work from home.

- Password-protect your computer and use a screensaver.
- Make sure your network is secured.
- Lock your computer away when you leave it at home for any length of time.
- Don't work on highly sensitive pieces of work in public places.
- Pack away all files out of sight before anyone comes to stay or you leave home for an extended period. Have a set of lockable drawers, so this isn't a big deal.
- Be extremely careful when traveling with your computer and try not to take confidential hard-copy documents on the road unless you really need them. (Laidlow, 2009).

Confidentiality and Interpreting

Another concern about confidentiality that may arise for case managers is working with a client(s) and an interpreter. Interpreters are individuals who possess specific knowledge and skills to enable cross-cultural communication by converting one language to another. Often referred to as foreign language interpreters or sign language or visual language interpreters, they function as a "language conduit," allowing persons who do not speak or understand spoken English to participate in court proceedings, medical matters, education, and other activities.

What can a case manager expect from an interpreter? Certification is one expectation. A certified interpreter is one who has demonstrated through evaluation, testing, professional development, and other means that he or she possesses the necessary knowledge and skills to function in this capacity (Davis, 2005). Many interpreters become team members along with the other professionals who may be working with a client. The case manager should also expect an interpreter to speak clearly in a manner that reflects the tone and emotions of the speaker, to remain impartial, and to have no unnecessary discussions with any participants. All participants, both helping professionals and clients, should be able to trust the interpreter to be accurate and to acknowledge when he or she is not able or qualified to interpret something (Davis, 2005).

Finally, all interpreters are guided by codes of ethics that provide guidance to professional behavior that helps ensure quality of service. In Virginia, for example, the Code of Professional Responsibility for Interpreters Serving Virginia Courts (2004) reads as follows:

> CANON 5: Interpreters shall protect the confidentiality of all privileged and other confidential information.

This brief but significant statement requires interpreters in court settings to uphold the attorney-client privilege and not repeat any confidential information obtained in the course of employment. Should the interpreter become aware of information that suggests harm to an individual or relates to a crime being committed, then that information must be disclosed to an appropriate authority in the judiciary. With this exception, violating confidentiality in this setting or any other setting is unprofessional, unethical, and in some cases, illegal (Davis, 2005). It may also be grounds for revoking professional interpreting certification.

The cultural dimensions of delivering case management services to clients who do not speak English are considerable and represent issues beyond confidentiality. As the number of non-native speakers continues to increase (Association of Latino Administrators and Superintendents, 2011), agencies delivering case management services need interpreters to participate in the helping process. The National Standards on Culturally and Linguistically Appropriate Services (CLAS) provide guidelines for delivering services to non-native speakers (Case Management Monthly, 2010). The guidelines support agency policy and services in five areas: interpreter services; written material, signage and way finding; notice of language access services to clients, and community involvement. Because language presents large cultural barriers to providing care to individuals from a non-Western or non-U.S. culture, a new definition of interpreters is emerging. These interpreters are Interpreter Cultural Mediators (ICM) responsible for providing liaison work between clients, patients, and health care and social service institutions. Community House Calls in Seattle, Washington, is an agency providing this service. The specific goals of this program are "to create a common fund of knowledge between medical and ethnic cultures, decrease language barriers to care, change institutional practices that promote satisfaction for non-English-speaking families, and improve cross cultural health education for providers, and enhance efficient utilization of resources by high risk/high needs families" (Community House Calls, 1998).

Duty to Warn

A case manager works in a nursing home for the elderly. One client, Mrs. Eddy, constantly expresses anger toward her husband. He lives at home and comes to visit her twice a day. He appears devoted to her and is her only contact with the world outside the nursing home. Today Mrs. Eddy says that she has a gun and plans to shoot her husband. The case manager is unsure whether to believe her. All the rooms are cleaned three times a week, and it is doubtful that a gun would remain unnoticed. Also, Mrs. Eddy is bedridden and has severe arthritis in her hands and fingers.

The **duty to warn** arises when a helping professional must violate the confidentiality that has been promised to a client in order to warn others that the client is "a threat to self or to others" (Remley & Herlihy, 2010). The *Tarasoff* court decisions in 1974 and 1976 established that mental health practitioners have a duty to warn not only anyone who works with the client and the police, but also the intended victim. The *Tarasoff* rulings govern the law of most states. It is a difficult judgment for a case manager to break confidentiality and invoke the decision to warn because confidentiality is such a fundamental responsibility. Violation of confidentiality is otherwise considered unacceptable practice in most instances.

The law and professional codes of ethics provide guidance in this matter. Further court rulings have clarified the duty to warn foreseeable victims, as first prescribed in the *Tarasoff* decision. However, state laws on professional confidentiality may conflict with the duty to warn. Also, professional codes of ethics do not always provide clear guidelines in this matter. Since case managers do not have their own professional code, they use the code of whatever discipline they hold credentials for. For instance, the *Code of Professional Ethics for Rehabilitation Counselors* states,

> Rehabilitation counselors will take reasonable personal action, or inform responsible authorities, or inform those persons at risk, when the conditions or actions of clients indicate that there is clear and imminent danger to clients or others after advising clients that this must be done. Consultation with other professionals may be used where appropriate. The assumption of responsibility for clients must be taken only after careful deliberation, and clients must be involved in the resumption of responsibility as quickly as possible. (Woodside & McClam, 2011)

Members of the helping professions are encouraged to use codes as initial guidelines, but ethical codes are not written to cover every possible situation. They are established as principles, which professionals should use to guide their behavior. Such principles sometimes conflict with one another and often do not address new issues that emerge.

The duty to warn is especially difficult for the case manager. He or she must consider the legal implications in case someone is hurt. In addition, there are complications in dealing with this issue in the context of the team.

Because the case manager sometimes provides direct services and also coordinates the treatment plan, the following questions may arise.

When do I give information that is related to duty-to-warn issues to other professionals who work with the client?

Do I tell individual professionals on the team on a need-to-know basis, or do I provide the information to the entire team?

What standards do these other professionals have concerning the duty to warn and confidentiality?

How do I involve the client in determining the answers to the preceding questions?

What guidelines does my agency have for confidentiality and duty to warn? Do these conflict with my professional code of ethics?

What are the guidelines of the other professional organizations represented on the interdisciplinary team? Are these in agreement with those of the professional organizations I belong to?

What legal standards are applied to the duty to warn and confidentiality in my state?

Because this issue is so complicated, the following recommendations guide aspects of informed consent.

Make informed consent, duty to warn, and confidentiality part of the introduction to services. Many case managers make it standard practice to discuss the concept of confidentiality with clients at the outset. Furthermore, many case managers now ask the client to sign an agreement stating that the case manager will report any information suggesting that the client may harm self or others.

Be informed of agency policy and procedures. The case manager must thoroughly research this issue before working with clients and discuss it with the supervisor and other agency officials familiar with relevant policy and procedure (Remley & Herilhy, 2010). The issue should also be discussed with the team to raise everyone's level of awareness and develop approaches for addressing these situations.

Develop plans of action. The case manager must think through exactly how he or she will handle a client who threatens harm to another and who claims to have a weapon available (in the home, office, or car). Conducting a risk assessment or contracting with clients may be part of the plan.

Obtain professional liability insurance. Many helping professionals either have their own professional liability insurance or are covered through the agency's liability policy. It is important to know if the agency has such a policy; consult with supervisors about this.

Include the client in the duty-to-warn process. There are advantages and disadvantages to involving the client if the case manager believes that the situation warrants a duty to warn. One way to involve the client is to remind him or her of your duty to violate confidentiality when there is a threat to harm self or others. The case manager can ask the threatening client to warn the intended victim. Although involving the client in this way can serve to maintain the trust of the relationship, it may also enrage the client and put the helping professional in danger. The client may feel cornered and become even more determined to carry out the destructive act.

Continue to learn more about the client. The case manager's development of a detailed client history should continue throughout the process. As information is acquired, special attention should be paid to any indications of past acts of violence and evidence of impulsivity or anger. Look for other clues or warnings of danger, such as threats, a plan, or a weapon.

Document in writing. The case manager should document any indications of violence, threats, or expressed anger. Such notes are to be written as objectively as possible.

Welfel (2013) provides guidelines for counselors to use when dealing with clients who represent a threat to themselves or others. They are relevant for case managers, too.

- Remind the client of your ethical and legal obligation to warn.
- Invite the client to participate in the process if possible.
- If possible, develop a plan with the client to surrender any weapons.
- Inform your supervisor, your attorney, law enforcement personnel, the local psychiatric hospital, and the intended victims.
- Keep the client committed, or have him or her committed for treatment.
- Keep careful records of all actions taken.

Working in the Managed Care Environment

Today many case managers work with managed care organizations to determine the services for which their clients are eligible. The context of managed care is based upon clearly defining client problems, determining the most effective and efficient service to provide, and measuring outcomes. Managed care organizations operate within a business atmosphere where the ultimate goal is to accruing the financial foundation needed to remain in business. Many case managers, who are committed to act in the best interest of their clients, encounter ethical dilemmas as they work with managed care. Two of these challenges are providing "best" services to clients and advocating for the client. There are times, in the opinion of the case manager, when the client may be harmed by the recommended services.

Managed care organizations are making decisions for treatment that balance the high cost of services with the needs of the client. This dilemma is well documented throughout the helping professions; many of these professions have responded by creating guidelines for working with managed care organizations (Reid & Silver, 2002). In addition, the rights of patients within the managed care system emerged as support for patients seeking quality care (Foundation for Tax Payer & Consumer Rights, 2005). The following vignette exemplifies in a simple, but dramatic way, the difficulties that occur when clients are denied treatment they need.

Zhewei, diagnosed with schizophrenia at the age of fifteen, received treatment from various individuals and social service agencies for twenty years. By the time she was thirty-five, she was taking medication, holding a part-time job, and receiving weekly counseling. She had been stable for the past fourteen months. Her insurance paid for her weekly counseling sessions; the case manager at the local community health center coordinated her care. Team members included her counselor, her physician, her employer, the case manager, and one of her parents. Zhewei had decided that she did not want to attend treatment team meetings about four years ago. During the last month, individual members of the treatment team had noticed small signs that Zhewei's behavior was changing, and her condition was deteriorating. All agreed that she needed additional intervention. A new plan was developed based upon these signs. The new plan, which required additional individual counseling sessions, a physical exam, and individual sessions with her psychiatrist, was rejected by the managed care organization.

In this situation, the case manager, as the leader of the team, receives the report from the managed care organization. It is the case manager's responsibility to advocate for the client. The case manager talks with the managed care assessment coordinator, writes an appeal, and visits with the agency. These activities are all in response to the ethical commitment to support what is best for the client. The case manager also works with the client and the family to help appeal the decision. What makes this situation difficult is its prevalence. Case managers today spent a good deal of time working with managed care organizations on behalf of their clients (Mullahy, 2010).

The second situation related to decisions made by managed care organizations expands the first challenge, that of providing to the client the services needed. The dilemma occurs when managed care offers treatment that the case manager believes will be harmful to the client. It is possible that in the process of a review of recommendations made by the case manager on behalf of the treatment team, the managed care organization will not just turn down the recommendation, but will recommend an alternative solution. The following short vignette illustrates this dilemma.

Suzie, an eight-year-old, is attending a public school presently. For the first three years, she attended a school for children with special needs. Her primary diagnosis was mild autism. A team has worked with Suzie's parents for the past three years. One of the goals of the team has been to reduce the medication that Suzie is taking. The team, with full support of the parents, would like to substitute weekly counseling and group therapy for the meds. Since Suzie has recently begun to attend public school and is involved in a "mainstreaming" effort, the case manager believed it was a good time to suggest this more drug-free approach. The managed care review turned down the request and suggested heavier doses of medication.

This decision made by the managed care organization places the case manager in the advocacy role, since the treatment team and the parents believe that more medications will decrease Suzie's abilities to function well in the public schools. The medication suggested by the managed care organization will sedate Suzie. Attending school in a sedated state does not allow the professionals to help Suzie to develop her socialization and coping skills. The case manager decides to appeal the decision and encourages the parents to appeal the decision. The case manager may also collect professional literature that reflects the current "standard of care" for children with mild autism. The psychologist on the team also will write a letter for support for the current plan.

Managed care is a part of the health and mental care system. Case managers have a responsibility to work with these organizations if they are to serve their clients effectively. Several suggestions support the work of the case manager as an important link between the client and the managed care organization.

Understand that there are limits on the amounts and types of treatment available for clients   The purpose of managed care organizations is to provide the minimum services that will meet the client's needs. If the case manager understands that this is the focus of the managed care organization, then the case manager can help these organizations by making a strong and documented case for services that help meet client's needs.

Work with the system   If the case manager establishes relationships with professionals who work in the managed care environment, it is easier to work with them than against them. A good relationship allows case managers to place phone calls to their "colleagues" and to talk with them about how they might use the system to promote better care for their clients.

Help clients understand about managed care   Many clients are not aware of their rights concerning managed care, their ability to appeal decisions, and their rights to request a review of their cases. Teaching clients how to effectively advocate for themselves helps clients and case managers speak with one voice.

Write clear, well-documented treatment plans   Many managed care decisions are based solely upon the documentation sent forward by the case manager. Plans and requests should be well written, and needs for services should be well documented.

Following these guidelines should help the case manager develop a relationship with the managed care professionals and advocate for clients. At times, using the voice of the treatment team strengthens the case for a request for services.

 # Autonomy

Client autonomy is a fundamental value in the case management process. As discussed in Chapter 1, case managers must be committed to the principles of client empowerment and participation, as well as to the ultimate goal of the process—that clients become able to manage themselves. Clients are to be involved in the process as much as possible, and treatment plans must support client choice and promote self-sufficiency. At the same time, the case manager has the obligation to provide the client with quality services and to act in his or her best interests. Sometimes, the case manager may believe that what the client prefers is not in his or her interest; this situation can entail an ethical conflict.

In an effort to encourage client participation, case managers can regard **autonomy** in a broader sense of increasing the range of choices a client can make. *Positive autonomy* means that the case manager works to broaden and strengthen the client's autonomy. If client preferences will result in danger to the client or others, the case manager must find a way to make those preferences more appropriate.

There are several instances in which client autonomy is not an absolute priority; for example, when client preferences interfere with other clients or other helping professionals, when the client is not competent to make decisions, and when clients need protection from their own decisions. In such situations, the case manager must have a clear conception of the reasons that autonomy should be restricted. The following case exemplifies the issue.

Mrs. Zeno is married and has three children. She has been a client of Child and Family Services for the past two years. She is now going to school, studying for an associate's degree. She intends to work full time after she completes her education. According to Mrs. Zeno, her husband takes all the money that she receives from Child and Family Services and from welfare. He gives her a small allowance, from which she buys groceries and pays for the rent and utilities, but there is never enough money for clothes for her children or educational supplies for herself. Mrs. Zeno is reluctant to challenge her husband. She does not want to disrupt her home, and she needs her husband's support if she is to finish school. In her last meeting with her case manager, Mrs. Zeno firmly stated that she would do nothing about her husband's behavior. The case manager believes that Mrs. Zeno should, at the very least, talk with her husband about the problem.

In dealing with issues of autonomy and self-determination, asking the following questions can help the case manager and guard the client's autonomy.

What are the facts of the case?   Before any autonomy issues can be thoroughly understood, the case manager must gather relevant information.

Can you understand the history of the case and the current situation, and view it from the client's perspective?   It is important to understand the client's perspective, which determines his or her preferences and choices.

Is the client competent?   It is difficult to determine competence. Clients may be able to make good decisions at certain times but not at other times. Also, they may demonstrate competence in one area of decision making but not in other areas. Only with great caution should a client be declared incompetent.

What does the client want?   Learn exactly what clients desire. Inform them fully of all the alternatives available, and help them understand the consequences (positive and negative) of each alternative.

What are the barriers to what the client wishes?   For many reasons, clients may not be able to get what they want. Sometimes, policies and regulations restrict the case manager's ability to support client choices. Often there are resource problems: the services are not available, or funds are insufficient to pay for them.

Can these obstacles be overcome?   It is the responsibility of the case manager to break down barriers to client autonomy. When making any decision that violates client autonomy, the case manager must inform him or her of the reasons for making that decision.

What are the risks involved in allowing client autonomy?   In determining whether to violate client autonomy, one must calculate the risks to the client and others. If a client's decision would cause harm to self or others, the case manager must overrule him or her.

What is the advice of the treatment team?   The opinions of the treatment team members must be sought and considered; these professionals have much at stake in their own work with the client. If any of the professionals involved do not regard client autonomy as a high priority, this issue must be addressed early in the case management process.

How can the case manager guard the values of the client?   The values of the client should guide the planning of services. If the client has conflicting values, the case manager should work with him or her to sort through the priorities.

Client Preferences: One Component of Autonomy

To develop the care plan, case managers take client preferences into account. In every part of the plan, client preferences receive priority, and everyone involved must respect those preferences. Sometimes this creates problems, as illustrated in the following examples.

The case manager who is developing a plan for Ms. Toomey has limited funds, but part of the money available is designated for health and recreation activities three days a week. Ms. Toomey hates any mention of good health; she especially dislikes exercise and fresh air. Ms. Toomey really loves to go to the movies; she does not watch them at home. The case manager is trying to decide whether she should use the resources available to pay for an attendant to take Ms. Toomey to the movies three days a week (which is Ms. Toomey's preference) or to have the attendant walk with her in the park at least once a week. The case manager is not sure that she can justify movies three times a week to the funding source.

Mr. Krutch needs child care for his baby while he works during the evening at a factory. The available child care workers are almost all Latino and speak little English, but they work well with children. Mr. Krutch refuses to have a Latino person in his home or caring for his child.

The case manager should consider the following guidelines when faced with issues involving client preferences.

Ask what the client needs.   The case manager must have a clear idea of the client's needs. The client's perspective is important in identifying these needs.

Ask why the client has a particular need.   For each of the needs listed, articulate clearly why the client needs the proposed services or support. The client and the case manager then understand why each of the needs is important. During this process, the cultural background of the client must be considered (Williams, 1993). Too often, case managers do not realize the preferences of their clients, particularly when they come from different cultures.

Can each of the needs be addressed?   It is possible that not all the client's needs can be addressed. Sometimes resources are limited, but the client usually has some choices available. The resources available may determine which needs can be met. The case manager must tell the client what is available, and he or she must also promote client interests by advocating for the introduction of services that are not currently available.

How are client preferences met?   Respect for client autonomy means that the case manager works to meet the preferences of the client. Creative advocacy and problem solving can help in finding ways of meeting client needs. By advocating institutional adjustments to client preferences, the case manager builds trust and strengthens the relationship. Within a good relationship, clients are more willing to make their preferences known.

When is a preference not legitimate?   The case manager has a responsibility to follow the agency's policies; for example, one client is not entitled to an undue share of the case manager's time or of any other resource at the agency's disposal.

Let's look at how to apply these guidelines to the cases presented earlier. With Ms. Toomey, the case manager faces a dilemma. Ms. Toomey does not want to take part in any activities involving exercise or good health. She has made it clear that she only wants to go to the movies. The funding agency insists that its money not be used to pay for three movies a week. The agency will only fund one movie per week; its criteria for funding eligibility require that there be a variety of activities and that they be related to physical health. In this case, the treatment team agrees with the funding agency. It suggests that an exercise program and an activity that will challenge her intellect will better support the goals set for Ms. Toomey. With this information, the case manager can discuss options with Ms. Toomey. The case manager is confident that she and Ms. Toomey will find a solution. She also knows that she has limited time to spend on the matter because she is responsible for thirty-five other cases.

The service provider working with Mr. Krutch to find child care is in a difficult position. When Mr. Krutch expressed his disdain for individuals of Latino origin, the care coordinator listened carefully. Two issues emerged. First, most of the available child care workers are Latino. They are very able and have excellent recommendations. Second, the agency has a carefully stated commitment not to discriminate in hiring based on race, gender, religion, or national origin. On the other hand, the care coordinator working with Mr. Krutch realizes that it is important for him to have a good relationship and to trust whomever cares for his child. In this case, the care coordinator decides to try to find a non-Latino child care worker, but she also gives Mr. Krutch a realistic picture of what candidates are likely to be available. She explores Mr. Krutch's reasons for his reluctance to hire a Latino worker, and she prepares him for the possibility that he may have to give such a worker a try. One criterion she used when deciding to seek a non-Latino worker was to ask what harm hiring a Latino worker might do to the client. Mr. Krutch was vehemently against such a hire, and going against his wishes would not have been helpful to the goals of the care process.

Autonomy and End-of-Life Issues

Earlier in this chapter we discussed the ethical issues of decision making that relate to patients who are dying and incompetent, unable to make their own decisions about their own end-of-life care. Today case managers are becoming increasingly involved in the care of competent individuals at the end of their lives as they work with elderly persons in long-term care facilities, with terminally ill inmates in correctional facilities, with AIDS patients, with people who have chronic illnesses, and with children and families caring for those who are dying. To give their clients more autonomy regarding their end-of-life decisions, case managers are helping clients consider establishing advance directives that will guide end-of-life care.

According to the Partnership for Caring (2005), an **advance directive** "is a general term that refers to your oral and written instructions about your future medical care, in the event that you become unable to speak for yourself. Each state regulates the use of advance directives differently. There are two types of advance directives: a living will and a medical power of attorney." The **living will** often defines what type of medical treatment the client wants at the end of life and what medical intervention the client does not want used, such as ventilator, heart/lung machine, nourishment and hydration, or resuscitation when sustaining life is futile. The client's right to accept or refuse medical treatment is protected by law (Partnership for Caring, 2005). A **medical power of attorney** is a document that records the name of the individual designated to decide about medical care if the individual is unable to do so.

It is the case manager's role to explain to the client advance directives or to refer the client to someone who can. Case managers help clients consider end-of-life issues and they discuss preferences with members of their families. In the absence of family members, the case manager can help the client identify a friend or professional to make these final decisions if the client is unable to do so. This is a form of advocacy and empowerment, helping clients ensure that their wishes are known. The client has the final choice about whether or not to have an advance directive and to communicate what those directives might be.

Aging with Dignity (2005) has developed a format, *Five Wishes*, for case managers and family members to use to guide the decision making with regard to **end-of-life care**. The Five Wishes program centers the discussion on five requests.

1. The person I want to make care decisions for me when I can't.
2. The kind of medical treatment I want or don't want.
3. How comfortable I want to be.
4. How I want people to treat me.
5. What I want my loved ones to know.

Helping clients think about and make their end-of-life decisions before their final days supports client autonomy and helps avoid the ethical dilemmas that do occur when these decisions have not previously been made.

 Breaking the Rules

One of the most difficult dilemmas that a case manager can face is whether to comply with laws, regulations, and rules of practice when these do not appear to meet client needs. Two roles of the case manager collide—representing the agency versus serving as the client's advocate.

Ms. Dimatto is a case manager at an AIDS center in a small community. One of her clients, Mr. Sams, is dying. Mr. Sams is living with his partner, and his family resides in a nearby town. He has a good deal of family support. Mr. Sams is bedridden now, and his partner works full-time and cares for him at night. His family shares the responsibility for his care during the day. His mother stays with him three days a week, and his father three days a week. There is one day when Mr. Sams is by himself. During the other days, his parents read to their son, talk with him, give medications, and feed and bathe him. Mr. Sams's partner takes part in a self-help partners' group sponsored by the AIDS center. The center also works with the family, helping with any crises that arise and finding needed resources. In addition, the agency gives the parents a small stipend for the care that they give their son during the day.

The family is facing a problem. Next month, the mother is having bypass surgery, and she will need her husband to help her during her recovery. Neither she nor her husband will be available to provide care to their son. Mr. Sams

> *knows that he needs daily care, so he has consented to hire an aide to stay with him and give him the help he needs. According to agency policy, however, his parents cannot receive any agency support for any care they give him for six months after the aide is hired.*

To protect the client, the agency has established a rule that members of the family who receive in-kind or monetary support for client care must be providing such care on a continuous basis. If the care is interrupted for more than fourteen days, the family cannot be supported by the agency as caregivers for six months. This rule works for stability of care and discourages families from providing help only when it is convenient for them. Another policy states, "The assistance must be provided by an individual whose sole responsibility is the care of the client." Thus, in this case, the father may not be reimbursed if he cares for both his wife after her surgery and for his son.

Ms. Dimatto, the case worker, has a conflict to resolve. On the one hand, in accordance with the rules, she could hire an attendant to stay with Mr. Sams until his death. His parents could then visit him as they wish, but they would not receive any compensation for the care they give their son, nor could the father receive a stipend for caring for his wife. Ms. Dimatto could also suggest that the parents return after fourteen days, asking the father to care for both his wife and his son. This solution would violate another policy—that the care of Mr. Sams be the father's sole responsibility—since the father would also be taking care of the mother during the postoperative period.

This case exemplifies the tension between the case manager's responsibilities as an employee of an agency and as an advocate for client interests. One way to approach this issue is to determine if the welfare of the client is more important than following the rules. Regulations established by governments and agencies are made to apply to everyone and every situation, but the reality is that few rules cover all situations. The obligation to follow regulations must be considered on a case-by-case basis. Here are some guidelines to use when considering these issues (Kane, 1993):

What is the purpose of the rule, and how does it help the client?   It is assumed that the rules have been developed to keep the client from harm. In Mr. Sams's case, the rule that encourages continuity of family care and penalizes any break in that service is written to discourage families from supporting the client only when it is convenient. The rule supports the importance of stability in care, especially for those who are ill and dying. The regulation that requires the attendant to have this client's care as his or her sole responsibility is likewise formulated for good reason—to ensure that the client receives maximum attention.

Who made the rule? What are the consequences of violating it?   Rules originate from many sources: federal, state, and local governments; local associations; and individual agencies. Each carries authority and imposes consequences for violating its rules. The case manager must know the source of the rule and the consequences of violating it. In Mr. Sams's case, the rule was instituted by the agency two years ago because families were taking agency stipends but were not providing good care for their relatives, who were the agency's clients. The agency finally decided that it was more important that clients receive quality care than that the care be given by family members.

What does the client lose and what does the client gain if the rule is followed?   The case manager carefully articulates the client's gains and losses in the event that a rule is broken. The weighing of advantages and disadvantages helps the case manager think through the issue and place client welfare at the center of consideration.

Is this a life-or-death situation for the client?   The answer to this question often gives perspective to consideration of the rule's effect on client welfare. If the answer to this question is yes, the case manager has an argument worth considering for bending or breaking the rule.

Asking for help   In breaking or ignoring rules for the client's benefit, the case manager must keep in mind three considerations: good decision making, use of supervision, and asking for exceptions. He or she must ask the preceding questions to have a firm basis for decision. When using supervision or asking for an exception, violation of the rules may be only one of several concerns. When asked for help, the supervisor may respond in one of several ways. First, he or she may tell the case manager not to violate the rule. At that point the case manager must decide whether to obey the rule and the supervisor or to break the rule and ignore instructions; there are negative consequences for both actions. Second, the supervisor may give the case manager permission to violate the rule. If the case manager does so, both the case manager and the supervisor are liable for the consequences. The supervisor may also ask the case manager not to act until an appeal for an exception has been filed.

Appealing for an exception   Appealing for an exception to a rule is a very different action from just bending a rule or violating a policy. The issue becomes more public. There is open dialogue about the rule and its purpose, the possible precedent of granting an exception, and the actual issues of the case for which the exception is requested. An appeal involves a foray into the realm of political activity; it is advocacy at a different level.

Petitioning for an exception can provide an impetus for change   It helps those who make the rules see that rigid enforcement may violate the reason for which the rule was made in the first place. In some situations, there is no way exceptions can be made. It can be useful to appeal for an exception even when there is no clear channel for such an appeal. Authorities may then understand the need to establish appeal procedures, since rules are not appropriate in every situation that arises.

· · · · · · · · · ·   **Deepening Your Knowledge: Case Study**

Consuela is a 24-year-old woman in the master's of human services program at the University of Nebraska. Her bachelor's degree is in English and she plans to pursue licensure as a human service professional upon completion of her program. She is currently meeting her internship requirement at the Children First Center in Lincoln, Nebraska. She spends ten hours each week shadowing different administrators, social workers, and counselors as they provide therapy for children and their families who have been referred for incidences of abuse and neglect.

During the first half of her internship, Consuela enjoyed all that she was learning and received high marks on her midterm evaluation from her site and faculty supervisors. She described her experience as "significantly increasing" her knowledge of child abuse, family therapy, and counseling skills for children and adolescents. On a particularly busy day, just after her midterm evaluation, Consuela's site supervisor is absent from work. One of the other therapists, Jessica, with whom Consuela does not have a strong relationship and by whom she often feels intimidated, is in charge of providing her duties and responsibilities for the day. She asks Consuela to co-facilitate a parents' group, work with Tonya in the file room, and then check back in with her for further assignments.

After completing these tasks, Consuela finds Jessica, who describes how "swamped" they are with the state officers coming in tomorrow. She explains to Consuela that an aunt has just brought a little girl into the office and the aunt thinks she may have been sexually abused. She says that the aunt just wants to talk with someone before they check the girl out, but "our protocol requires an evaluation for any suspicion of sexual abuse." Jessica explains, "you never know who is telling the truth," but that they need to start the paperwork, complete an evaluation and then conduct an initial interview. Jessica gestures toward the other employees, saying, "there's no one else that's free to do it but you and I," and asks Consuela if she has shadowed a physical evaluation. Consuela says that she has and Jessica responds "Great! Then I will start the paperwork while you take the girl for the evaluation and we can meet together so you can shadow me on the initial interview."

Consuela starts to express reservation but before she can speak Jessica pats her on the shoulder and says, "I really appreciate this and won't forget it on your next evaluation. Days like this make me wonder why I ever get out of bed." Confused by the mix of emotions surrounding the afternoon and concerned about the impact of expressing hesitancy toward the assignment, Consuela conducts the physical evaluation.

Morgan, C. (2012). Unpublished manuscript, Knoxville, Tennessee. Used with permission.

Discussion Questions

1. What key legal issues are present in this scenario?
2. How might this scenario present problems with family disagreement?
3. In working with this type of population, what other forms of family disagreement would the case manager likely need to navigate?
4. What options does Consuela have in this situation? Who might be among her first choices for consultation before following through with the physical evaluation?

Chapter Summary

Numerous ethical and legal issues surround the case management process. The issues arise within a context that makes decision making complex. Factors such as the increasing involvement of family members in client care and the reduction of resources available to assist clients increase the number of ethical issues case managers confront.

Family disagreements are one set of issues that case managers often face. Because families and friends are involved in client care, they expect to have a voice in the planning and treatment process. Frequently, there are strong disagreements among families. Case managers need to consider guidelines they will use when these situations might arise.

Another issue relevant to the process of case management is working with potentially violent clients. Violence is much more likely to occur in some settings, and there are ways that individuals, agencies, and organizations can address the prevention of violence. Agreeing on intervention goals once the client has become violent, teaching case managers intervention skills to defuse violent situations, and developing team responses to violence are all good strategies to address a very difficult issue.

Confidentiality presents ethical dilemmas for case managers regardless of settings and client populations. Standards of confidentiality must be upheld, such as keeping confidences unless the information concerns harm to self or others. Some major decisions relative to confidentiality that the case manager makes are what should be included in a permanent record, what information should be shared with professionals in other agencies, and what information should be discussed with family and friends of the client. Technology has made the keeping of confidences more difficult, but strategies exist to make client information more secure. Confidentiality issues also arise in situations that require an interpreter who facilitates the communication between the client and the case manager.

The duty to warn is an exception to keeping client confidences. This concept, clarified by the *Tarasoff* court case, states that if the case manager knows that the client is intending harm to self or others, the professional must alert appropriate individuals. Many situations have no clear-cut answers, and the case manager must make a judgment about the true intention of the client.

Many case managers work directly with professional staff in managed care organizations. Because the managed care organizations make decisions about client treatments, several dilemmas may arise with regard to treatment. It is important for case managers to develop relationships with members of the managed care network and know how to present well-documented requests for services.

Autonomy represents a basic value of case management and a goal of the case management process. Case managers are committed to the principles of client empowerment and participation. We present a series of questions that guide the case manager when determining how autonomous the client can be. "Is the client competent?" "What does the client want?" "What are the barriers to what the client wishes?" A special case of autonomy that is becoming increasingly important is end-of-life decision making. Case managers are working with more clients who are competent but are dying and feel the need to determine their fate in their last days.

A case manager often has dual responsibilities to the client and to the agency. Codes of ethics, the law, agency policies and procedures, and professional values and judgment are not all in agreement. The case manager must determine what the rule or guideline for practice is and, in some circumstances, whether or not it should be broken.

Chapter Review

Key Terms

Violence	*Advance directive*
Confidentiality	*Living will*
Duty to warn	*Medical power of attorney*
Autonomy	*End-of-life care*

Reviewing the Chapter

1. What pressures make it difficult for case managers to confront ethical issues?
2. Describe the main issue that arises when families disagree about treatment for clients.
3. List ways that case managers can include family members in the case management process.
4. Why is violence so prevalent in human services today?
5. List the steps case managers can follow to confront potential violence.
6. Explain the importance of confidentiality in the case management process.
7. How does confidentiality affect recordkeeping?
8. Describe how technology has affected confidentiality.
9. How do interpreters influence confidentiality?
10. How does the duty to warn relate to confidentiality?
11. Describe how the case manager can experience conflict between the duty to warn and the requirement of confidentiality.
12. List the considerations for the case manager when giving duty-to-warn information to the interdisciplinary team.
13. List the dilemmas case managers face when working with managed care organizations.
14. Why can it be difficult to grant client preferences?
15. Relate the case of Ms. Toomey to the guidelines for thinking about issues of autonomy.
16. Why is a discussion of end-of-life issues important?
17. How does the case manager determine when it is permissible to break the rules?

● *Questions for Discussion* ....

1. Why do ethical issues arise from the involvement of families in the case management process?
2. Confidentiality involves some special difficulties. Discuss them. What conclusions can you draw?
3. Do you think that it is a good idea always to grant client preferences? When is it not appropriate?
4. What advice would you give a new case manager about breaking a rule to meet a client's need?

● *References* ....

Aging with Dignity. (2005). *Five wishes.* Tallahassee, FL: Author.

American Counseling Association. (2005). *ACA code of ethics and standards of practice.* Retrieved from www.counseling.org/resources/codeofethics.htm

Aslam, F., Astab, O., & Janjua, N. Z. (2005). Medical decision-making: The family-doctor-patient triad. *PLoS Med, 2*(6), e120. Retrieved from http://www.journal.pmed.0020129.pdf

Association of Latino Administrators and Superintendents. (2011). *Transforming education: Breaking language barriers to achieve accurate student assessments.* White paper: February 18, 2011. Retrieved from http://www.alasedu.net/resources/1/Publications/White%20Papers/White%20Paper%20-%20White%20Paper%20-Transforming%20Education%20-%20Breaking%20Language%20Barriers%20to%20Achieve%20Accurate%20Student%20Assessment.pdf

Case Management Monthly. (2010). *Interpreters help overcome linguistic and cultural barriers.* Retrieved from http://www.hcpro.com/CAS-247244-2311/Interpreters-help-overcome-linguistic-and-cultural-barriers.html

Community House Calls. (1998). *Beyond medical interpretation.* Retrieved from http://ethnomed.org/about/related-programs/community-house-calls-program/icm-manual98.pdf

Corey, G., Corey, M. S., & Callanan, P. (2003). *Issues and ethics in the helping professions.* Pacific Grove, CA: Brooks/Cole.

Costa, L., & Altekruse, M. (1994). Duty to warn guidelines for mental health counselors. *Journal of Counseling and Development, 72*(4), 346–350.

Davis, J. E. (2005). Working with sign language interpreters in human service settings. *Human Service Education, 25,* 41–52.

Department of Health and Human Services. (2003). *Fact sheet: Protecting the privacy of patients' health information.* Retrieved from http://www.hhs.gov/news/facts/privacy.html.

Foundation for Tax Payer & Consumer Rights. (2005). *California patient rights.* Retrieved from http://www.calpatientguide.org/

Kane, R. A. (1993). Uses and abuses of confidentiality. In R. A. Kane & A. L. Caplan (Eds.), *Ethical conflicts in the management of home care: The case manager's dilemma* (pp. 147–158). New York: Springer.

Laidlaw, G. (2009). *Confidentiality in your home office.* WebWorkerDaily. Retrieved from http://gigaom.com/collaboration/confidentiality-in-your-home-office/

Lang, F., & Quill, T. (2004). Making decisions with family at the end of life. *American Family Physician, 70*(4), 719–723.

Occupational Health and Safety Administration (OHSA) (2004). *Guidelines for preventing workplace violence for health care and social service workers.* Retrieved from http://www.osha.gov/Publications/OSHA3148/osha3148.html

Mid-Ohio Regional Planning Commission (2012). *Social media policy of a government agency.* Retrieved from http://shinydoor.com/govt_social_media_policy

Mullahy, C. (2010). *The case manager's handbook* (4th ed.). Boston: Bartlett and Jones.

Partnership for Caring. (2005). *Advance directives: Living wills, durable power of attorney for health care.* Retrieved from http://www.choices.org/ad.htm

Pecchioni, L., & Sparks, L. (2007). Health information sources of individuals with cancer and their family members. *Health Communication, 21,* 1–9.

Reid, W. H., & Silver, H. B. (2002). *Handbook of mental health administration and management.* London: Taylor and Francis.

Remley, T. P. (2011). *Ethics and the helping professions in a technological world.* Unpublished paper presented in Knoxville, TN, November 4, 2011.

Remley, T. P., & Herlihy, B. (2010). *Ethical, legal, and professional issues in counseling* (3rd ed.). Upper Saddle River, NJ: Prentice Hall.

Sparks, L., (n.d.). *Family decision-making.* Retrieved from http://media.myfoxla.com/mentalhealth/documents/Family-Decision-Making.pdf

U.S. Department of Education (n.d.). *Family educational rights and privacy act (FERPA).* Retrieved from http://www.ed.gov/policy/gen/guid/fpco/ferpa/index.html

Virginia's Judicial System. (2004). *Code of Professional Responsibility for Interpreters Serving in Virginia Courts.* Retrieved from http://www.courts.state.va.us/interpreters/code.html

Welfel, E. (2013). *Ethics in counseling and psychotherapy* (5th ed.). Pacific Grove, CA: Brooks/Cole/Cengage.

Williams, O. J. (1993). When is being equal unfair? In R. A. Kane & A. L. Caplan (Eds.). *Ethical conflicts in management of home care: The case manager's dilemma* (pp. 206–208). New York: Springer.

Woodside, M., & McClam, T. (2011). *An introduction to human services* (7th ed.). Pacific Grove, CA: Brooks/Cole/Cengage.

CHAPTER FIVE

The Assessment Phase of Case Management

Thriving and Surviving as a Case Manager

Sara

When I got a new admission I actually completed all of the assessments. The first assessment was basically a psychosocial assessment. It usually meant spending about an hour or an hour and a half with the child or adolescent and gathering some preliminary information about mental health and school, family, peers. Most of this is historical information. From that I would make a referral to our intensive outpatient program—for individual therapy always, and for family therapy if they had an appropriate family member. I would also contact human services right after he or she was admitted and request a child and family team meeting upon the individual's admission to our program. The purpose of that was to ensure that we were doing what we were supposed to be doing when working with their permanency plan and following the steps that we needed to. It had to fit into the psychosocial assessment. We also have a school on campus. The school gets the same admission paperwork and they receive my psychosocial assessment. That means I am doing the assessment for case management, for the school, and for human services.

Another assessment is directly related to schools. Our school would immediately try to get previous school records; until we receive the previous school assessments the parents and the child are our primary sources of educational information. It can be difficult with the population that we work with to get information from the schools—some of the children haven't been in school very consistently or they've been on the run, or they have lived in multiple placements and have attended several schools.

For assessments, families are so hard for us to engage. We take mainly kids from across the state who are not from our county. Because of this, contacts with families are very, very rare. Phone contact is a little better but even that can be hard at times.

—Permission granted from Sara Bergeron, 2012, text from unpublished interview

The first time I encounter the client is when I receive the referral. Next I prepare the groundwork for screening the family to see if their needs match our program. I schedule a home visit; other times the family visits with me in my office and then I go to their home. My first job is to certify the HIV/AIDS status. I get this information from one of their medical providers. I then conduct a needs assessment and strengths assessment of the family and determine which needs are being met and which are not. I follow through to make sure all needs are met.

—Case manager, family service center, New York, NY

The first time we hear from a potential client is by phone. Kids call and tell us about themselves; what they tell is all that we know . . . we don't have a way to check out the information, to know if it is true or not . . . once they come to our shelter, we get their permission to talk with their parents, school officials, and other helpers to verify their story and their situation. Using the phone as a first source of screening and getting to know the client, especially since he or she is a child, is difficult for us. We try to find out if they need our shelter and our services and also if they are a danger to self or others.

—Hotline case manager, youth services, St. Louis, MO

In our facility we have an emergency shelter for men. Clients can come in the evening and get a change of clothes and take a shower. We combine this service with our chapel time. We tell the men that they can change and we try to provide a message of hope. We also get men coming for our services through referrals from other local agencies and some who are self-referred. We allow persons to enter the program who want to change, for whatever reason. To stay in the program they must commit to change. For one week they can stay and learn about the program. We conduct an initial assessment but with limited results. When men are using drugs or alcohol, it is difficult to talk with them about their issues and current status.

—Caseworker, rescue mission, Miami, FL

Assessment means appraisal or evaluation of a situation, the person(s) involved, or both. As the initial stage in case management, assessment generally focuses on identifying the problem and the resources needed to resolve it. Focusing on the people who are involved includes attention to client strengths, and that can be a valuable resource for encouraging client participation and facilitating problem solving. Specifically, the case manager identifies the initial presenting problem and makes an eligibility determination. As the three opening quotations show, data are gathered and assessed at this phase to show the applicant's problem in relation to the agency's priorities. Identifying possible actions and services and determining who will handle the case are also part of the assessment phase. These activities occur differently at each agency. The case manager at the family service center does a preliminary screening after she has received a referral. At the rescue mission, the initial contact is a pre-application or intake interview. At youth services, the initial contact is a hotline call.

This chapter explores the assessment stage of case management: the initial contact with an applicant for assistance, the interview as a critical component in data gathering, and the case record documentation that is required during this phase. The assessment phase concludes with the evaluation of the application for services. For each section of the chapter, you should be able to accomplish the following objectives.

APPLICATION FOR SERVICES
- List the ways in which potential clients learn about available services.
- Compare the roles of the case manager and the applicant in the interview process.
- Define *interview*.
- Describe a strengths assessment.
- Distinguish between structured and unstructured interviews.
- State the general guidelines for confidentiality.
- Define the case manager's role in evaluating the application.
- List the two questions that guide assessment of the collected information.

CASE ASSIGNMENT
- Compare the three scenarios of case assignment.

DOCUMENTATION AND REPORT WRITING
- Distinguish between process recording and summary recording.
- List the content areas of an intake summary.
- State the reasons for case or staff notes.

Application for Services

Potential clients or **applicants** learn about available services in a number of ways. Frequently, they apply for services only after trying other options. People having problems usually try informal help first; it is human nature to ask for help from family, friends, parents, and children. Some people even feel comfortable sharing their problems with strangers waiting in line with them or sitting beside them.

A familiar physician or pastor might also be consulted on an informal basis. On the other hand, some people avoid seeking informal help because of embarrassment, loss of face, or fear of disappointment. At times the decision to seek help or view help in a positive way is rooted in cultural expectations and values.

Previous experiences with helping agencies and organizations also influence the individual's decision to seek help. Many clients have had positive experiences with human service agencies, resulting in improved living conditions, increased self-confidence, the acquisition of new skills, and the resolution of interpersonal difficulties. Others have had experiences that were not so positive, having encountered helpers who had different expectations of the helping process, delivered unwanted advice, lacked the skills needed to assist them, were inaccessible, or never understood their problems. Increasingly, clients may also encounter local, state, and national policies that may make it difficult to get services. An individual's prior experiences also play a role in the decision regarding whether to seek help. Finally, there may be a financial cost of help that the individual or family cannot afford.

An individual who does decide that help is desirable can find information about available services from a number of sources. Informal networks are probably the best sources of information. Family, friends, neighbors, acquaintances, and fellow employees who have had similar problems (or who know someone who has) are people we trust to tell us the truth about seeking help. Natural helpers such as barbers, hairdressers, and bartenders may also refer. Other sources of information are professionals with whom the individual is already working, the media (posters, public service announcements, and advertisements), the telephone book, or Web-based sources. Once people locate a service that seems right, they generally get in touch on their own (self-referral).

Other individuals may be referred by a human service professional if they are already involved with a human service organization but need services of another kind. They may be working with a professional, such as a physician or minister, who also makes referrals. These applicants may come willingly and be motivated to do something about their situation, or they may come involuntarily because they have been told to do so or are required to do so. The most common referral sources for mandated services are courts, schools, prisons, protective services, marriage counselors, and the juvenile justice system. These individuals may appear at the agency but ask for nothing, or even deny that a problem exists.

The words of case managers at four different locations illustrate the various ways referral can happen. The first case manager works at a community-based agency.

Creswell Center receives referrals every day of the week all year long. One of the services we provide is outreach; we use vans and bicycles and some outreach workers walk on the streets; counselors make contact with people who are homeless, offering them assistance and services. We may see over 200 clients per day for services. The length of stay to receive services might be a month or even a year, usually not longer than a year. We know the length of time they stay in our program, some stay a month or so . . . sometimes as long as a year. (Case manager (outreach), Creswell Center, urban area, Missouri)

The second quote is from a caseworker at an institution that serves the elderly in a large metropolitan city.

> Our referrals come from everywhere; some clients hear about us through word of mouth, others are referred by other professionals who know about our work in the community. We have a large network of professionals in the community involved in discharge planning; we get referrals from them. Most of our clients are referred from the Services for the Blind. The good news in this city is that by law, if a client needs services related to disabilities, then agencies must refer to appropriate agencies. This helps us gain clients who can benefit from our services. (Director and case manager, home for the aged, New York, NY)

The third case manager works at a community mental health center that coordinates services for clients with chronic mental illness. At this agency, client records are available at time of referral. These provide important information about the client and previous services.

> Once I receive a referral I begin my research about their client. The first test is if the client, on paper, meets the criteria for our agency. Then, if the answer to this question is "yes," we go to see the client. Usually the client is in the hospital. The hospital visits are really important because we can talk with professional staff, medical social workers, the client, and often the family. All of the sources we need are in one place. Also, it is a good place to meet the clients for the first time, and when we see them in the community after their release, we already have begun to establish a relationship with them. They are reluctant to talk with us once they leave the hospital. (Case manager, intensive case management, Los Angeles, CA)

The final example is from a program for the homeless in Brooklyn, New York

> We provide housing to homeless individuals. We offer our housing to them once they leave the shelter. To screen individuals for our program and housing, we go to the homeless shelters. We also hope that workers at the shelters will make a referral. We see ourselves as providing bridge services between a shelter, employment, and permanent housing. (Social worker case manager, shelter to home Program, Brooklyn, NY)

These examples illustrate the different ways in which a referral occurs. In an institutional setting, referrals are made by other professionals and departments in the institution. In other settings, referrals can also occur in house or come from other agencies or institutions. Sometimes the referral procedure may include an initial screening by phone **(Figure 5.1)** or on the Internet, a committee

---

PROJECT LIVE INTAKE FORM

DATE _____

NAME _____ AGE _____

PHONE _____

ADDRESS AND DIRECTIONS _____

_____

_____

REFERRAL SOURCE (AND REASON OR RELATIONSHIP) _____

_____

PROBLEM _____

_____

_____

OTHER _____

Figure 5.1   Phone screening form

deliberation, an individual interview, or the perusal of existing records or reports, or both. Usually, the individual who receives the referral makes sure that the necessary paperwork is included.

Not all applicants who seek help or are referred for services become clients of the agency. **Clients** are those who meet eligibility criteria to be accepted for services. An intake interview is the first step in determining eligibility and appropriateness.

The Interview

An **interview** is usually the first contact between a helper and an applicant for services, although some initial contacts are by telephone or letter. The first helper an applicant talks with may be an intake worker who only conducts the initial meeting, or he or she may in fact be the case manager or service provider. If the first contact is an intake worker, the applicant (if accepted for services) will be assigned to a case manager, who will coordinate whatever services are provided. In this text, we will assume that the intake interviewer is the case manager.

The initial meeting with an applicant takes place as soon as possible after the referral. The interview is an opportunity for the case manager and applicant to get to know one another, define the person's need or problem, and give some structure to the helping relationship. These activities provide information that becomes a starting point for service delivery, so it is important for the case manager to be a skillful listener, interpreter, and questioner (skills that you will read about in Chapter 7). First, we will explore what an interview is, the flow of the interview process, and two ways to think about interviews that you may conduct as part of the intake process.

Interviewing is a critical tool for communicating with clients, collecting information, determining eligibility, and developing and implementing service plans—in all, a key part of the case management process. Primary objectives of interviewing are to help people explore their situation, to increase their understanding of it, and to identify client resources and strengths. The roles of the applicant and the case manager during this initial encounter reflect these objectives. The applicant learns about the agency, its purposes and services, and how they relate to his or her situation. The case manager obtains the applicant's statement of the problem and explains the agency and its services. Once there is an understanding of the problem and the services the agency offers, the case manager confirms the applicant's desire for services. The case manager is also responsible for recording information, identifying the next steps in the case management process, informing the applicant about eligibility requirements, and clarifying what the agency can legally provide a client. There are several desirable outcomes of the initial interview. First, rapport establishes an atmosphere of understanding and comfort. Second, the applicant feels understood and accepted. Third, the applicant has the opportunity to talk about concerns and goals. Of course, cultural considerations guide preparation for and conducting this interview.

Exactly what is an interview?   In case management, an interview is usually a face-to-face meeting between the case manager and the applicant; it may have a number of purposes, including getting or giving information, resolving a disagreement, or considering a joint undertaking.

An interview may also be an assessment procedure. It can be a testing tool in areas such as counseling, school psychology, social work, legal matters, or employment applications. One example of this is testing an applicant's mental status. We can also think of the interview as an assessment procedure in which one of the first tasks is to determine why the person is seeking help. An assessment helps define the problem, and the resulting definition then becomes the focus for intervention.

Another way to think about defining the interview is to consider its content and process (McAuliffe & Lovell, 2006). The content of the interview is *what* is said, and the process is *how* it is said. Analyzing an interview in these terms gives a systematic way of organizing the information that is revealed in the interaction between the case manager and the applicant. It also facilitates an understanding of the overall picture of the individual. This is particularly relevant in case management, since many clients have multiple problems. You will read more about process and content in the next chapter.

Ivey, Ivey, and Zalaquett (2010) caution that although the terms *counseling* and *interviewing* are sometimes used interchangeably, interviewing is considered the more basic process for information gathering, problem solving, and the giving of information or advice. An interview may be conducted by almost anyone— business people, medical staff, guidance personnel, or employment counselors. Counseling is a more intensive and personal process, often associated with professional fields such as social work, guidance, psychology, and pastoral counseling. Interviewing is a responsibility assumed by most case managers, whereas counseling is not always the job of the case manager.

How long is an interview?   An interview may occur once or repeatedly, over long or short periods of time. However, some helpers limit the use of the word *interview* to the first meeting, calling subsequent meetings *sessions*. In fact, the actual length of time of the initial meeting depends on a number of factors, including the structure of the agency, the comprehensiveness of the services, the number of people applying for services (an individual or a family), and the amount of information needed to determine eligibility or appropriateness for agency services.

A case manager from a community action agency recalled that she had conducted an intake interview in fifteen minutes. "In school I would read books about helping and it seemed to me I would need about an hour and a half for my first interviews. I thought I would have all of the time I needed. Boy was I wrong!" When I was in school reading textbooks, I thought I would always have an hour and a half for every interview. And we would just ask questions to get all the information about this person. Not true!" (Case manager, family health program, Knoxville, TN).

Where does the interview take place?   Interviews generally take place in an office at an agency, school, hospital, or other institution. Sometimes, however, they are held in an applicant's home. In such cases, the case manager has the distinct advantage of observing the applicant in the home, which gives information about the applicant that may not be available in an office setting. An informal location, such as a park, a restaurant, or even the street, can also serve as the scene of an interview. Whatever the setting, it is an important influence on the course of the interview.

What do all helping interviews have in common?   There are commonalities to all interviews in the human services that should occur in any initial interview. First, there must be shared or mutual interaction: Communication between the two participants is established, and both share information. The case manager may be sharing information about the agency and its services, while the applicant may be describing the problem. No matter what the subject of their conversation is, the two participants are clearly engaged as they develop a relationship.

A second factor is that the participants in the interview are interdependent and influence each other. Each comes to the interaction with attitudes, values, beliefs, and experiences. The case manager also brings the knowledge and skills of helping, while the person seeking help brings the problem that is causing distress. As the relationship develops, whatever one participant says or feels triggers a response in the other participant, who then shares that response. This type of exchange builds the relationship through the sharing of information, feelings, and reactions.

The third factor is the interviewing skill of the case manager. He or she remains in control of the interaction and clearly sets the tone for what is taking place. The knowledge and expertise of the case management process distinguishes the case manager from the applicant and from any informal helpers who have previously been consulted. Because the helping relationship develops for a specific purpose and often has time constraints, it is important for the case manager to bring these characteristics to the interaction, in addition to providing information about the agency, its services, the eligibility criteria, community resources, and so forth.

THE INTERVIEW PROCESS

The interview's structure refers to the arrangement of its three parts: the beginning, the middle, and the end. The beginning is a time to establish a common understanding between the case manager and the applicant. The middle phase continues this process, through sharing and considering feelings, behaviors, events, and strengths. At the end, a summary provides closure by describing what has taken place during the interview and identifying what will follow. Let's examine each of these parts in more detail.

The beginning   Several important activities occur at the beginning of the interview: greeting the client, establishing the focus by discussing the purpose, clarifying roles, and exploring the problem that has precipitated the application for services. The beginning is also an opportunity to respond to any questions that the applicant may have about the agency and its services and policies. Clients raise several questions during the initial interview. Most of these questions relate to how the service is provided.

- How often do you want me to come to see you?
- Will you come to my home?
- What happens if I have a question or need you and the office is closed?
- If I have an appointment and cannot make it, what do I do?
- Tell me again about confidentiality, please.
- What if I have an emergency?
- When do we complete our work?
- What will I be charged for services?
- Do I need insurance?

Answering these questions can lead to a discussion of the applicant's role and his or her expectations for case management.

The middle   The next phase of the interview is devoted to developing the focus of the relationship between the case manager and the applicant. Assessment, planning, and implementation also take place at this time. Assessment occurs as the problem is defined in accordance with the guidelines of the agency. Often assessment tends to be problem focused, but a discussion that focuses only on the client's needs or weaknesses, or both, can be depressing and discouraging. Spending some of the interview identifying strengths can be energizing and can result in a feeling of control. Assessment also includes consideration of the applicant's eligibility for services in light of the information that is collected. All these activities lead to initial planning and implementation of subsequent steps, which may include additional data gathering or a follow-up appointment.

The end   At the close, the case manager and the applicant have an opportunity to summarize what has occurred during the initial meeting. The summary of the interview brings this first contact to closure. Closure may take various forms, including the following scenarios. (1) The applicant may choose not to continue with the application for services. (2) The problem and the services provided by the agency are compatible, the applicant desires services, and the case manager moves forward with the next steps. (3) The fit between the agency and the applicant is not clear, so it is necessary to gather additional information before the applicant is accepted as a client.

A technique that some case managers find successful as an end to the initial interview is a homework assignment that once again turns the applicant's attention to strengths. Individuals may be asked to complete a self-assessment instrument such as the Strengths Self-Assessment Questionnaire, a forty-question self-report instrument designed to involve the client and significant others in identifying strengths (McQuaide & Ehrenreich, 1997). Although this type of questionnaire can be administered at any time in the case management process, the value in such a technique at this time is the movement that might occur from problem-oriented vulnerability to problem-solving resilience. It may also solidify the client's intent to return and promote his or her involvement in the case.

STRUCTURED AND UNSTRUCTURED INTERVIEWS

Interviews may be classified as structured or unstructured. A brief overview of these two types of interviews is given here; the next chapter will provide more in-depth information about the structured interview in connection with the discussion of skills for intake interviewing.

Structured interviews are directive and focused; a form or a set of questions that elicit specific information usually guides them. The purpose is to develop a brief overview of the problem and the context within which it is occurring. It can range from a simple list of questions to solicitation of an entire case history.

Agencies often have application forms that applicants complete before the interview. If the forms are completed with the help of the case manager during

Voices from the Field: Strengths Assessment

Ragan and Sekar (2006) provide a model of strengths-based assessment and intervention that supports positive interventions and goals for the client and family. As discussed earlier in Chapter 1, the strengths perspective is a necessary aspect of client empowerment and the move to self-sufficiency. Using this approach solidifies the sense of collaboration and partnership the client has with the case manager. Strengths assessment provides an opportunity for clients and families and friends to tell the client's stories. Content from these stories provides the focus for the case manager activities to follow. Dennis Saleebey (2002), an expert in strengths-based assessment and intervention, outlines questions that guide the case manager. He presents survival questions, support questions, exception questions, and esteem questions. Case managers may find this guide helpful as they think about the initial interview.

Dennis Saleebey's Five Types of Questions to Assess Strengths

Survival Questions

- How have you managed to survive (or thrive) thus far, given all the challenges you have had to contend with?
- How have you been able to rise to the challenges put before you?
- What was your mindset as you faced these difficulties?
- What have you learned about yourself and your world during your struggles?
- Which of these difficulties have given you special strength, insight, or skill?
- What are the special qualities on which you can rely?

Support Questions

- What people have given you special understanding, support, and guidance?
- Who are the special people on whom you can depend?
- What is it that these people give you that is important?
- How did you find them or how did they come to you?
- What did they respond to in you?
- What associations, organizations, or groups have been especially helpful to you in the past?

Exception Questions

- When things were going well in life, what was different?
- In the past, when you felt that your life was better, more interesting, or more stable, what about your world, your relationships, your thinking was special or different?

- What parts of your world and your being would you like to recapture, reinvent, or relive?
- What moments or incidents in your life have given you special understanding, resilience, and guidance?

Possibility Questions

- What now do you want out of life?
- What are your hopes, visions, and aspirations?
- How far along are you toward achieving these?
- What people or personal qualities are helping you move in these directions?
- What do you like to do?
- What are your special talents and abilities?
- What fantasies and dreams have given you special hope and guidance?
- How can I help you achieve your goals or recover those special abilities and times that you have had in the past?

Esteem Questions

- When people say good things about you, what are they likely to say?
- What is it about your life, yourself, and your accomplishments that gives you real pride?
- How will you know when things are going well in your life—what will you be doing, who will you be with, how will you be feeling, thinking, and acting?
- What gives you genuine pleasure in life?
- When was it that you began to believe that you might achieve some of the things you wanted in life?
- What people, events, and ideas were involved?

From Saleebey, D. (2002). *The strengths perspective in social work practice* (3rd ed.). New York: Allyn & Bacon.

the interview, the interaction is classified as a structured one. Of course, this is a good way to establish rapport and to identify strengths, but case managers must be cautious. The interview can easily become a mere question-and-answer session if it is structured exclusively around a questionnaire. Asking yes/no questions and strictly factual questions limits the applicant's input and hinders rapport.

The intake interview and the mental status examination are two types of structured interviews, each of which has standard procedures. Generally, an agency's **intake interview** is guided by a set of questions, usually in the form of an application. The **mental status examination** (which typically takes place in psychiatric settings) consists of questions designed to evaluate the person's current mental status by considering factors such as appearance, behavior, and general intellectual processes. In both situations, the case manager has responsibility for the direction and course of the interview, even though the areas to be covered are predetermined. A further look at the mental status examination illustrates the structured interview process.

A mental status examination is a simple test that assesses a number of factors related to mental functioning. Sometimes agencies will use a predetermined list of questions. Others may use a test such as the *Mini Mental Status Examination* (MMSE). Both modalities structure the interview. Whatever the method, generally the focus is on four main components that may vary slightly in their organization: appearance and behavior, mood and affect (emotion), thought disturbances, and cognition (orientation, memory, and intellectual functioning). Some of these components may be assessed by observing the client, while others require asking specific questions and listening closely to responses. For example, appearance and behavior are observable and may target dress, hygiene, eye contact, motor movements or tics, rhythm and pace of speech, and facial expressions. On the other hand, thought disturbances are based on what a client tells the examiner and his or her perception of things: "Do your thoughts go faster than you can say them?" "Do you hear voices that others cannot hear?" and "Have you ever seen anything strange you could not explain?" A final word to the examiner: When testing a client, be sure to take into consideration factors such as a person's country of origin, language skills, and educational level. Terms like "high school" and tasks like naming former U.S. presidents may not be appropriate for all examinees.

As you read the following report, determine the results of the assessment of the four components of a mental status examination:

"Pops" Bellini arrived for his appointment on time. He was dressed in slacks and a shirt but appeared disheveled with wrinkled clothing and uncombed hair. He stated that he has been homeless for six months and has not bathed in four days. He wants to work. He rocked forward when answering a question and tightly clasped his hands in his lap. He reports that he is on no medication at the present time although he has taken "something" in the past for "nerves." His mood was anxious and tense, and he wondered if he would be able to find some work. Speech was slow, clear, and deliberate. Initially he responded with a wide-eyed stare. Mr. Bellini often asked that a question be repeated; upon repetition, he provided responses that were relevant to the question. Cognition seemed logical and rational although slow. Mr. Bellini is aware of the need to find work to support himself, but is concerned about his ability to perform at a job. He seems most worried about "pressures" that cause him to "freeze."

In contrast to the structured interview, the **unstructured interview** consists of a sequence of questions that follow from what has been said. This type of interview can be described as broad and unrestricted. The applicant determines the direction of the interaction, while the case manager focuses on giving reflective responses that encourage the eliciting of information. Helpers who use the unstructured interview are primarily concerned with establishing rapport during the initial conversation. Reflection, paraphrasing, and other responses discussed in Chapter 7 will facilitate this.

Assessing the interview   We suggest that it is helpful for the case manager to conduct an informal self-assessment of his or her own competence at the beginning and the conclusion of each interview. It is also important for case managers to conduct formal assessments of their skills development. The Strengths-Based Restorative Justice Assessment Tools for Youth (2004) manual provides ideas about evaluating interviews with clients. The following evaluation questions focus on cultural competence, justice, and strengths-based work. We inserted the term case manager to replace "counselor/probation officer" in the questions.

Related to cultural competence and justice

- "Was your case manager helpful?
- How fair do you feel the case manager was?
- How sensitive was the case manager to your family's background or to experiences you have had because of your race, ethnicity, gender, sexual orientation?" (Strengths-Based Restorative Justice Assessment Tools for Youth, 2004, p. 34)

Related to strengths assessment

- "Asks about strengths
- Points out positives
- Uses strengths
- Encourages youth/family involvement
- Moves toward a positive plan
- Focuses on the future
- [Uses] Individualized planning
- Encourages community connection
- Encourages development of youth's healthy identity" (Strengths-Based Restorative Justice Assessment Tools for Youth, 2004, p. 40–41)

CONFIDENTIALITY

Our discussion of the initial meeting would be incomplete if we did not address the issue of **confidentiality**. Human service agencies have procedures for handling the records of applicants and clients and for maintaining confidentiality; all case managers should be familiar with them. An all-important consideration is access to information.

All communication during an interview should be confidential in order to encourage the trust that is necessary for the sharing of information. Generally, human service agencies allow the sharing of information with supervisors, consultants, and other staff who are working with the applicant. The client's signed consent is needed if information is to be shared with staff employed by other organizations. The exception to these general guidelines is that information may be shared without consent in cases of emergency, such as suicide, homicide, or other life-threatening situations.

Applicants are frequently concerned about who has access to their records. In fact, they may wonder whether they themselves do. Legally, an individual does indeed have access to his or her records; the Federal Privacy Act of 1974 established principles to safeguard clients' rights. In addition to the right to see their records, clients have the right to correct or amend the records.

There are two potential problem areas related to confidentiality. First, it is sometimes difficult in large agencies to limit access of information to authorized staff. Support staff, visitors, and delivery people come and go, making it essential that records are secure and conversations confidential. The second problem has to do with **privileged communication**, a legal concept under which certain communications between clients and professionals may not be used in court without client consent (Corey, Corey, & Callanan, 2011). State laws determine which professionals' communications are privileged; in most states, human service workers are not usually included, but therapists are. Thus, the helper may be compelled to present in court any communication from the applicant or client. Sometimes it can be a challenge for the case manager to explain this limitation to an applicant while trying to gather essential information. It can also be perplexing to the applicant.

Evaluating the Application for Services

During this phase, the case manager's role is to gather and assess information. In fact, this process may actually start before the initial meeting with the applicant, when the first report or telephone call is received, and continue through and beyond the initial meeting. The initial focus on information gathering and assessment then narrows to problem identification and the determination of eligibility for services. Guidelines and parameters established by the agency or by federal or state legislation influence this process to some extent. At this point in the process, the case manager must pause to review the information gathered for assessment purposes.

Part of the assessment of available information is responding to the following questions:

- Is the client eligible for services?
- What problems are identified?
- Are services or resources available that relate to the problems identified?
- Will the agency's involvement help the client reach the objectives and goals that have been established?
- What types of services might be needed to meet unique issues presented by the client, family, or circumstances?

Reviewing these questions helps the case manager determine the next steps. To answer the questions and evaluate the application for services, the case manager engages in two activities: a review of information gathering and an assessment of the information.

Review of information gathering   Usually the individual who applies for services is the primary source of information. During the initial meeting, the case manager forms impressions of the applicant. The problem is defined, and judgments are made about its seriousness—its intensity, frequency, and duration. As the case manager reviews the case, he or she considers these impressions in conjunction with the application for services, the case notes summarizing the initial contact with the client, and any case notes that report subsequent contacts. The case manager learns more about the applicant's reasons for applying for services; his or her background, strengths and weaknesses; the problem that is causing difficulty; and what the applicant wants to have happen as a result of service delivery. The case manager also uses information and impressions from other contacts. Other information in the file that may contribute to an understanding of the applicant's situation comes from secondary sources, such as the referral source, the client's family, school officials, or an employer. Information from secondary sources that can be part of the case file might be medical reports, school records, a social history, or a record of services that have previously been provided to the client.

An important part of the review of information gathering is to ascertain that all necessary forms, including releases, have been completed, and signatures obtained where needed. It is also a good idea at this point to make sure that all necessary supervisor and agency reviews have occurred and are documented.

Assessing information   Once the case manager has reviewed all the information that has been gathered, the information is assessed. Many case managers have likened this part of case management to a puzzle. Each piece of information is part of the puzzle; as each piece is revealed and placed in the file, the picture of the applicant and the problem becomes more complete. A case manager working in New York City described it as "lots of information that comes in many forms from many individuals, professional and otherwise, that you have to go through and figure out, 'What seem to be the salient issues and challenges? What facts do we know?'" This case manager added that helpers "need the special skills of collecting lots of information, making sense of it, reducing it to a manageable size, and then do[ing] something with it." As a social worker case manager in Houston, Texas, suggests,

> One way that the case manager understands information is through how it feels, what is happening beyond the details. When you deal with multiple cultures—here there are covens and witchcraft, African Americans, immigrants, and the like—there are different cultures and different needs. I think you become part of the community and part of the culture. It is important that you are aware of what is happening in the community and in the various cultures. Take alternative religious perspectives: some would think a person suffers from mental illness that believes in shamans and spells. Sometimes the client needs a better spell. Frankly, I have seen voices cleared by a shaman.

After learning what information is in the file, the case manager's task shifts to assessing the information. Two questions guide this activity. Is there sufficient information to establish eligibility? Is additional information necessary? In addition to answering these questions, the case manager also evaluates the information in the file, looking for inconsistencies, incompleteness, and unanswered questions that have arisen as a result of the review.

Is there sufficient information to establish eligibility?   To answer the question of sufficient information, the case manager must examine the available data to determine what is relevant to the determination of eligibility. The quantity of data gathered is less important than its relevance. Human service organizations usually have specific criteria that must be met to find an applicant eligible. The data must correspond to these criteria if the applicant is to be accepted for services.

The criteria for acceptance as a client for vocational rehabilitation services are a good example. Vocational rehabilitation is a state and federal program whose mission is to provide services to people with disabilities so as to enable them to become productive, contributing members of society. Essentially, the criteria for acceptance for services are the following: The individual must have a documented physical or mental disability that is a substantial handicap to employment, and there must be a reasonable expectation that vocational rehabilitation services will render the applicant fit for gainful employment.

During the assessment phase, a vocational rehabilitation counselor assesses the information gathered to determine whether the applicant has a documented disability. The next step is to document that the disability is a handicap to employment. Does the disability prevent the applicant from returning to work? Or, if the person has not been employed, does the disability prevent him or her from getting or keeping a job? If the answers are yes, the counselor's final task is to find support for a reasonable expectation that, as a result of receiving services, the applicant can be gainfully employed. This brings us to the second main question in the information assessment activity.

Is additional information necessary in order to determine eligibility for services?   If the answer is no, the case manager and the client are ready to move to the next phase of case management. If the answer is yes—that is, additional information is necessary to establish eligibility—a decision is then made about what is needed and how to obtain it. In the vocational rehabilitation example, the counselor examines the file for the documentation of a disability. Specifically, the counselor is looking for a medical report from a physician or specialist that will establish a physical disability or a psychological or psychiatric evaluation that will establish a mental disability. If the needed information is not in the file, the helper must make arrangements to obtain the necessary reports.

Establishment of eligibility criteria is not the sole purpose of this phase. Data gathered at this time may prove helpful in the formulation of a service plan. Certainly the case manager does not want to discard any information at this point; neither does he or she want to leave unresolved any conflicts or inconsistencies. Relevant, accurate information is an important part of the development of a plan. Ensuring relevance and accuracy at this point in the process saves time and effort and allows the helper and the client to move forward without delay.

 ## Case Assignment

Once eligibility has been established and the applicant accepted for services, there are three possible scenarios, depending on the particular agency. In all three, the applicant becomes a client who is assigned a case manager to coordinate services.

In many instances, the case manager is the same person who handled the intake interview and determination of eligibility. In some agencies, however, there are staff members whose primary responsibility is conducting the intake interview. After a review, the case is assigned to a case manager—the helping professional who assumes primary responsibility for the case and is accountable for the services given to the client, whether provided personally, by other professionals at the agency, or by helpers at a different agency.

A second scenario involves the *specialized worker*, a term that may refer to either level specialization or task specialization. Level specialization has to do with the overall complexity and orientation of the client and the presenting problem. Is the case under consideration simple or complex? Is it a case of simply providing requested information, or is it a multiproblem situation? Task specialization focuses on the functions needed to facilitate problem resolution. Does the case require highly skilled counseling, or is coordination sufficient?

The third scenario occurs most often in institutions where a team of professionals is responsible for a number of clients. For example, in a facility for children who are developmentally disabled, clients interact daily with a staff that includes a teacher, a nurse, an activity coordinator, a cottage parent, and a social worker. These professionals work together as a team to provide services to each client.

 ## Documentation and Report Writing

Documentation and report writing play critical roles in the assessment phase of service coordination. The main responsibilities facing the case manager in this phase are identifying the problem or problems and determining the applicant's eligibility for services. Documentation of these two responsibilities takes the form of **intake summaries** and **staff notes**. Most agencies have guidelines for the documentation of information gathered and decisions made. In this section, we discuss the forms of documentation, their purposes, and how to prepare them. Before we discuss intake summaries and staff notes, let's distinguish between process recording and summary recording.

Process Recording and Summary Recording

A **process recording** is a narrative telling of an interaction with another individual. In the assessment phase of case management, a process recording shows what each participant has said by an accurate account of the verbal exchange, a factual description of any action or nonverbal behavior, and the interviewer's analysis and observations. The person making the recording should imagine that a tape recorder and a camera are taking in everything that is heard or seen. Of course, because records are required to be brief and goal oriented, the helper would not attempt an exhaustive description, but this approach helps focus the recorder's attention on accuracy and impartiality.

Process recording is a useful tool for helping professionals in training, especially those who are learning to be case managers. Tape recorders, video cameras, and VCRs are readily available today, but many agencies and organizations don't have them, and case managers may not have time or authorization to use the equipment. Process recording is still an effective way to hone one's skills of direct observation.

Process recording is most often used with one-on-one interviews. It includes the following elements.

- Identifying information: names, date, location, client's case number or identifying number, and the purpose of the interview.

Paulette Maloney saw the client, Rosa Knight, for the first time on Monday, November 5, at the agency. Ms. Knight is applying for services, and the purpose of the interview was to complete the application form and inform her of the agency's services.

- Observations: description of physical and emotional climate, any activity occurring during the interaction, and the client's nonverbal behavior.

Ms. Knight appeared in my office on time, dressed neatly in a navy dress. I asked Ms. Knight to come in, introduced myself, shook her hand, and asked her to sit down. Although there was little eye contact, she smiled shyly with her head lowered. In a soft voice, she asked me to call her Rosa. Then she waited for me to speak.

- Content: an account of what was said by each participant. Quotes are helpful here, to the extent that they can be remembered.

I explained to Rosa that the agency provides services to mothers who are single parents, have no job skills, and have children who are under the age of 6. She replied, "I am a single parent with two sons who are 18 months old and 3 years old. I have worked briefly as a domestic." I asked her how long she had worked and where. She replied that she worked about five months for a woman a neighbor knew. She quit "because the woman was always canceling at the last minute." She related that one time when she went to work, the house was locked up and the family was out of town. During this exchange, Rosa clasped her hands in her lap and looked up.

- Recorder's feelings and reactions: Sometimes this is called a *self-interview,* meaning that the recorder writes down feelings about and reactions to what is taking place in the interview.

I was angry about the way Rosa had been treated in her work situation. She appears to be a well-mannered, motivated young woman who genuinely wants a job so that she can be self-supporting. Her shyness may prevent her from asserting herself when she needs to. Her goal is to get out of the housing project. I wonder if my impressions are right.

- Impressions: Here the recorder gives personal impressions of the client, the problem, the interview, and so forth. It may also be appropriate to make a comment about the next step in the process.

Rosa Knight is a 23-year-old single mother of two sons, ages 18 months and 3 years. She has worked previously as a domestic. Particular strengths seem to be motivation and reliability. Her goal is to receive secretarial training so that she can be self-supporting and move away from the projects. Based on the information she has provided during this interview, she is eligible for services. I will present her case at the next staff meeting to review her eligibility.

The other style of recording, **summary recording**, is preferred in most human service agencies. It is a condensation of what happened into an organized presentation of facts. It may take the form of an intake summary or staff notes (both discussed later in this section). Summary recording is also used for other types of reports and documentation. For example, a diagnostic summary presents case information, assesses what is known about the client, and makes recommendations. (See the report for juvenile court in Chapter 9.) A second example is *problem-oriented recording*, which identifies problems and treatment goals. This type of recording is common in an interdisciplinary setting or one with a team structure. (An example is the psychological evaluation, also in Chapter 9.)

Summary recording differs from process recording in several ways. First, a summary recording gives a concise presentation of the interview content rather than an extensive account of what was said. The focus remains on the client, excluding the case manager's feelings about what transpired. Summary recordings usually contain a summary section, which is the appropriate place for the writer's own analysis. Finally, summary recording is organized by topic rather than chronologically. The case manager must decide what to include and omit under the various headings: Identifying Information, Presenting Problem, Interview Content, Summary, and Diagnostic Impressions.

Summary recording is less time-consuming to write, as well as easier to read. It is preferred for these reasons, not to mention the fact that it uses less paper, thereby reducing storage problems. Where computerized information systems are used, a standard format makes information easy to store, retrieve, and share with others. A reminder is in order here: Agencies often have their own formats and guidelines for report writing and documentation. Generally speaking, however, the basic information presented here applies across agency settings.

Intake Summaries

An intake summary is written at some point during the assessment phase. It is usually prepared following an agency's first contact with an applicant, but it may also be written at the close of the assessment phase. For purposes of illustration, assume that it is written after the intake interview. After the case manager conducts the intake interview, he or she assesses what was learned and observed about the applicant. This assessment takes into account the information provided by the applicant, the mental status examination, if appropriate, forms that were completed, and any available information about the presenting problem. Client strengths are also identified at this point. The case manager also considers any inconsistencies or missing information. While integrating the information, the case manager also considers the questions that are presented in the previous section. Is the applicant eligible for services? What problems are identified? Are services or resources available that relate to the problems identified? Will the agency's involvement improve the situation for the applicant?

This information is organized into an intake summary, which usually includes the following data.

- Worker's name, date of contact, date of summary
- Applicant's demographic data—name, address, phone number, agency applicant number
- Sources of information during the intake interview
- Presenting problem
- Summary of background and social history related to the problem
- Previous contact with the agency
- Diagnostic summary statement
- Treatment recommendations

Figure 5.2 is a sample intake summary from a treatment program for adult women who are chemically dependent. To qualify for the residential program, applicants must have a child who is three years of age or younger and has been exposed to drugs. The day program is available to mothers who have a child older than three years. The applicant whose intake is summarized in Figure 5.2 is applying to the residential program, which lasts one year. During this time

INTAKE SUMMARY
OAK HILL CENTER

CLIENT'S NAME __Katie Dunlap__   ADMISSION DATE _____

INTERVIEWER _____   DATE __8-11-XX__

Katie Dunlap was admitted to Oak Hill Center August 11, XXXX. The client's date of birth is May 22, XXXX. She is from Nashville. This is her first admission to this center. Prior to admission the client resided with her sister at 1010 Western Avenue in Memphis, TN. Her drug of choice is crack cocaine. The client started using at age 14 and the date of last use was August 9, XXXX. She has been in four rehabilitation centers; however, she has not graduated from any of them. The client has 2 daughters: Candy, who is 4 years of age, and Kristy, who is 2. They have been in foster care for one year. The younger child was chemically exposed. The children will be returned to the client one month from the admission date. The client's father and mother were also chemically dependent. Her father died when she was 11. The client stated that her mother had mental health problems and that they do not have a good relationship. The client has not been treated for being emotionally abused by her mother and physically abused by the father of her children. Currently she has no means of income, nor does she have health insurance. A psychological evaluation is being completed at present. She stated that she views herself as happy, easygoing, and appreciative.

_____   _____
INTAKE WORKER'S SIGNATURE   DATE

Figure 5.2   Intake summary

children live with their parent in one of ten agency apartments. To graduate from the program, the client must be employed or enrolled in school and be free of substance abuse. Some individuals enter the program voluntarily, and others are ordered to come as part of their probation. On admission, a staff member conducts a thirty-minute interview, which is written up and placed in the

Want More Information?

● ● ● ● ● ● ● ● ● ● ● ●

West Bridge Organization offers services to those suffering with mental illness. On their Web site, they describe their availability and philosophy of the intake interview.

A Fresh Style of Intake Interview

Did you know that WestBridge will travel anywhere in the country to complete an intake interview? In our ongoing effort to live our mission of flexibility, mobility, and responsiveness, WestBridge will go to a home, hospital, or otherwise convenient location to meet with potential participants and their family, introduce ourselves, and determine if WestBridge is the right fit.

It is no secret that relationships are the foundations of our lives. They are also the cornerstone to long-term recovery from mental illness and substance use. In that spirit, WestBridge believes in beginning the relationship process by sending our staff (admissions counselors and/or clinical staff members) to meet with potential participants and their family. This enables the potential participant to meet some of our staff, hear about our program and get a feel for the work we do. This service aids our admission process by our being able to have face-to-face conversations with potential participants and families. Our objective is to determine goals, strengths and if WestBridge offers what is needed. Often, this process helps build trust and clinical rapport that can ease the transition to treatment, and can be the catalyst of a strong recovery process.

If the participant and his or her family are interested in pursuing an admission and WestBridge decides that our programming fits the needs of the participant and family, there are many ways to facilitate the admission from this point.

SOURCE: http://www.westbridge.org/TheBridgeNewsletter/fall_07_Newsletter.pdf. Reprinted by permission.

client's file within two days. The client then receives an orientation to the program and is assigned a care coordinator in charge of that case. The care coordinator and the client then have a more extensive interview, lasting approximately ninety minutes.

Staff Notes

Staff notes, sometimes called *case notes*, are written at the time of each visit, contact, or interaction that any helping professional has with a client. Staff notes usually appear in a client's file in chronological order. They are important for a number of reasons.

- Confirming a specific service. The helper wrote, "John missed work 10 of 15 days this month. He reported that he could not get out of bed. I referred him to our agency physician for a medication review."

- Connecting a service to a key issue. The helper may write, "I observed the client's interactions with peers during lunch," or "I questioned the client about his role in the fight this morning."
- Recording the client's response. "Mrs. Jones avoided eye contact when asked about her relationship with her family," "Janis enthusiastically received the staff's recommendation for job training," or "Joe resisted the suggestion that perhaps he could make a difference."
- Describing client status. Case notes that describe client status use adjectives and observable behaviors: "Jim worked at the sorting task for 15 minutes without talking," and "Joe's parents were on time for the appointment and openly expressed their feelings about his latest arrest by saying they were angry."
- Providing direction for ongoing treatment. Documenting what has occurred or how a client has reacted to something can give direction to any treatment. "During our session today, Mrs. Jones said she felt angry and guilty about her husband's illness in addition to the feelings of sadness she expressed at our last meeting. These feelings will be the focus of our next meeting."

The format of case notes depends on the particular agency, but they are always important. For instance, one substance abuse treatment facility uses a copy of its form for each client every day. A worker on each of the three shifts checks the behavior observed and makes chronological case notes on the other side. These notes allow the case manager and others working with a client to stay up to date on treatment and progress and provide the means of monitoring the case.

There are a number of other case note models, such as data, assessment, and plan (DAT), functional outcomes reporting (FOR), and individual educational programs (Cameron & Turtle-Song, 2002). These are all variations of the original **SOAP** format, originally developed by Weed (1964). SOAP is an acronym for subjective, objective, assessment, and plan. Developed to improve the quality and continuity of client services by enhancing communication among professionals, SOAP supports the identification, prioritization, and tracking of client problems so they can receive attention in a timely fashion.

Subjective and objective are both parts of data collection. Subjective refers to information about the problem from the client's perspective or that of other people: "reports difficulty getting along with her coworkers" or "mother complains about client losing control and striking younger sister." The helper's observations and external written materials make up the objective component, which is written in quantifiable terms: "client seemed nervous as evidenced by repeatedly shifting in chair, chewing nails, rocking, and looking down." The assessment section combines both previous sections for interpretation or a summarization of the counselor's clinical thinking regarding the problem. This is often stated as a psychiatric diagnosis based on the *DSM-IV-TR* (see Chapter 9). The last section of the SOAP notes is the plan that generally includes both action and prognosis: "next appointment is 5/25/05 @ 10 p.m.; prognosis is good due to motivation and interest in changing anger behavior. Also referred to anger management group for weekly meetings."

The notes presented below, from a residential treatment facility for emotionally disturbed adolescents, illustrate another format for case notes. These notes are in the file of a fifteen-year-old female client who has been diagnosed with borderline personality disorder. At this facility, a staff person from each shift is required to make a chart entry; the abbreviations *DTC*, *ETC*, and *NTC* refer to day, evening, and night treatment counselors. Any other staff member who has contact with the client during the day (such as the therapist, teacher, nurse, or recreation specialist) also writes a staff note. The word *Level* refers to the number of privileges that a client (Ct.) has. For example, a client at Level 1 has a bedtime of 9 p.m.; a client at Level 3 would go to bed at 10.

1/22/05, 8:15 A.M. DTC: Ct. awoke on time this morning. She was showered, dressed, and ready for school on time. She had positive interactions with her peers this morning. Ct. received one cue for being loud. Ct. completed her chores in a timely manner and responded appropriately to all staff requests. Ct. remains on Level 2 and under constant supervision. _____ Jan Allen

---

1/22/05, 2:45 P.M. TEACHER: Ct. completed all of her classroom assignments in a timely manner. She received one C for cursing. She accepted the C appropriately and maintained a good attitude. Ct. had positive interaction with peers throughout the day. She had a slight confrontation with one peer, but the two of them worked it out in a positive manner. Ct. has been very talkative and cheerful today. Ct. responded appropriately to all staff requests. Ct. remains on Level 2 and under constant supervision. _____Carlos Chaney

---

1/22/05 4:30 P.M. THERAPIST: Ct. requested and received her Level 3 today. She appeared very excited and stated that she has worked very hard for this and feels that she deserves it. We discussed the fact that being on Level 3 requires displaying Level 3 behavior. Ct. was receptive to this and seems to want to make the effort to do so. We also mutually decided that her new goal will be to take responsibility for her own actions and not blame other people. She understands that this is essential to her treatment and was receptive to doing so. _____Kim Stuck

---

1/22/05 9:30 P.M. ETC: Ct. participated appropriately in all group activities this evening. Went outside for structured activity time and participated appropriately. Interacted positively with her peers. Ct. participated actively and appropriately in group. Discussed her new goal and how much she desires to achieve it. Gave positive and appropriate feedback in group. Ct. completed her chore in a timely manner without being prompted by staff. She discussed with staff how much she wants to go home and how she wants to work through the program as quickly as possible so that she can go home. She seemed a little quieter than usual this evening. Ct. required no cues for the evening and responded appropriately to all staff requests. Ct. remains on Level 3 and under constant supervision. _____*B. Greer*

1/23/05 5:00 A.M. NTC: Ct. slept through the night with no problems and was present for each 15-minute bed check. _____*Phil Thress*

This residential treatment facility gives its staff members strict guidelines for charting and consistent abbreviations: Ct. to mean client, cue to mean a warning, and C to signify a consequence. Also, there must be no blank lines or spaces in a case note, so the recorder puts in the line and signature at the end of each entry. Note also that no names of other clients or staff members appear in any entry. The case manager routinely reviews notes such as these.

You will also note that the word *appropriate* appears often. In this facility, it is an important word because clients with borderline personality disorder often behave and interact inappropriately. It is common for them to act out sexually, exhibit interpersonal difficulties, and rebel against any rules or authority figures.

 # Deepening Your Knowledge: Case Study

Lonnie and Dorothy live in the southern part of town in a low-income area. Lonnie, age sixty-eight, works at a nursing home as a janitor and Dorothy, age sixty-six, does not work due to health problems (though she has worked as a nurse in the past). They have been married for thirty-nine years but do not get along very well. They basically coexist together without much substantial interaction.

Lonnie's job is rough on him because he works seven-day shifts and then has three days off. He drives forty-five minutes to work in his 1985 pickup (which he still makes payments on). He spends all day on his feet and is exhausted when he gets home. He has talked of finding better work but, because they live from paycheck to paycheck, they cannot afford for him to take any kind of pay cut, even temporarily. Lonnie spends most of his spare time in front of the television.

This is hard on Dorothy. Due to heart problems, she is on disability and rarely gets away from the house, often spending days alone. If she does get away, it is only to go to the grocery store or the pharmacy. She also has a tendency to be paranoid. For instance, if her disability check does not come on the exact day it did last month, she spends the afternoon on the phone talking to the disability office and the post office positive that someone has "made a mistake" or has "taken her check." Last month, Lonnie started to have some medical problems himself and has had to go the doctor more frequently than usual.

After receiving information during her disability appointment about financial planning and counseling services from a local non-profit agency, Dorothy and Lonnie decide to make an appointment. They arrive at the office and meet with Sandra, a case manager who handles between fifteen and twenty cases at any given point in time. Sandra welcomes Dorothy and Lonnie and describes the first steps in the case management process. She asks Dorothy and Lonnie if they would be willing to take a mental status exam after a discussion of their situation and needs. Although Lonnie expresses reluctance, both agree to the examination and discussion. Sandra also asked Lonnie and Dorothy to describe their strengths and weaknesses before concluding with a discussion of confidentiality and the scheduling of a follow-up appointment.

Morgan, C. (2012). Unpublished manuscript, Knoxville, Tennessee. Used with permission.

Discussion Questions

1. Did Lonnie and Dorothy's case manager conduct a formal interview, informal interview, or both? What would be the purposes of these interviews as they relate to Lonnie and Dorothy?
2. After the initial meeting, Sandra sits down to assess the case of Lonnie and Dorothy. What are some of the key questions she should ask in this process?
3. When they left the interview, Lonnie and Dorothy discussed their thoughts and feelings about the interaction with Sandra. What could Sandra have done to help this discussion take place with all three participants present so that she could understand their opinions more clearly?

 # Chapter Summary

The assessment phase of case management includes the initial contact with the individual or individuals who need or desire services, the intake interview to gather data, and the documentation that is required. Evaluating the application for services concludes the phase.

Applicants learn about services in different ways, which often result in a referral. The first step in determining eligibility for services is the intake interview, which may be structured or unstructured. The intake interview is an opportunity to establish rapport, explore the need or problem, understand client strengths, and provide some structure to the relationship. Culture is an important consideration in structuring this interview.

Case assignment follows acceptance for services and may occur in one of three ways. First, the case manager who conducts the intake interview may be assigned the case. Second, the case may be assigned to a specialized worker who may provide either complex services (level specialization) or specific services (task specialization). Finally, a team of professionals may be responsible for a number of clients.

Documentation and report writing are important parts of the assessment phase of case management. Two different types of documentation are process and summary recording. Intake summaries and case notes are examples of the documentation that occurs in this phase.

Chapter Review

Key Terms ....

Applicants
Client
Interview
Structured interview
Intake interview
Mental status examination
Unstructured interview

Confidentiality
Privileged communication
Intake summary
Staff notes
Process recording
Summary recording
SOAP

Reviewing the Chapter ....

1. How does a person's previous experience with human service agencies influence his or her decision to seek help?
2. In what different ways do people learn about available services?
3. Distinguish between the terms *applicant* and *client*.
4. What purposes does the interview serve?
5. Describe the roles of the helper and the applicant during the initial meeting.
6. Discuss different ways to define *interview*.
7. What is the difference between interviewing and counseling?
8. List the three factors common to all interviews.
9. Describe what takes place in each of the three parts of an interview.
10. Give three examples of issues that applicants may raise during the initial interview.
11. Why are intake interviews and mental status examinations structured interviews?
12. What does a helper need to know about confidentiality in the assessment phase?
13. Discuss the two problem areas related to confidentiality.
14. What questions guide the assessment of the information gathered during this first phase of service coordination?
15. What might a strengths assessment look like?
16. Describe case assignment.
17. Compare process recording and summary recording, and provide an example of each.
18. What purposes do staff notes (case notes) serve?
19. Describe the SOAP format and its components.

Questions for Discussion ....

1. Why do you think the interview is an important part of the case management process?
2. What evidence can you give that a good assessment is vital to the case management process?

3. Speculate on what could happen if a client's confidentiality were violated.
4. Develop a plan to interview a client who has applied for public housing.

● *References* ....

Cameron, S., & Turtle-Song, I. (2002). Learning to write case notes using the SOAP format. *Journal of Counseling and Development, 80,* 286–292.

Corey, G., Corey, M. S., & Callanan, P. (2011). *Issues and ethics in the helping professions* (8th ed.). Pacific Grove, CA: Brooks/Cole/Cengage.

Ivey, A. E., Ivey, M. B., & Zalaquett, C. P. (2007). *Intentional interviewing and counseling: Facilitating client development in a multicultural society* (6th ed.). Pacific Grove, CA: Brooks/Cole/Cengage.

McAuliffe, G., & Lovell, C. (2006). The influence of counselor epistemologies on the helping interview: A qualitative study. *Journal of Counseling Development, 84,* 308–317.

McQuaide, S., & Ehrenreich, J. H. (1997). Assessing client strengths. *Families in Society,* 78(2), 201–212.

Rangan, A. M., & Sekar, K. (2006). *Strengths perspective in mental health: Evidence-based case study.* Retrieved from http://www.strengthbasedstrategies.com/PAPERS/16RanganFormatted.pdf

Saleebey, D. (2002). *The strengths perspective in social work practice* (3rd ed.). New York: Allyn & Bacon. Retrieved from http://www.uwgb.edu/newpart/PDF/PostTraining/104/TOL15%20-%20Five%20Types%20of%20Questions%20to%20Assess%20Strengths%20HO28.pdf

Strengths-based restorative justice assessment tools for youth: Final project report. (2004). NPC Research. Retrieved from http://www.npcresearch.com/Files/Strengths%20Final%20Report%20SS%2052104.pdf

Weed. L. L. (1964). Medical records, patient care and medical education. *Irish Journal of Medical Education, 6,* 271–282.

. . . .

CHAPTER SIX

Effective Intake Interviewing Skills

Thriving and Surviving as a Case Manager

Ellen

There is more than one way for clients to enter our system of services. Sometimes the state determines the level of the services and so we are assigned based on the criteria of the grant that we are working through. For clients, like those at Adult Community Support, or if someone wants to go into counseling services, there is a number to call that's advertised on the website. If people drop in and want to get engaged in services, we will direct them through the intake line. They will call the intake line and describe what's going on for them. The intake specialist will gather some initial information about eligibility through Medicare, Medicaid, those sorts of things. Then we will assign that person to a case manager who is capable of doing intakes. If a person comes in, they sit with the intake specialist to schedule the intake time. If I were to do an intake, the client would come into the intake appointment and then we would sit down and go through the required legalities and the Duty to Warn and all of those sorts of documents. I'd have the client sign the informed consent, make sure we've got billing information correct, and do all those sorts of things. Then I'd start the intake process as far as gathering information. Depending upon what I found out during intake, there were times that I actually had to turn clients away. This didn't happen very often, but I remember once in particular, a young woman came in, very intelligent, college degree; she had a whole lot of anxiety and had a fair amount of insight. She was describing symptoms that she called panic attacks but what she was describing was actually neurological and so we had to determine that she did not meet criteria for any particular diagnosis, which is what needs to happen in order for someone to receive services. Therefore I couldn't get her in, but was able to give her some resources so she could go through some neurological testing to find out what was happening.

—Permission granted from Ellen Carruth, 2012, text from unpublished interview

Intake in our agency occurs on a regular basis, every Monday morning. We get clients from lots of sources, from the courts primarily. These are parolees or sometimes they are clients on probation and they are about to be released. Several of us conduct the intake; it is one of my favorite parts of my job. In my setting, many clients come in with handcuffs on and with armed guards if they are still in the custody. So my job in the intake is to assess attitudes for release, health, communicable diseases, and their history of violence. Because of the underlying danger to self and others, this intake is critical.

—Director, homeless shelter and recue mission, Miami, Florida

The intake interview is important because here you set the stage for future work and represent the policy, process, and atmosphere of the agency. In the first contacts with the family, you need to gain information from the family but you also need to be credible as a helper. Some of our clients are really smart, very educated, while others are unsure of what their problems are. They don't know what they need and cannot articulate this. We try to gain a holistic picture of the family and the situation; we want to gather as much information as we can.

—Director and case manager, children's services, New York, New York

I like to listen carefully to what the individuals being interviewed say about themselves. Listening is the key here. I don't think that I can tell clients as much as they can tell me. Once clients begin talking, they even surprise themselves, because they actually know so much about their situations. For instance, they will ask a question and then they will answer it. They are the experts in their own lives. Just be with them. That is important.

—Caseworker, family services, Bronx, New York

Interviewing is described in the previous chapter as directed conversation or professional conversation. Many helpers consider it an art as well as a skilled technique that can be improved with practice. In case management, the intake interview is a starting point for providing help. Its main purpose is to obtain an understanding of the problem, the situation, and the applicant. A clear statement of the intentions of the interview helps the case manager and the client reach the intended outcomes.

The chapter-opening quotations illustrate some important skills that are needed during the interviewing process. At the homeless shelter and rescue mission in Florida, the interviewer screens by seeking specific information about attitude, communicable diseases, and violent behavior. The director and caseworker at the children's services agency begins her contact with families by using questions to grasp the big picture. She makes a distinction between questions in general and the *right* questions. The caseworker at family services in the Bronx emphasizes listening as a critical skill in the interviewing process. It is her belief that clients will tell their problems if given an opportunity. Each of these professionals describes interviewing in a different way, but a common thread is respect for the considerable skills involved in using the interview to gain an understanding of the client's situation.

A number of factors influence interviewing in the helping professions. Some factors apply directly to the interviewer, such as attitudes, characteristics, and communication skills. In addition, factors such as gender, race, ethnicity, and power influence the process. Others are determined by the agency under whose auspices the interview occurs: the setting, the purpose of the agency, the kinds of information to be gathered, and recordkeeping. This chapter explores many of these factors.

The intake interview is usually the first face-to-face contact between the helper and the applicant. In some agencies, the person who does the intake interview will be the case manager; other agencies have staff members whose primary responsibility is intake interviews. Interviews are also a part of the subsequent case management process, and some of the skills used in the intake interview apply there, too. This chapter uses the term *case manager* to refer to the helping professional who is conducting the interview.

This chapter is about effective interviewing in case management: the attitudes and characteristics of interviewers, the skills that make them effective interviewers, how these skills are used in structured interviews, and the pitfalls to avoid when interviewing. For each section of the chapter, you should be able to accomplish the following objectives.

ATTITUDES AND CHARACTERISTICS OF INTERVIEWERS
- List two reasons why the attitudes and characteristics of the case manager are important to the interview process.
- Describe four populations of clients that may require the case manager's approach to be culturally sensitive.
- Name five characteristics that make a good interview.
- Describe a physical space that encourages positive interactions between the client and the case manager.
- List barriers that discourage a positive interview experience.

ESSENTIAL COMMUNICATION SKILLS
- List the essential communication skills that contribute to effective interviewing.
- List three interviewing skills.
- Support the importance of listening as an important interviewing skill.
- Offer a rationale for questioning as an art.
- Write a dialogue illustrating responses that a case manager might use in an intake interview.

INTERVIEWING PITFALLS
- Name four interviewing pitfalls.
- Describe each of these pitfalls.

 # Attitudes and Characteristics of Interviewers

The case manager's attitudes and characteristics as an interviewer are particularly important during the initial interview because this meeting marks the beginning of the helping relationship. Research supports the view that the personal characteristics of interviewers can strongly influence the success or failure of helping (Sommers-Flanagan & Sommers-Flanagan, 2009). In fact, they concluded after a review of numerous studies that these personal characteristics are as significant in helping as the methods that are used.

One approach to the attitudes and characteristics of interviewers is a framework that looks at two sets of critical attitudes: one related to self and the other related to how one treats another person. Consistent research confirms this approach (Brammer & McDonald, 2003; Combs, 1969; Sommers-Flanagan & Sommers-Flanagan, 2009). Those related to self include self-awareness and personal congruence, whereas respect, empathy, and cultural sensitivity are among the attitudes related to treatment of another person. Elsewhere in the literature, other perspectives on helping attitudes and characteristics have as common themes the ability to communicate, self-awareness, empathy, responsibility, and commitment (Woodside & McClam, 2012).

The case manager communicates helping attitudes to the applicant in several ways, including greeting, eye contact, facial expressions, and friendly responses. The applicant's perceptions of the case manager's feelings are also important to his or her impression of the quality of the interview. Communicating warmth, acceptance, and genuineness promotes a climate that facilitates the exchange of information, which is the primary purpose of the initial interview. The following dialogue illustrates these qualities.

INTERVIEWER: (*stands as applicant enters*) Hello, Mr. Johnson (*shakes hands and smiles*). My name is Clyde Dunn—call me Clyde. I'll be talking with you this morning. Please have a seat. Did you have any trouble finding the office?

APPLICANT: No, I didn't. My doctor is in the building next door, so I knew the general location.

INTERVIEWER: Good. Sometimes this complex is confusing because the buildings all look alike. Have you actually been to the Hard Rock Cafe in Cancun (*pointing to the applicant's shirt*)?

APPLICANT: No, I haven't. A friend brought me this T-shirt. I really like it.

INTERVIEWER: They certainly are popular. I see them all over the place. Well, I'm glad you could come in this morning. Let's talk about why you're here.

The case manager communicates respect for the applicant by standing and shaking hands. It is also easy to imagine that Clyde Dunn is smiling and making eye contact with Mr. Johnson. Clyde takes control of the interview by introducing himself, suggesting how Mr. Johnson might address him, and asking him to have a seat. His concern about Mr. Johnson finding the office and his interest in the T-shirt communicate warmth and interest in him as a person. Clyde also reinforces Mr. Johnson's request for help in a supportive way. All these behaviors reflect an attitude on Clyde's part that increases Mr. Johnson's comfort level and facilitates the exchange of information.

The positive climate created by such a beginning should be matched by a physical setting that ensures confidentiality, eliminates physical barriers, and promotes dialogue. It is disconcerting to the applicant to overhear conversations from other offices or to be interrupted by phone calls or office disruptions. He or she is sharing a problem, and such events may lead to worries about the confidentiality of the exchange. Physical barriers between the client and the case manager (most commonly, desks or tables) also contribute to a climate that can interfere with relationship building. As much as the physical layout of the agency allows, the case manager should meet applicants in a setting where communication is confidential and disruptions are minimal. It is preferable to have a furniture arrangement that places the case manager and the applicant at right angles to one another without tables or desks between them and that facilitates eye contact, positive body language, and equality of position.

A sensitive case manager is also cognizant of other kinds of barriers, such as **sexism**, **racism**, **ethnocentrism**, **ageism**, and attitudes towards **sexual orientation**. Problems inevitably arise if the case manager allows any biases or stereotypes to contaminate the helping interaction. To help you think about your own biases and stereotypes, indicate whether you believe each of the following statements is true or false.

T	F	Boys are smarter than girls when it comes to subjects like math and science.
T	F	Men do not want to work for female bosses.
T	F	Mothers should stay home until their young children are in school.
T	F	Women cannot handle the pressures of the business world.
T	F	Asians are smarter than other ethnic groups.
T	F	People on welfare do not want to work.
T	F	People who do not attend church have no moral principles.

T	F	A mandatory retirement age of 65 is necessary because people at that age have diminished mental capacity.
T	F	The older people get, the lower their sexual interest and ability becomes.
T	F	Gays are incapable of commitment in relationships.

How did you respond to these statements? Each statement reflects an unjustified opinion that is based solely on a stereotype of gender, race, age, or attitude toward sexual orientation.

Sensitivity to issues of ethnicity, race, gender, age, and sexual orientation is important for the case manager when conducting interviews. Many clients and families have backgrounds very different from that of the case manager. In the United States today, an increasing number of the population originates from non-European backgrounds, a large number of clients are women, and the population proportion of elderly people is increasing rapidly. For many people in these populations, life is difficult, and they have few places to turn for help. Many of them live in poverty, have inadequate education, have a disproportionate chance of getting involved in the criminal justice system (either as a victim or a perpetrator), possess few useful job skills, are unemployed, and suffer major health problems at a disproportionate rate (Factline, 2012).

Case managers should ask themselves, "How do I become sensitive to my clients and relate to them in a way that respects and supports their race, culture, gender, age, and sexual orientation?" Although we introduce aspects of multicultural perspective in Chapter 5, the following points may be helpful when considering the intake interview.

Each client is unique   It is easy to stereotype cultural, racial, gender, or age groups, but clients cannot be understood strictly in terms of their particular culture. For example, poverty-stricken, homeless clients share values and experience similar life events, but they are not all the same. During interviews, case managers must take special care to get to know each individual client rather than categorizing him or her as a member of one particular group. For example, one case manager at a housing development for the homeless in New York City explained how she struggles to see each individual as unique: "I work on seeing individuals as unique every day. It is easy to see one client and then see a whole host of clients that have the 'same' issue or problem. I try to look for what makes a person unique."

Language has different meanings   Do not assume that words mean the same thing to everyone who is interviewed. When the case manager asks interview questions, clients sometimes do not understand the terminology. Likewise, words or expressions that clients use may have a very different meaning for the interviewer. For example, questions about family and spouse are familiar subjects in an intake interview. When clients talk about "partners" or "family," these terms can have various meanings, depending on the cultural background and life experiences of the individual being interviewed. For example, in the Native American culture, the *family* is an extended one that includes many members of the clan. For gay men and lesbian women, the word *partner* has the special meaning of "significant other."

Another example of language having different meanings arises when working with a client who is deaf. One general rule of thumb is to avoid idioms and figurative language, such as "Cat got your tongue?" Someone who is hearing impaired may respond, "Where is the cat?" after interpreting the phrase literally. A second general rule is to be aware of words with multiple meanings. For example, *hard* may mean difficult—or it could mean rigid or unyielding. Words with multiple meanings are difficult for individuals with hearing impairment.

Discussed earlier, and still relevant, is the barrier of language for non–English-speaking individuals. Frequently, interpreters are not available. Even with interpreters present, clients indicate they still do not understand what information is provided. They do not feel understood (Factline, 2012). This hinders how those interviewed ask questions, answer questions, and engage with the intake interview process.

Explain the purpose of the intake interview and the case manager's role   Clients may show up for the interview without understanding its purpose or the role of the interviewer in the helping process. Confidentiality may also be an important issue for them—sharing information about themselves and others may be contrary to the rules of their culture. For example, for many people raised in Asian cultures, to describe a problem to someone who is not in the family implies making the matter public, which is considered to bring shame to the family.

Clients may be different from you   It is easy to make the mistake of expecting the clients we serve to be like us. We begin the interview process wanting to find similarities as a way of building a bridge to the client. When clients prove to be very different, or we cannot understand them, we often want them to change so that they will be easier to "manage." In the United States, we often like to think of our country as a melting pot in which all cultures mix together and lose their original identities. When individuals do not want to lose their own culture, there is a tendency to blame them for being difficult. Case managers must take special care in the interview process to let clients know that you respect their differences.

We have assembled some suggestions for developing sensitivity in interviewing individuals with certain cultural backgrounds (Atkinson, 2004; Choudhuri, Santiago-Rivera, & Garrett, 2012; Gilligan, 1982; Slattery, 2004; Sue & Sue, 2008). These are meant to be guidelines and points of awareness; they should be consulted with discretion. As we mentioned earlier, individuals seldom exhibit all the characteristics of their cultural group.

Interviewing clients of Native American origin   In many Native American cultures, sharing information about oneself and one's family is difficult. It is important not to give others information that would embarrass the family or imply wrongdoing by a family member. Listening behaviors such as maintaining eye contact and leaning forward are considered inappropriate and intrusive in some Native American cultures. For many Native Americans, trust increases as you become more involved in their lives and show more interest in them. Making home visits and getting to know the family can significantly improve an interviewer's chances of getting relevant information. Native Americans tend not to make decisions quickly. The slowness of the process could influence how soon the client is willing to share information or make judgments.

Native American cultures sometimes incorporate a fatalistic element—a belief that events are predetermined. During the initial stages of the process, the client may not understand how his or her responses and actions can influence the course of service delivery.

Interviewing clients with a common background of Spanish language and Hispanic customs   Individuals living in the United States who are of Mexican, Central and South American, or Caribbean ancestry are often referred to as Hispanic, Latino, or Chicano. There is actually little agreement on the appropriate term for identification across groups and even within subgroups. Although they share some commonalities, they may differ in appearance, country of origin, date of immigration, location and length of time in the United States, customs, and proficiency in English. Case managers should be sensitive to terminology and avoid stereotypes.

Many cultures with this common background view informality as an important part of any activity, even the sharing of information. Taking time to establish rapport with the client before direct questioning begins is helpful.

Some people of this origin may be perceived as submissive to authority because they appear reticent or reluctant to answer questions. Their behavior, in fact, may be shyness or the natural response to a language barrier.

The father may be seen as aloof as he performs his roles of earning a living for the family and establishing the rules. The mother and other members of the family tend to assume more nurturing roles. Questions that do not take these roles into consideration may be misinterpreted by the clients or may suggest to them that the interviewer is an outsider incapable of understanding their culture or of helping them.

Fatalism often plays a role in these cultures. These clients may not see any point in discussing the future, preferring to talk about the present.

Interviewing African Americans   Many African Americans do not believe that they receive the same treatment from social service agencies and professionals as white Americans. Reactions to this belief include a distrust of the human service delivery system, anger about discriminatory treatment, or both. This distrust may result in a reluctance to share information during the intake interview. During the intake interview, it is important to focus on concrete issues that can be connected to services. This approach shows respect for the client's right to expect fair treatment and quality services (Sue & Sue, 2008).

When being interviewed by a white professional, an African American may feel powerless or believe that his or her input does not matter.

Consideration of cultural values such as family characteristics, extended family and friends, educational orientation and experiences, spirituality, and racial identity may help demonstrate to the client that his or her input does matter (Sue & Sue, 2008).

Interviewing women   Many women do not know how to talk about the difficulties that they are experiencing, and they may not know how to respond to the questions they are asked. Some have had few opportunities to discuss their problems and may believe they do not have the right to complain. Listening carefully is very important.

Anger may play a part in the initial interview. Many women come to the helping process frustrated, either because their efforts have been unrecognized or because they believe that others expect them to be perfect. Often this anger must be expressed before any information can be gathered.

Women often feel powerless and do not expect the bureaucracy to serve them well. They may be reluctant to communicate and doubtful that the interview or the process as a whole can make a difference.

Women may also fill different roles in their lives that may conflict or cause confusion. When interviewing about client strengths, women from some traditional cultures in the United States may defer to males and elders and subordinate their own individuality, yet at work and at school, they may be assertive and confident. Without exploration, these differences may be perceived as weaknesses while in fact, the flexibility and role shifts may be strengths. Learning about roles and demands contributes to an understanding of the client's situation.

Women may be overly dependent as clients and assume that the case manager will take complete control of the interview. They may want the interviewer to be the one to identify problems and possible goals. In such cases, care should be taken to give the woman opportunities and encouragement to respond more fully.

Interviewing elderly clients   In this society, elderly people are often disregarded and devalued. During the interview, the case manager must show respect for the elderly client's answers and opinions about the issues discussed. Such a client needs to be assured that his or her responses are important and have been heard by the helper.

Pay special attention to the elderly client's description of support in his or her environment. Many live in an environment of decreasing support (changing neighborhood, death of friends) and with decreasing mobility. Others live with limited family support. These clients may not realize how their environment has changed.

Elderly clients may be reluctant to share their difficulties for fear of losing much of their independence. They may understate their needs or overstate the amount of support they have, hoping to avoid changes in their living conditions, such as being removed from their homes or relinquishing their driving privileges.

Interviewing individuals with disabilities   Although individuals with disabilities are not traditionally considered a cultural group, it is important to develop a sensitivity to the issues these individuals may encounter. Attitudes toward people with disabilities are often based on the amount of information and education about disabilities and on the amount of contact a person has had with individuals with disabilities (Atkinson, 2004). These factors are also the best predictors of positive attitudes toward people with disabilities. Case managers working with this population need to know about mental, physical, and emotional disabilities; the onset of disability; acceptance of disability; disabilities as handicaps; accommodations; and treatment.

A major source of information about a disability is the client. As with other clients, establish the helping relationship by building rapport and trust. Then address the disability or condition: Is it the problem? If not, does it affect the problem? Is it even related to the reason the client is seeking services? Don't make assumptions about why the individual is there or about the disability, and don't generalize. Each person is unique. Case managers also need to increase self-awareness about their own attitudes and knowledge. Know your limits and control your reactions. Increase your knowledge by learning from your client about a particular disability, the difficulties faced, and the environmental situations that are problematic.

Interviewing sexual minorities   In this society, discrimination against members of the gay, lesbian, bisexual, and transgender (GLBT) community is practiced in religious, legal, economic, and social contexts. This discrimination exists as an obvious external practice as well as a more subtle internal practice. To begin work with individuals who are part of the GLBT community, case managers need information about lifestyle issues and challenges these individuals confront. Issues include understanding the effects of prejudice, developing a positive identity, and becoming more aware of community resources. Those in the GLBT community also experience a loss of support from family and friends and are often victimized and harassed. One way case managers can establish a positive atmosphere is by using nonheterosexist language such as "partners" instead of "husband and wife." Intake forms can be revised to use nonheterosexist language (Sue & Sue, 2008). It is also important to conduct intake interviews that focus on the special issues described earlier that this community experiences.

These are only a few of the differences that helpers may encounter during the intake interview with individuals of various ethnic, racial, gender, and age groups. In several ways, case managers can continue to learn more about how to interview culturally diverse clients. Among them are becoming knowledgeable about other cultures, reading professional articles that focus on ways to modify the interviewing process to meet the needs of certain client groups, and talking with other helpers whose own cultural origins give them insight into cultural barriers. Gaining an understanding of diversity is a process that continues throughout the professional life of every effective helper. Such an understanding enhances the interviewing environment for both parties.

Essential Communication Skills

Communication forms the core of the interviewing process. In interviewing, communication is the transmission of messages between applicant and helper. As the first face-to-face contact, the interview is a purposeful activity for both participants. In many cases, the motivation is a mutual desire to decide whether the applicant is in the right place for the needed services. This is a negotiation that is facilitated by **effective communication skills.**

An important skill that promotes the comfort level of the applicant and lays the foundation for a positive helping relationship is using language the person understands. This means avoiding the use of technical language. For example, terms such as *eligibility*, *resources*, and *Form 524* may not mean much to an applicant who has not become familiar with the human service system. Another example is to imagine that the interviewer is discussing the benefits of taking a vocational or interest test. Rather than going into detail about the validity or reliability of the test, the case manager should discuss how it might help the applicant establish a vocational objective. Using language or words the applicant does not understand tends to create distance and disengagement.

Congruence between verbal and nonverbal messages is another way to facilitate the interaction between an applicant and a case manager. A major part of the meaning of a message is communicated nonverbally, so when conflict is apparent between the verbal and nonverbal messages, the applicant is likely to believe the nonverbal message. A common example of this is the person who says, "Yes, I have time to talk with you now," while dialing the phone or looking through her desk drawer for a folder. The lack of eye contact or any other encouraging nonverbal message communicates to us that the person is indeed busy or preoccupied with other matters.

Another skill that facilitates the interview process is **active listening**—making a special effort to hear what is said, as well as what is not said. An interviewer who is sensitive to what the applicant is communicating, nonverbally as well as verbally, gains additional information about what is really going on with the individual. This ability is particularly helpful in situations in which the presenting problem may differ from the underlying problem and in interviews with individuals from other cultures. Later in the chapter, we present a more detailed discussion of listening as it relates to the intake interview.

A popular way to elicit information is by asking questions. **Questioning** is an art as well as a skill. Unfortunately, case managers don't often develop their professional questioning skills, relying instead on questioning techniques that have served them well in informal or friendly encounters. Typically, this means asking questions that focus on facts, such as "What happened?" "Who said that?" "Where are you?" or "Why did you react that way?" Questions such as these usually lead to other questions, placing the burden of the interview on the case manager and allowing the applicant to settle into a more passive role. The applicant's participation is then limited to answering questions, so the interview may begin to feel like the game "Twenty Questions." Skillful questioning combined with effective responding helps elicit information and keep the interaction flowing. Appropriate questioning and responding techniques are introduced later in this chapter.

Patterns of communication vary from culture to culture, according to religion, ethnicity, gender, and lifestyle differences. In the dominant culture in the United States, it is effective to use a reflective listening approach when feelings are important. Many of the techniques that are useful in this approach are not appropriate for all cultures. For example, eye contact is inappropriate among some Eskimos. The sense of space and privacy is different for Middle Easterners, who often stand closer to others than Americans do. Some people from Asian cultural backgrounds may prefer more indirect, subtle approaches of communication. Thus, a single interviewing approach may have different effects on people from various cultural backgrounds. The skillful and sensitive case manager must be aware of these differences.

Both spoken language and body language are expressions of culture. Many helpers work with clients from several cultures, each with their own assumptions and ways of structuring information. Both talking and listening provide many occasions for misunderstanding. A director and case manager at a comprehensive community service center in St. Louis, Missouri, talks about the necessity of understanding various cultural backgrounds. "We serve individuals and families from a variety of cultures. For example, thirty of our clients are male and the rest are female. Fifty-five percent are African Americans, seventeen percent are white, and the rest are Asian and Hispanic . . . we also have a few individuals and families from the Middle East and a few Native Americans." Their staff is continually learning about the meaning of words and expressions used by a diversity of clients. When staff members do not understand or are lost in translation, they ask the clients to help them understand.

Assigning great significance to any single gesture by the applicant is also risky, but a pattern or a change from one behavior to another is often meaningful (Sielski, 1979). Once again, the key is the case manager's awareness during the interview process.

Now that you have read about general guidelines for essential communication, let's focus on the specific skills of listening, questioning, and responding.

Interviewing Skills

Interviewing skills aim to enhance communication, which involves both words and nonverbal language. Spoken language varies among individuals and cultures. Spoken language is challenging to understand because it is always changing, it is usually not precise, and it is ambiguous. Body language, which is also important and challenging to understand, includes body movement, posture, facial expression, and tone of voice. Knowing the ways in which body language varies culturally can help the interviewer fathom the thoughts and feelings of the applicant.

In talking with an applicant, the case manager must strive for effective communication, making sure that the receiver of the message understands the message in the way the sender intended. In the intake interview, the case manager listens, interprets, and responds. To understand the applicant's problem as fully as possible, the case manager constantly interprets the meanings of behaviors and words. He or she should always have a "third ear" focused on this deeper interpretation.

At the same time, the applicant is interpreting the words and behaviors of the case manager. A case manager who is an effective interviewer can help the applicant make connections and interpretations. Also contributing to correct interpretations and connections is a good working relationship between the two of them, good timing, and sensitivity to whether the material being discussed is near the applicant's level of education or experience.

A caseworker at the family services agency in New York City describes the initial meeting at her agency. She is sensitive to how clients are treated this first visit; she expresses concern that without sensitive treatment, the clients may not return.

> You have to be very careful with each client who comes to your agency for the first time. For most individuals, asking for help is a major step in their lives. You cannot create barriers. Several times, I have been busy and have told clients to come back in an hour or two. I remember one time in particular. A homeless man brought his young daughter to our door. We asked if he would like to wait, but he said "no." He and the daughter never returned.

A local police detective in California who works with gang-related crimes and gang members explains his approach to relationships.

> I approach gang members in a casual way. They know I am a cop. I talk to them about how things are in their world, family, friends, the neighborhood, fun, and school. I share some things about myself and the neighborhood where I grew up. Once I have them engaged in conversation, I think they begin to see me as a person, not as a cop, and I honestly try to see them as a person, not just a member of a gang. All in all I want them to tell me what is happening with them, the gang, and the neighborhood. I always ask them what they would do if they were not in a gang.

Both of these helping professionals are experienced at intake interviewing. They value the helping relationship and recognize its importance in the service delivery that is to follow. To establish the relationship, they use communication skills, such as listening, questioning, and responding. These are discussed and illustrated next, with excerpts from intake interviews.

LISTENING

Listening is the way most information is acquired from applicants for services. The case manager listens to the applicant's verbal and nonverbal messages. "Listening with the eyes" means observing the applicant's facial expressions, posture, gestures, and other nonverbal behaviors, which may signal his or her mood, mental state, and degree of comfort. Verbal messages communicate the facts of the situation or the problem and sometimes the attendant feelings. Often, feelings are not expressed verbally, but nonverbal messages provide clues. A good listener should be sensitive to the congruence (or lack of it) between the client's verbal and nonverbal messages. The case manager must pay careful attention to both forms of communication.

Good listening is an art that requires time, patience, and energy. The case manager must put aside whatever is on his or her mind—whether that is what to recommend for the previous client, the tasks to be accomplished by the end of the day, or making a grocery list—to focus all attention on the applicant. The case manager must also be sensitive to the fact that his or her behavior gives the applicant feedback about what has been said. During the interview, the case manager must also recognize cultural factors that play into the interpretation of body language. For example, the proper amount of eye contact and the appropriate space between case manager and applicant may vary according to the cultural identity of the applicant. As you can see, listening is indeed complicated. What behaviors characterize good listening? How are attentiveness and interest best communicated to the applicant?

Attending behavior, *responsive listening*, and *active listening* are terms that indicate ways in which case managers let applicants know that they are being heard. The following five behaviors are a set of guidelines for the interviewer (Egan, 2009, pp. 68–70). They can be easily remembered by the acronym SOLER.

- **S:** Face the client *squarely*; that is, adopt a posture that indicates involvement.
- **O:** Adopt an *open posture*. Crossed arms and crossed legs can be signs of lessened involvement with or availability to others. An open posture can be a sign that you're open to the client and to what he or she has to say.
- **L:** Remember that it is possible at times to *lean* toward the other. The word *lean* can refer to a kind of bodily flexibility or responsiveness that enhances your communication with a client.
- **E:** *Maintain good eye contact*. Maintaining good eye contact is a way of saying, "I'm with you; I'm interested; I want to hear what you have to say."
- **R:** Try to be relatively *relaxed*. Being relaxed mean two things. First, it means not fidgeting nervously or engaging in distracting facial expressions. Second, it means becoming comfortable with using your body as a vehicle of personal contact and expression.

Attending behavior is another term for appropriate listening behaviors. Eye contact, attentive body language (such as leaning forward, facing the client, facilitative and encouraging gestures), and vocal qualities such as tone and rate of speech are ways for the interviewer to communicate interest and attention (Ivey, Ivey, & Zalaquett, 2009). Attending behavior also means allowing the applicant to determine the topic.

Epstein provides additional guidelines for good listening (1985, pp. 18–19).

1. Be attentive to general themes rather than details.
2. In listening, be guided by the purpose of the interview in order to screen out irrelevancies.
3. Be alert to catch what is said.
4. Normally, don't interrupt, except to change the subject intentionally, to stop excessive repetition, or to stop clients from causing themselves undue distress.
5. Let the silences be, and listen to them. The client may be finished, or thinking, or waiting for the practitioner, or feeling resentful. Resume talking when you have made a judgment about what the silence means, or ask the client if you do not understand.

A skillful listener also hears other things that may help him or her understand what is going on. A shift in the conversation may be a clue that the applicant finds the topic too painful or too revealing, or it might indicate that there is an underlying connection between the two topics. Another consideration is what the applicant says first. "I'm not sure why I'm here" and "My probation officer told me to come see you" give clues about the applicant's feelings about the meeting. Also, the way the applicant states the problem may indicate how he or she perceives it. For example, an applicant who states, "My mother says I'm always in trouble," may be signaling a perception of the situation that differs from the mother's. Concluding remarks may also reveal what the applicant thinks has been important in the interview. The skilled interviewer also listens for recurring themes, what is not said, contradictions, and incongruencies.

Good listeners make good interviewers, but as you have just read, listening is a complex activity. It requires awareness of one's own nonverbal behaviors, sensitivity to cultural factors, and attention to various nuances of the interaction. It is further complicated by the fact that people seeking assistance don't always say what they mean or behave rationally. However, the use of good listening skills always increases the likelihood of a successful intake interview.

QUESTIONING

Questioning, a natural way of communicating, has particular significance for intake interviews. It is an important technique for eliciting information, which is a primary purpose of intake interviewing. Many of us view questioning as something most people do well, but it is in fact a complex art. This section elaborates on questioning skills, introduces the appropriate use of questions, identifies problems that should be considered, and explores the advantages of open inquiry as one way to elicit information.

Questioning is generally accepted by some as low-level or unacceptable interviewer behavior (Egan, 2009). Others view it as a complex skill with many advantages (Ivey, Ivey, & Zalaquett, 2009). Let's explore its complexity and its advantages.

There are several reasons questioning is a complex skill: Questioning may assist *and* inhibit the helping process; it can establish a desired as well as an undesired pattern of exchange; and it can place the client in the one-sided position of being interrogated or examined by the helper.

For these reasons, we may consider questioning an art form. The wording of a question is often less important than the manner and tone of voice used to ask it. A counselor serving also as a care coordinator at a family-focused agency in Tucson, Arizona says this:

> As a counselor and as a care coordinator, I have to think and act like a detective . . . this is especially true when I make a home visit. I enter the home as a stranger and I take in all of the information I can with all of my senses. . . . I go way beyond the formal five-page assessment that I have to fill out about each potential client and family.

A case manager at the School for the Deaf in Knoxville concurs:

> [The] way that I learn about my clients, I have to be curious about who they are, where they are, and why they are what they are. To do this I must appear interested but not too nosy.

Too many questions will confuse the applicant or produce defensiveness, whereas too few questions place the burden of the interview on the client, and may lead to the omission of some important areas for exploration. The pace of questions influences the interview, too. If the pace is too slow, the applicant may interpret this as lack of interest, but a pace that is too fast may cause important points to be missed. A delicate balance is required.

What are the advantages of questioning? One is that questioning saves time. If the case manager knows what information is needed, then questioning is a direct way to get it. Questioning also focuses attention in a particular direction, moves the dialogue from the specific to the general as well as from the general to the specific, and clarifies any inaccuracies, confusion, or inconsistencies. Let's examine some examples of the appropriate uses of questioning. After each example, you are asked to provide two relevant questions.

To begin the interview Could you tell me a little about yourself? What would you like to talk about? Could we talk about how I can help?

You work at the county office on aging. A woman comes in with her elderly mother. List two questions that you might use to begin the interview.

—————————————————————————————————

—————————————————————————————————

—————————————————————————————————

To elicit specific information How long did you stay with your grandmother? What happens when you refuse to do as your boss asks? Who do

Voices from the Field: Counseling Intake Interview Questionnaire

The following intake interview, posted on the Internet, reflects the approach one counselor uses to gain initial information about clients. She asks clients to fill out the following form. This form provides just one example of the questions case managers can ask during the intake interviewing process.

Intake Interview

The following questions may help me to better assist you in the counseling process. If for any reason you do not want to answer the questions, you do not have to. You may leave any question blank. During the course of therapy, I may ask follow-up questions based upon your answers. Again, these questions and the answers you provide me may likely increase my ability to be helpful.

Client Name _____

Date _____

Age _____

Who suggested you come to see me?

_____

_____

OK to thank referral? _____Yes _____No

In your own words, why are you seeking counseling at this time?

_____

_____

_____

_____

_____

_____

_____

_____

_____

_____

Have you had counseling in the past? _____Yes _____No

Who did you see?

_____

Approximate dates?

_____

Was it a good experience? _____Yes _____No

Why or why not?

_____

_____

_____

_____

_____

_____

_____

_____

Please list any medications you are currently taking.

_____

_____

_____

_____

_____

_____

_____

_____

_____

(continues)

(continued)

Have you ever gone to the hospital for mental health reasons?

_____Yes   _____No

Have you ever gone to the hospital for substance abuse reasons?

_____Yes   _____No

Approximate dates?

_____

_____

Where did you go?

_____

_____

In your own words, what do you hope to gain from counseling?

_____

_____

_____

_____

_____

_____

_____

_____

_____

What do you see as your strengths? (For couples, please identify relationship strengths)

_____

_____

_____

_____

_____

_____

_____

_____

_____

If you or any of your family has a history of any of the following, please indicate and describe *briefly.*

Health problems
Mental health problems
Substance abuse problems
History of abuse (physical, emotional, or sexual)
Legal problems
Economic problems
Occupational problems
Housing problems

Thanks for taking the time completing this form. We can talk about your reactions and any concerns in our next session.

© Semmler, P. (n.d.). Intake Interview. Retrieved from http://pamsemmler.com/forms.php

you think is pressuring you to do that? Can you give me an example of a time when you felt that way?

A client tells you about mistreatment by her boss at her new job. She claims that she is being sexually harassed. What two questions would you ask to help you understand what happened?

_____

_____

_____

To focus the client's attention   Why don't we focus on your relationship with your daughter? What happens when you do try to talk to your husband? Of the three problems you've mentioned today, which one should we discuss first?

A client facing surgery is worried about how the surgery will go, who will care for her children while she is in the hospital, and whether she will be fired for missing so much work. She wrings her hands and seems ready to burst into tears. What are two questions you could use to focus her attention?

_____

_____

_____

To clarify   Could you describe again what happened when she left? How did you feel about that conversation compared with others you have had with him? What is different about these two situations?

A young man shares his anguish over his mother's death a year ago. You notice that he is smiling, and you are confused about what he is really saying. Write two questions that would help you clarify what is going on.

_____

_____

_____

To identify client strengths   What is a current problem you have also faced in the past? Can you now use the same resources to solve your current problem? What did you do to keep the problem from turning into a crisis?

A young woman with a disability is questioned to assess functioning level and suitability for a program that requires her to ride public transportation. Write two questions that would help you identify her strengths.

_____

_____

_____

These scenarios are examples of interview situations in which the case manager might legitimately use questions. In all of them, the general rule of questioning applies: Question to obtain information or to direct the exchange into a more fruitful channel.

Although questioning may seem to be the direct path to information, sometimes this strategy can have negative effects. Long, Paradise, and Long (1981) suggest that interviewers not rely on questions to carry the interaction or interview. This is particularly problematic for beginning helpers because people generally have a tendency to ask a question whenever there is silence (discussed later in this section). Questions may also be inappropriate when the case manager does not know what to say. Asking questions nervously may lead to more questions, which can put the case manager in the position of focusing on thinking up more questions rather than listening to what the client is saying. Prematurely questioning to assess client strengths during the interview can also be problematic and may be viewed as rejection by the client.

An overreliance on questioning can create other problems for both interviewers and clients. For the client, too many questions can limit self-exploration, placing him or her in a dependent role in which the only responsibility is to respond to the questions. A client may also begin to feel defensive, hostile, or resentful at being interrogated. Using too many questions may place the case manager in the role of problem solver, giving him or her most of the responsibility for generating alternatives and making decisions. In the long term, overreliance on questioning leads to bad habits and poor helping skills. Using questions to the exclusion of other types of helping responses eventually results in the withering of these other skills (as discussed in the next section).

In conclusion, questioning is an important strategy for effective interviewing, but it is more than a strategy for obtaining information. Because of the subtleties of questioning, the matter of its appropriate uses in interviewing, and the potential problems, questioning is an art that requires practice. The skillful case manager who uses questioning to best advantage knows when to use open and closed inquiries to gather information during the intake interview. These types of questions are discussed next.

Closed and open inquiries   The questions used in intake interviews can be categorized as either open or closed inquiries. Determining which one to use depends on the case manager's intent. If specific information is desired, closed questions are appropriate: "How old are you?" "What grade did you complete in school?" "Are you married?" If the case manager wants the client to talk about a particular topic or elaborate on a subject that has been introduced, open questions are preferred: "What is it like being the oldest of five children?" "Could you tell me about your experiences in school?" "How would you describe your marriage?"

Closed questions elicit facts. The answer might be yes, no, or a simple factual statement. An interview that focuses on completing a form generally consists of closed questions like those in the previous paragraph. However, the interviewer must be cautious, for a series of closed questions may cause the client to feel defensive, sensing an interrogation rather than an offer of help. One approach is to save the form until the end of the interview, review it, and complete the unanswered questions at that time. If the completion of an intake form is allowed to take precedence in the interview, the case manager misses the opportunity to influence the client's attitudes toward the agency, getting help, and, ultimately, service provision. Perhaps just as importantly, information that could be acquired through listening and nonverbal messages may be missed if the interviewer is focused on writing answers on the intake form.

Open inquiries, on the other hand, are broader, allowing the expression of thoughts, feelings, and ideas. This type of inquiry requires a more extensive response than a simple "yes" or "no." The exchange of this type of information contributes to building rapport and explaining a situation or a problem. Consider the following example.

FATHER: I'm having trouble with the oldest boy, William. He's in trouble again at school.
INTERVIEWER 1:   How old is William?
INTERVIEWER 2:   Could you tell me more about what's going on?

Interviewer 1's response is a closed question that asks for a simple factual answer. Interviewer 2's response is an open inquiry that asks the father to elaborate on what he thinks is happening with William. This allows William's father to determine what he wishes to tell the interviewer about the situation. Such an open inquiry emphasizes the importance of listening—to what the individual says first, how he or she perceives the problem, and what is considered important.

You now should understand how valuable open inquiries can be in intake interviewing. They also provide an opportunity for the clients to introduce topics, thereby putting them at ease by allowing discussion of their problems in their own way and time. Besides providing the information that the case manager needs, open inquiries encourage the exploration and clarification of the client's concerns.

Four methods are commonly used to introduce an open inquiry (Evans, Hearn, Uhlemann, & Ivey, 2010). Each is presented here with an example of a client statement, the interviewer's response, and the kind of information that the client might volunteer in response to the open inquiry.

- "What" questions are fact oriented, and elicit factual data.

  MR. CAGLE: I'm here to get food stamps. Here's my application.
  INTERVIEWER: Let's review it to make sure you've completed it correctly. What's your income?
  MR. CAGLE: Well, I make minimum wage at my job, and my wife don't make much either. We have three children and we live in a low-income apartment.

- "How" inquiries are people oriented, encouraging responses that give a personal or subjective view of a situation.

  TAMISHA: My boyfriend doesn't like my parents, and when we are all together, nobody agrees with anyone about anything.
  INTERVIEWER: How do you feel about that?
  TAMISHA: I hate it. Everyone is so uncomfortable. I want everyone to get along, but I dread the times we have to be together. Sometimes I feel like somebody will yell at someone else or even hit somebody.

- "Could," "could you," or "can you" are the kinds of open inquiries that offer the client the greatest flexibility in responding. These inquiries ask for more detailed responses than the other types.

  JUAN: I hate school. My teacher doesn't like me. She's always on my case about stuff.
  INTERVIEWER: Could you describe a time when she was on your case?
  JUAN: Well, I guess. Like yesterday, she was mad at me because I was late to class ... but I was only five minutes late. Then she called on me to answer a question. Well, I hadn't read the stuff because I lost the book, so how could I answer the question? I mean, give me a break.

The fourth type of open inquiry is the "why" question, which experienced interviewers often avoid because it may cause defensiveness in clients. Examples of "why" questions that may do this are "Why did you do that?" and "Why did you think that?" Phrased this way, these questions may be perceived as judgments that the client should not have done something, felt a certain way, or had certain thoughts. Less-risky "why" questions are those phrased less intrusively: "Why don't we continue our discussion next week?" "Why don't we brainstorm ways that you could handle that?"

In what follows, we analyze some excerpts from an intake interview that occurred at juvenile court. Tom Rozanski is the case manager who was assigned to court on that particular day. In some such cases, the juvenile is remanded to state custody that very day; that is, he or she can leave the courthouse only to go to a local or state detention facility. The juvenile in this case, Jonathan Douglas, has been charged with breaking and entering. He has a history of substance abuse and school truancy and is well known to the judge, who finds him guilty and remands him to state custody. The case then comes under the jurisdiction of an Assessment, Care, and Coordination Team (ACCT), which takes responsibility for assessing the case, developing a plan of services, and coordinating the needed services among the agencies that are involved with the plan. Tom finds on this day that court is very crowded. Once Jonathan Douglas has been remanded to state custody, Tom asks him to follow him into the hall, and the initial intake interview occurs there. Jonathan's parents also join them, as do two officers, who suspect that Jonathan will run if he gets the chance. They stand together in the hall for a brief interview so that Tom can gather enough information to arrange a placement that afternoon. Here's what happens.

TOM:	Jonathan, my name is Tom. (*Shakes hands*) I work for the Assessment, Care, and Coordination Team. We are responsible for assessing your case and planning services for you.
JONATHAN:	(*Limply shakes hands and looks everywhere but at Tom*)
TOM:	Jonathan, are you listening? Please look at me. Are you on any drugs right now?
JONATHAN:	(*Unintelligible response*)

Tom realizes that it is futile to try to talk with Jonathan now and hopes that in a few hours he will be down from whatever drugs he has taken.

TOM:	Mr. and Mrs. Douglas, I am Tom Rozanski, a case manager for the Assessment, Care, and Coordination Team in this county. Let me review for you what has happened. The judge found Jonathan guilty of breaking and entering. Because of his prior record, he is in state custody, and it is my job to find a place for him to stay while we evaluate his case. I need some basic information right now. Can you help me?
MRS. DOUGLAS:	Yes, we want to help him any way we can.
TOM:	Does Jonathan live with either of you?

MRS. DOUGLAS:	He stays with me once in a while, but mostly he stays with his dad.
TOM:	Mr. Douglas, could you describe his behavior when he stays with you?
MR. DOUGLAS:	Well, I guess he goes to school sometimes. Leastways, when I leave for work, I try to get him up. I don't know if he goes, though. Sometimes he's here when I get home and sometimes he isn't. He's a big boy now, and I can't do much with him, so I just let him be.
TOM:	Do either of you have health insurance?
BOTH PARENTS:	No.

The interview lasts approximately five more minutes, and Tom obtains some key information about the family situation. He has very little time and needs specific information, so he hurriedly asks closed questions. "What is your address, Mr. Douglas?" "What grade is Jonathan in?" "Has he had a medical examination recently?" Finally, Tom has enough information to complete most of the intake form. That afternoon, he meets with Jonathan and makes another attempt to talk with him. He is relieved to find Jonathan more communicative at this meeting. Here's an excerpt; note Tom's use of open inquiries.

TOM:	Jonathan, I would like to talk with you about what's going to happen. I'd also like you to tell me your side of what's going on.
JONATHAN:	(*Looks at Tom but makes no comment*)
TOM:	When we finish talking, Deputy Johnston will take you to Mountainview Hospital, where you will spend the next two weeks. During that time, we will talk again, you will take some tests, and you will meet with a group of young people who are your age. At the end of that time, we will develop a plan of services for you. Now, could you tell me about yourself?
JONATHAN:	Well, I'm fifteen. I don't like school and I don't get along with either of my parents. My mother doesn't want me since she moved, and my dad doesn't care if I'm at home or not.
TOM:	This is the first time you have been in trouble for breaking and entering. What happened?
JONATHAN:	Well, I was with these guys and we needed money for some weed. It looked easy. I think I made a mistake.
TOM:	Yeah. It seems so. Let's talk about what you can do now. What kind of changes would you like to see?
JONATHAN:	Well, I don't want to go to jail and I don't want to go to Red River [a juvenile correctional facility]. I can't stay home though. They don't care about me and I don't care about them.
TOM:	How would you describe your relationship with your parents?
JONATHAN:	We don't have a relationship. They don't care about me. Sometimes I stay with my mom, but she's looking for

	another husband and she doesn't want me around. My dad, he just doesn't want to be bothered.
TOM:	Hmm. Sounds as though you're not sure if there's a place for you with them. What changes would you like to see in your relationship with your parents?
JONATHAN:	I wish they … wish … I wish they liked me.
TOM:	I see. Could you give me an example of what they would do if they liked you?
JONATHAN:	I don't know.
TOM:	Can you describe a time when you did something they liked?
JONATHAN:	My mom likes it when I come in early. My dad, he doesn't care.
TOM:	What have you done to please your mom?
JONATHAN:	(*Pauses*) I cleaned up the kitchen once.

In this excerpt, Jonathan mentions his family and school in his first response. Tom picks up on the family situation and decides to explore it with Jonathan. He has talked with the parents, and although they are not living together, he senses that both are interested in Jonathan and willing to help him but don't seem to know what to do, and they feel that Jonathan rebuffs any overtures they make. Tom is trying to discover what kind of support may be available to Jonathan from his parents and how receptive he would be to it. Tom uses open inquiries in his conversation with Jonathan to elicit the boy's thoughts and feelings about this issue. The use of "what" questions gets at factual information, and the "how" questions are aimed at people-oriented information.

In summary, case managers who are good interviewers use both open and closed inquiries, although open inquiries are preferred whenever possible. They are also careful to ask one question at a time and to avoid asking consecutive questions of a kind that might create the feel of a cross-examination. What other types of responses do interviewers use? The next section suggests other ways of responding to clients in an interview situation.

RESPONDING

A case manager might use various kinds of responses during the course of an intake interview. Of course, the type of response depends on the intent at that particular point. Let's review some of the most common responses. In the following material, each response is followed by an example of its use. Joe Barnes, a recent parolee, has returned home and is having a difficult time with his wife. His parole officer, sensing that the relationship is in trouble, suggests that Joe see a counselor at the Family Service Center.

Minimal responses   Sometimes called **verbal following,** minimal responses let the client know that you are listening. "Yes," "I see," "Hmm," and nodding are minimal responses. Using them is important when getting to know the applicant.

> JOE BARNES: I'm here because my probation officer thought it would be a good idea for me to talk with someone about things at home. Things haven't been very good since I came home.
>
> MIKE MATSON: I see.

Paraphrase   This response is a restatement (in different words) of the main idea of what the client has just said. It is often shorter and can be a summary of the client's statement. Paraphrasing lets the client know that the case manager has absorbed what was said.

> JOE: I just don't know what the trouble is. I was glad to get home, and I thought my wife would be glad to have me there. But we fight about everything—even stuff like when to feed the dog. I don't know what to do.
>
> MIKE: You don't know what's happening between you and your wife since you got home. Sounds like it's pretty unpleasant for both of you … and you're wondering what to do about it.

Reflection   Sometimes people get out of touch with their feelings, and reflection can help them become more aware. The feelings may not be named by the individual but, rather, communicated through facial expression or body language. For example, a flushed face or a clenched fist may show anger. The case manager's reflective response begins with an introductory phrase ("You believe," "I gather that," "It seems that you feel") and then clearly and concisely summarizes the feelings the case manager perceives.

> JOE: Yes. I don't know how we can continue to live like this. I know she is really angry about me getting in trouble with the law, but I've paid my dues, learned my lesson. I don't plan to ever get in that mess again.
>
> MIKE: I gather that you really do feel bad about what you did, but you would like to put the past behind you and focus on the future and how to make your marriage work.

Reflection is a response that facilitates a discussion of the client's feelings, particularly when he or she may feel threatened by such a discussion. It is also helpful as a way to check and clarify the case manager's perception of what was said during the interview.

Clarification   Clarifying helps the case manager find out what the client means. When the case manager is confused or unsure about what has taken place, it is more productive to stop and clarify at the time than to continue.

> JOE: I got so angry last week because she wouldn't listen to me and she didn't seem to care that I was home. I was yelling, she was yelling, she threw a bowl at me, and I almost hit her.

Want to Know More: Reducing the Blind Dimension

Rehabilitation counseling field services introduced case managers with responsibility for seeing clients through the entire service delivery process. In the ERIC archives, the authors found a document prepared by Rubin and Farley in 1980 designed to train rehabilitation counselors in the case management process. Particularly intriguing is how the article describes one objective of the intake interview as "decreasing three dimensions." It indicates that one of the goals of the intake interview is inform both client and helper and to gather information unknown to both. Page 10 of Rubin and Farley's participant workbook explains how to decrease the blind, hidden, and unknown dimensions.

Decreasing the Blind Dimension

During the intake process, the client needs much information. Extensive client participation and involvement is desired throughout the rehabilitation process but is very crucial during the intake process. The client needs information about the rehabilitation agency, its role and function, services that are offered, the objectives of those services, eligibility requirements, etc, to decide if the rehabilitation agency can meet his/her needs. Your role and the client's rights and responsibilities are other areas to be discussed with the client very early. You are more likely to practice effective information dissemination by knowing the information needed by the client and the most effective and efficient way to communicate that information.

Decreasing the Hidden Dimension

The client has much information needed by the counselor. You are more involved with the task of collecting information during the intake process than at any other time during the rehabilitation process. The hidden dimension is more likely to be diminished when you are aware of all of the information collection areas that should be explored and when you are an effective facilitator of free client expression and self-revealing behavior.

Decreasing the Unknown Dimension

The blind dimension is decreased via effective information dissemination. The hidden dimension is decreased via effective information collection. During this information exchange process, additional information known neither by the client or the counselor becomes known. The unknown dimension is further decreased via information from external sources, i.e., medical evaluation, psychosocial evaluation, vocational evaluation, etc.

Rubin, S. E., & Farley, R. C. (1980). Intake interview skills for rehabilitation counselors: A trainer's guide. Fayetteville, AR: Rehabilitation Research and Training Center. Retrieved from http://www.eric.ed.gov/PDFS/ED218535.pdf

MIKE: Sounds to me like you got so angry and frustrated that you were almost out of control.

Summarizing   With this response, the interviewer provides a concise, accurate, and timely summing up of the client's statements. It also helps organize the thoughts that have been expressed in the course of the interview. Summarizing is used to begin an interview when there is past material to review. It is also useful during the interview when a number of topics have been raised. Summarizing directs the client's attention to the topics and provides direction for the next part of the interview.

From the summary, the client can choose what to discuss next. Summarizing is also useful when the client presents a number of unrelated ideas or when his or her comments are lengthy, rambling, or confused; a summarizing response can add direction and coherence to the interview. Finally, summarizing is a way to close the interview: The case manager goes over what has been discussed. Prioritizing next steps or topics becomes easier at this point.

JOE: I told her I didn't care what she thought. I'm sure she knew what I meant even though I didn't know what I meant. She won't give me a chance. I am trying hard, so what does it mean to her that I have been gone? She has no idea what I have been through.

MIKE: Let me see if I can summarize what we've talked about today. Returning home has been very difficult for you and you're confused about your relationship with your wife. She still seems angry about your trouble with the law, and the two of you just can't seem to communicate.

JOE: I guess that's about it.

MIKE: Let's focus on the communication problems at our next meeting.

The following is an excerpt from an intake interview that incorporates all the responses that you have just read about: open and closed inquiries, minimal responses, paraphrases, reflection, clarification, summarization, interpretation, confrontation, and informing. Notice how and when the case manager uses each response and the client's reaction to it.

Mathisa walked into the AIDS Community Center one Wednesday evening about 8 o'clock. She had come to talk to a counselor because she had just discovered that her best friend had AIDS. Her friend had told Mathisa and no one else, and Mathisa was scared. She did not know what to tell her friend, and she did not know what to do. Mathisa passes by the center on her way to school each morning, but she had barely noticed it. And now she was here.

A young man came up to her and introduced himself. She said "Hi," but did not want to tell him her name. In fact, she really did not want anyone to know that she was there. He asked her if she had come to talk and she nodded. He led her into a small room that had three comfortable chairs. He sat in one and pointed to one where she could sit.

The young man, Dean, started by telling Mathisa about the agency and about his job as a service coordinator. He also talked to her about the confidentiality policies of the agency.

DEAN: I'm glad you're here.

MATHISA: I'm not sure I'm glad to be here. I've never been in this place before.

DEAN: It's scary to be in a place for the first time. We're always glad to welcome newcomers and visitors. (Smiles) What's going on?

MATHISA: (Pauses) I'm here for a friend.

DEAN: Your friend is very lucky that you could come for him or her. How did you decide to come here?

MATHISA: Well, this is a place I pass every morning on my way to school. Sometimes I wonder what it's like here. And today I knew that I needed to come. Can I be sure that nobody will find out what I tell you?

DEAN: Yes, what you tell me stays between the two of us. Confidentiality is very important to you.

MATHISA: I have some information, and I don't want anyone else to know. I don't know what I can do.

DEAN: Umm … (Nods)

MATHISA: You need to know what before you can help, I guess.

DEAN: Could you describe the event that brought you here?

MATHISA: I'm just so scared and I don't know what to do.

DEAN: It's scary having information and not having any idea what to do with it. How do you think I can help you?

MATHISA: I don't know for sure. But I do know that you understand AIDS and you help people with AIDS. I only know what they taught us in school. (Mathisa is obviously in distress; she is almost in tears and is choosing her words carefully.)

DEAN: Your quiet voice and your tears let me know that the reason you came is very upsetting to you.

MATHISA: (Nods)

DEAN: (Silence)

MATHISA: My best friend just told me that she has AIDS. She got tested when she was on a trip a month ago. She went to a state that does not ask your real name. She just found out yesterday. She's really blown away by it. No one else knows—not even her parents.

DEAN: She told you and you don't know what to do.

MATHISA: I don't really know anything about it. I don't want her to die, and I don't want to die. Her boyfriend doesn't know, and I don't know what she'll tell her parents. She may even run away or kill herself, but what if the tests are wrong? And seeing the really sick people here makes me think that I don't want to live.

DEAN: *Mathisa, I'm not sure what you said just then; you said that you didn't want your friend to die, and then you said that you didn't want to live.*

In this interview, Dean promoted good rapport with Mathisa by providing a good physical setting. It was simple, without distractions; they sat in close proximity, with no barriers between them, in comfortable chairs. Perhaps most important, it was an environment that was private. Dean introduced himself and assured her of confidentiality so that Mathisa felt comfortable beginning to talk.

Dean used a combination of open inquiries and responses. His first open inquiry was "Can you tell me why you're here?" This was designed to elicit a fact from Mathisa. She did not elaborate, but she did give enough information to continue the conversation. Dean also used "how" and "could" questions to encourage Mathisa to provide more information.

Dean's responses included a paraphrase ("Confidentiality is very important to you") as well as reflection ("Your quiet voice and your tears let me know that the reason you came is very upsetting to you"). Both of these responses helped Mathisa understand that Dean was actively listening to her and had heard what she had said. He had also interpreted her nonverbal messages.

At the conclusion of this excerpt from the interview, Dean used clarification ("I'm not sure what you said just then …") to try to sort through the information that Mathisa had given. In the remainder of the interview, Dean will continue to find out more about the problem and its implications for Mathisa and her friend. When they finish talking, Dean will summarize what has transpired and perhaps suggest where the relationship can go at that point.

Interviewing Pitfalls

Clearly, interviewing requires a great deal of skill. An effective interviewer is one who listens attentively, questions carefully, and uses other helpful responses to elicit information and promote client understanding. However, caution is necessary. The desire to be helpful and the anxiety of conducting that first interview can lead to a number of pitfalls. Four of them will be discussed here.

Premature problem solving   This arises from a desire to be helpful to the applicant by removing the pain, the discomfort, or the problem itself as soon as possible. Unfortunately, if the interviewer suggests a change, strategy, or solution before the problem has been fully identified and explored, this may address a symptom of the presenting problem rather than the actual problem. Premature problem solving may cause the client to lose confidence in the case manager's knowledge and skills or to become impatient. Also, premature problem solving undermines the client's self-determination and can lead to false assumptions or misinterpretation of what the client says, and steer him or her in the wrong direction. In the case of mental illness, misdiagnosis can result.

Giving advice   In attempting to solve the problem or offer a solution, the case manager may mistakenly give advice. When given hurriedly and before the problem has been explored sufficiently, advice may be seen as indicating a lack of interest or thoroughness. The client may also feel misunderstood, or he or she may superficially agree, without intending to follow through. Advice giving also tends to diminish the client's level of responsibility, self-determination, and partnership in problem solving.

Overreliance on closed questions   The pitfall of overuse of closed questions has been discussed elsewhere in this chapter. Remember that closed inquiries are usually directive and focused on facts; they rarely provide the opportunity for exploration. A series of closed inquires might also make the client defensive. Once this feeling is established, it is difficult to overcome.

Rushing to fill silence   Because silence is often awkward in everyday social situations, beginning helpers as well as seasoned professionals are sometimes uncomfortable with pauses and rush to fill them, believing that silence indicates that nothing is happening. In fact, silence does have meaning. The client may be waiting for direction from the interviewer, thinking about what has transpired so far, or just experiencing an emotion. Constant dialogue can be a false signal that something is happening. Skillful case managers learn to listen to silence.

Deepening Your Knowledge: Case Study

Marcus works for Tribe Solutions, a provider of behavioral counseling services to adolescents on a Native American reservation in southern Florida. He started the position two months ago, and just last week received some training on intake interviewing, with the expectation that he would be able to conduct his own intakes by the end of the month.

After his second opportunity to shadow Marcia, a veteran counselor with over a decade of experience at the site, Marcus sat down to make notes of his observations about the intake interview process. Marcus has noticed that his office is very different than Marcia's. Because he has been so busy since starting the job, many of his belongings and files are disorganized. His office furniture consists of a large desk, a filing cabinet, an office chair, and a straight-back chair in the corner of the room. In contrast, Marcia's office is very inviting, with plants, artwork, and a warm, open space between three comfortable chairs. Marcus has wondered if these physical characteristics have any impact on the clients' experience.

Marcus also observed Marcia's friendliness and inviting style with her clients. Though she has to confront them on many occasions, she consistently conveys an attitude of caring and helpfulness. Even though the paperwork for the intake interview indicates a highly structured series of questions, Marcia often lets her clients lead for portions of the conversation before returning to the script. Marcia told Marcus that she believes her attitude and framing of client issues must focus on support and encouragement because Native American youth come from a marginalized culture and often view outsiders with skepticism and hostility. Marcus remains unsure of how to balance this approach with the program's expectation that delinquent youth must be held accountable for their actions, and thinks that Marcia's infrequent eye contact does little to help convey her stated attitude. Marcus is troubled by the indifference toward life circumstances and behavioral problems he has observed in Marcia's clients and wonders why Marcia does not give her clients more advice. Marcia is also careful in conversation with clients to be sure that they understand the content and meaning of their dialogues, which often seems unnatural to Marcus.

When considering Marcia's interactions with clients, Marcus has trouble understanding how she chooses between questioning and active listening. He has little confidence that he will be able to distinguish between the appropriate moments for these delicate skills. After taking these notes, Marcus realizes that he is very unsure about his suitability for the job. He decides to meet with Marcia the next day to discuss his questions and concerns.

Morgan, C. (2012). Unpublished manuscript, Knoxville, Tennessee. Used with permission.

Case Study Discussion Questions

1. Why is it important that Marcus try to model the physical space of his office more closely on what he has seen in Marcia's office?
2. How would you rate Marcia's cultural sensitivity toward her clients, as indicated by Marcus's statements? How should she approach this topic with Marcus to help him better understand the steps he needs to take?
3. If you were in Marcia's role of discussing these questions with Marcus, which issues would you see as most important, and how would you go about answering them for Marcus?

Chapter Summary

This chapter focuses on the interviewing process, especially the intake interview that initiates the helping process. Critical to this process are the helper's values and attitudes, because these can convey to the client how the helper feels about him- or herself and how the helper feels about the client. A positive interaction is more likely to occur if the helper demonstrates warmth and caring. Attention to the physical space can also facilitate the intake interview process. For example, talking with the client in a private area can signal that confidentiality is important. Barriers that discourage the client can be a lack of sensitivity to racial, religious, cultural, or gender issues.

Basic communication skills are important if the case manager is to establish a dialogue with the client. Effective communication includes demonstrating congruence between what is said verbally and nonverbally, engaging in active listening, and being sensitive to cultural differences. Listening is key to conducting an effective interview. This means imparting to the client that attention is being paid to what is said, which in turn conveys respect and the desire to learn about a client. Listening also includes responsive listening and the active listening described earlier. Questioning is both a skill and an art. Used appropriately, questioning is an effective way of helping people talk about themselves without asking direct, closed questions.

   Good interviewers not only develop communication skills, but they also learn to avoid pitfalls, including premature problem solving and giving advice. Case managers are often tempted to identify the problem too quickly or to seek an immediate solution. Both of these responses may focus more on the case manager than the client. Another pitfall is an overreliance on closed questions, which discourage clients from talking. The final barrier is rushing to fill silence rather than giving clients time to assume responsibility for part of the dialogue. These barriers can be replaced with other communication skills that encourage client participation.

Chapter Review

Key Terms ....

Sexism	*Active listening*
Racism	*Questioning*
Ethnocentrism	*Attending behavior*
Ageism	*Closed questions*
Sexual orientation	*Open inquiries*
Effective communication skills	*Verbal following*

Reviewing the Chapter ....

1. What attitudes and characteristics facilitate the development of a helping relationship?
2. Write a dialogue representing the beginning of an intake interview to illustrate desirable attitudes and characteristics of the helper.
3. Describe an office setting that facilitates relationship building.
4. Discuss the problems that are created by stereotypes based on gender, race, age, and sexual orientation.
5. Write general guidelines for essential communication skills in interviewing.
6. Discuss the importance of listening in the intake interview.
7. What is attending behavior (active listening)?
8. What are the five listening behaviors represented by the acronym SOLER?
9. Why are listening and questioning both complex skills and arts?

10. State the advantages and disadvantages of questioning.
11. Describe the five situations in which questioning is appropriate.
12. Distinguish between closed and open inquiries.
13. What are the four commonly used methods of introducing an open inquiry?
14. Name four pitfalls of interviewing and tell how each may be avoided.

● Questions for Discussion ....

1. Do you think that you will be able to conduct a good interview? What skills will you need to strengthen your competence as an interviewer?
2. What are the advantages and disadvantages of recording client information on the computer while conducting the intake interview? Speculate about how interviewing changes if computers are used in the process.
3. Discuss the kinds of activities that might help you practice your listening skills.
4. Do you believe that certain communication skills are essential to effective interviewing? If your answer is no, why not? If yes, what are they, and why are they important?

● References ....

Atkinson, D. R. (2004). *Counseling American minorities*. Boston: McGraw Hill.

Atkinson, D. R., & Hackett, G. (2004). *Counseling diverse populations*. Boston: McGraw Hill.

Brammer, L. M., & MacDonald, G. (2003). *The helping relationship*. Boston: Allyn & Bacon.

Carkhuff, R. R. (1969). *Helping and human relations: A primer for lay and professional helpers. Vol 1: Selection and training*. New York: Holt, Rinehart & Winston.

Choudhuri, D. D., Santiago-Rivera, A., & Garrett, M. (2012). *Counseling and diversity*. Pacific Grove, CA: Brooks Cole/Cengage.

Combs, A. W. (1969). *Florida studies in the helping professions*. Gainesville: University of Florida Press.

Egan, G. (2009). *The skilled helper: A problem-management and opportunity-development approach to helping*. Pacific Grove, CA: Brooks/Cole.

Epstein, L. (1985). *Talking and listening: A guide to the helping interview*. St Louis: Times Mirror/Mosby.

Evans, D. R., Hearn, M. T., Uhlemann, M. R., & Ivey, A. E. (2010). *Essential interviewing: A programmed approach to effective communication*. Pacific Grove, CA: Brooks/Cole.

Factline. (2012). *Tracking health in underserved communities*. Retrieved from http://www.mmc.edu/www.meharry.org/fl/access_to_health_care/barriers_to_care_for_immigrants.html

Gilligan, C. (1982). *In a different voice*. Cambridge, MA: Harvard University Press.

Ivey, A. E., Ivey, M. B., & Zalaquett, C. P. (2009). *Intentional interviewing and counseling: Facilitating client development in a multicultural society*. Pacific Grove, CA: Brooks/Cole.

Sielski, L. M. (1979). Understanding body language. *Personel and Guidance, 57*(5), 238–242.

Sommers-Flanagan, J., & Sommers-Flanagan, R. (2009). *Clinical interviewing.* Boston: John Wiley.

Sue, D. W., & Sue, D. (2008). *Counseling the culturally diverse.* Boston: John Wiley.

Woodside, M., & McClam, T. (2012). *Introduction to human services.* Pacific Grove, CA: Brooks/Cole.

Yuker, H. E., (1994). Variables that influence attitudes toward people with disabilities: Conclusions from the data. *Journal of Social Behavior and Personality, 9*, 3–22.

CHAPTER SEVEN

Service Delivery Planning

Thriving and Surviving as a Case Manager

Katie

I work with school-age children, and most of these children have behavior problems. During the planning for them we use a central file for all records. All of our records are electronic, so everything is on the computer. For every visit we have to do a case management progress note, which is just kind of an outline. It has subjective findings of what you and the client are working on and what comes out of that. It has objective findings where I might record the child's appearance and affect. There is also an assessment part; I just record my ideas and opinions about what's going on. In that section I have to talk about our SNAP goals, which refers to the client's strengths, needs, abilities, and preferences, and that's in every single assessment. I write something positive and also something they might need to work on. And then at the end we also talk about goals. I ask, "What are the goals for next time we meet?" So it's always based on the treatment plan. The treatment plan is developed by me as the case manager.

There is a treatment plan outline, which is an assessment of their living arrangements, school, and support; then there is the treatment plan where we actually outline the goals. They have to be in the client's words, because we want to work on what the client and the parents want to work on. So first we outline their goals, and then we describe the steps that we are going to take to reach their goals.

I also meet with a team of four people in our school-based services group. We also have a partner team with ten or twelve people. Once a week we have a team meeting where we talk about updates that are going on in the center. We do a resource highlight every week and talk about things that are going on, such as client concerns and referrals.

—Permission granted from Katie Ferrell, 2012, text from unpublished interview.

Although some agencies call the individuals they work with "clients," we use the term "consumers." We changed our language when we began to evaluate our services. These individuals actually come here for the services we provide. In other words, they actually "buy" our services. They could go elsewhere. Consumers choose agencies for a variety of reasons; we hope they see this agency as a welcoming and friendly place.

    —Case manager, family services, New York, New York

We use the Millon Adolescent Clinical Inventory for each of our new clients who come to live with us. We call our clients "residents." This inventory has 160 items. It is a questionnaire and addresses a range of issues including personality patterns, expressed concerns, and clinical syndromes. First we score the questionnaire, and then we identify issues of concern from the question and from the resident. Finally the resident and the case manager together make a treatment plan that addresses the areas of concern.

    —Case and intake worker, emergency shelter, St. Louis, Missouri

When I first meet clients during the interview, I help them understand what will happen if they join our program. We have some basic criteria, such as having a 9.5 in math and a 10.0 in reading on the TABE test. Plus they have to be able to type 30 words a minute and have a minimum understanding of computers. If they join the program, they must attend classes on a regular basis. They may only miss eight days within a six-month period. That is difficult for some of them.

    —Case manager, welfare-to-work and case management services, Knoxville, Tennessee

At this point in the process, the agency has determined that the applicant meets the eligibility criteria and the services are appropriate, so the person can now receive services. At the family services agency, the applicant becomes a *consumer*, whereas the welfare-to-work program uses the term *client*. An agency in South Dakota that serves adults with developmental disabilities calls the service recipients *individuals*, explaining that "they are not clients or consumers anymore. They are just people."

Other agencies or organizations use the term *customer*. The change in status from applicant to recipient of services marks the move into the second phase of case management: planning service delivery.

The quotes that introduce this chapter identify some of the activities that occur during this phase. At the emergency shelter in St. Louis, Missouri, client participation is important in planning. In fact, clients determine the goals. Client interests and expectations and use of test data are shared at the welfare-to-work program in Knoxville, Tennessee. One case manager in New York City summarizes this phase of case management: "You just have to read through the information several times and say, 'What stands out here? What issues should I pay attention to?' Then you say, 'Well, if these are the problems my clients and I think are important, what do we do about them?'"

A caseworker at a school for the deaf also notes the importance of gathering information in order to see the big picture: "One skill that I need in this job is to be able to talk with the different professionals, family, and friends of the children I work with. So many people have to be involved to address such complex problems. I search and search for all of the little pieces. But the job is still not done. I need to step back and see the big picture. This is the key to case management. Without a case manager or someone taking the case management approach, no one has the big picture."

This chapter explores the planning phase of case management, when the helper and the client together determine the steps necessary to reach the desired goal. The activities involved in this phase include reviewing and continuing to assess the problem, developing a plan, using an information system, and gathering additional information. Running through our discussion in this chapter are two critical components of the case management process—client participation and documentation.

For each section of the chapter, you should be able to accomplish the objectives listed here.

REVISITING THE ASSESSMENT PHASE
- List the two areas of concern that are addressed when reviewing the problem.

DEVELOPING A PLAN FOR SERVICES
- Identify the parts of a plan.
- Write a plan.

IDENTIFYING SERVICES
- Locate available services.
- Create an information and referral system.

GATHERING ADDITIONAL INFORMATION
- Compare interviewing and testing as data collection methods.
- Identify the types of interviews.
- Show how sources of error can influence an interview.
- Illustrate the role of testing in case management.
- Define *test*.
- Categorize a test.

- Identify sources of information about tests and the information that each provides.
- Analyze the factors to be considered when selecting, administering, and interpreting a test.

Revisiting the Assessment Phase

The next phase of case management begins with a review of the problems and strengths identified during the assessment phase. Before moving ahead with the process, the case manager will need to know if the problem has changed, if the same client resources are available, and if any shift in agency priorities has occurred. In order to complete the review quickly before moving into a planning mode, the case manager and the client examine two aspects of a case.

The first area of concern involves a review of the relevant facts regarding the problem. At this point, the case manager and the client revisit the identification of the problem. The initial question the helper asks can help determine whether the problem still exists. Working with people requires an element of flexibility; clients' lives change, just as ours do. Thus, the problem may have changed in some way; the client may have a different perspective on it; the participants may be different; or assistance may no longer be needed, appropriate, or wanted. Once the case manager has confirmed that the problem still exists and has documented any changes that have occurred, the problem itself is revisited. Is the problem an unmet need, such as housing or financial assistance, or is it stress that limits the client's coping abilities or causes interpersonal difficulties? Is the problem a combination of several factors? This activity is best accomplished by talking with the client and reviewing his or her file. The client is still considered the primary source of information and a critical partner in the case management process.

A second area of concern in the review of the problem requires an examination of available information to answer the following six questions.

- What do I know about the source of the problem?
- How does culture influence the client and environment?
- What attempts have been made previously (before agency contact) to resolve the problem?
- What are the motivations for the client to solve the problem?
- What are the interests and strengths of the client that will support the helping process?
- What barriers may affect the client's attempts to resolve the problem?

An important source of information is the client. Talking with the client can reveal what he or she has thought about doing, what has been tried, and some possible solutions. Exploring with the client his or her motivations, strengths, interests, and cultural considerations indicates that the process of case management continues to be a partnership between the client and the case manager.

Other techniques that are helpful in reviewing the problem are observations and documentation. In the course of receiving the application, conducting the intake interview, making a home visit, or all three, the case manager has opportunities to observe the client beyond the office setting. For example, these observations may be richer if they occur in the home or if the client is accompanied to the office by family members or a significant other. Information available from such observations includes the client's thoughts, feelings, behaviors, and relationships.

Documentation in the case file also provides facts and insights about the client. Case notes, reports from other professionals, and intake forms help the case manager pin down past occurrences and pertinent facts about the present situation. Case managers who have a long history in service delivery may call on knowledge and experience from the past to understand a current case. Sometimes, knowledge comes from a case manager's own perception, instinct, experience, or street know-how. Many case managers mention rapid insight they sometimes have about a client, the client's environment, possible difficulties, and creative approaches to the case management process. This insight is treated as just one piece of information and must undergo the same scrutiny as the other information collected.

Once the case manager has revisited the problem, confirmed its existence, documented any changes, and reaffirmed the client's desire for assistance, the two of them move to the next step of the planning phase, which addresses the need to determine the steps necessary to reach the identified goal or goals. This is the plan that will guide service provision.

Developing a Plan for Services

The **plan** is a document, written in advance of service delivery, which sets forth the goals and objectives of service delivery and directs the activities necessary to reach them. The plan also serves as a justification for services by showing that they meet the identified needs and will lead to desired outcomes. More specifically, a plan describes the service to be provided, who will be responsible for its provision, and when service delivery will occur. If there are financial considerations, the plan may also identify who will be responsible for payment. Sometimes financial support is available from outside sources, including the client and the family. Usually, the completed plan is signed by the client and the case manager as the representative of the agency. It may then be approved by someone else in the agency before the authorization to provide services is granted.

Clearly, the plan is a critical document, since it identifies needed services and guides their provision. How is it developed? What is included? What are goals and objectives? What factors might present planning challenges? You will learn answers to these questions as you read this section.

Plan development is a process that includes setting goals, deciding on objectives, and determining specific interventions. The process begins with the synthesis of all the available data. This information is scrutinized carefully to assemble as complete a picture of the case as possible. It is analyzed to identify inconsistencies, desirable outcomes, or both. It is also important to consider the veracity of the available data. For example, if substance abuse is a problem, how accurate is the client's report of the amount of alcohol consumed daily or the extent of with-drawal (sleeping disturbances, d.t.'s, blackouts, convulsions, hallucinations, etc.).

For the beginning case manager, the following method uses a step-by-step approach to synthesize data and integrate the information into a workable plan. Using the worksheet displayed in Table 7.1, the case manager can record his or her analysis.

- Reread the client file and fill in the following categories on the worksheet: sources of information, relevant facts.
- With this snapshot of the contents of the client's file, assess and record conclusions, contradictions, and missing information.
- Review this assessment with the client and make revisions according to his or her input and other new data gathered; fill in client motivations, strengths, interests, and cultural considerations with client input.
- Discuss desirable outcomes with the client.

In Roy Roger Johnson's case in Chapter 1, the information available at the time of plan development was derived from Roy's application for services, the intake interview, reports from his orthopedic surgeon, case documentation, a general medical examination report, a psychological evaluation, and a vocational evaluation report. When Roy and his counselor developed the plan of services, they reviewed and considered all this information using the steps listed in Table 7.1.

Roy had a back injury and needed assistance finding a job; he also met economic eligibility criteria. His service plan, reproduced in Figure 7.1, included a program objective and intermediate objectives. For each objective, a service was identified, as well as a method of checking progress toward the achievement of the objective. The form also provided space to describe any other client, family, or agency responsibilities or conditions. Because this agency values client participation, Roy's view of the program was also noted. Then both Roy and the counselor signed the plan.

Exactly what a plan looks like varies from agency to agency. However, if you are employed by an agency that provides case management or client services, you can be sure that a plan will guide your work. Let's examine the components of a plan of services.

Service plans are goal directed and time limited, so they should include both long-term and short-term goals. Long-term goals state the situation's ultimately desired state. Short-term goals aim to help the client through a crisis or some other present need. Whatever the time constraints, goals establish the direction for the plan and provide structure for evaluating it.

Goals are statements that describe a state, a condition, or an intent. For clients, a goal is a brief statement of intent concerning where they want to be at the end of the process; for example, "Learn daily living skills in order to live independently," "Acquire knowledge and skills for a career in business communications," or "Develop a support network for help in coping with phobias."

Having written goals helps us focus on what we are trying to accomplish before we take action or provide any services. Action is often easy, but sometimes

TABLE 7.1   INTEGRATING CLIENT INFORMATION

CLIENT WORKSHEET

Client Name: _____

Date: _____

Source of Information	Relevant Facts	Conclusions	Contradictions	Missing Information	Motivations of Client	Strengths of Client	Interests of Client

SERVICE PLAN

1. NAME ____JOHNSON, ROY_____ PROGRAM TYPE ☒ INITIAL ☐ AMENDMENT

2. YOU ARE ELIGIBLE FOR: ☒ VOCATIONAL REHABILITATION SERVICES ☐ EXTENDED EVALUATION SERVICES ☐ PAST EMPLOYMENT SERVICES

    BECAUSE: ☒ A. YOU HAVE A PHYSICAL OR MENTAL DISABILITY WHICH CONSTITUTES A SUBSTANTIAL HANDICAP TO EMPLOYMENT AND:

    ☒ B. YOU CAN REASONABLY BE EXPECTED TO BENEFIT IN TERMS OF EMPLOYABILITY FROM SERVICES.

    ☐ C. IT CANNOT BE DETERMINED WHETHER OR NOT YOU CAN BENEFIT IN TERMS OF EMPLOYABILITY FROM REHABILITATION SERVICES.

    ☐ D. POST EMPLOYMENT SERVICES ARE NEEDED FOR YOU TO MAINTAIN EMPLOYMENT.

3. PROGRAM OBJECTIVE: ___Business Communications____ ANTICIPATED DATE OF ACHIEVEMENT: MONTH__1_YEAR xx
    ESTIMATED DATES TO REACH OBJECTIVE & RECEIVE SERVICES

4. INTERMEDIATE OBJECTIVE, SERVICES METHODS OF CHECKING PROGRESS.

	RESPONSIBILITY	FROM	TO
OBJECTIVE To correct physical impairment so that client might reach vocational objective	Client	1/XX	1/XX
SERVICES Possible office visit with the doctor			

METHOD OF CHECKING PROGRESS ___Medical information____

	RESPONSIBILITY	FROM	TO
OBJECTIVE To provide background information and educational skills so that client might reach vocational objective	VR	1/XX	1/XX
SERVICES A. Tuition/		1/XX	1/XX
B. Miscellaneous Educational Expenditures		1/XX	1/XX

METHOD OF CHECKING PROGRESS ___R-II, Grade Reports____

	RESPONSIBILITY	FROM	TO
OBJECTIVE To follow client's progress and develop plan amendment if needed so that client might reach objective	Client	1/XX	1/XX
SERVICES A. Possible RP-B		1/XX	1/XX
B. Client/Counselor Contacts		1/XX	1/XX

METHOD OF CHECKING PROGRESS R-11

5. CLIENT OR FAMILY AND AGENCY RESPONSIBILITIES AND CONDITIONS: I. Client is responsible to maintain contact with counselor twice each semester by mail, phone or in person. II. Client is responsible to furnish VR Counselor with a copy of grades at the end of each term. III. Client is responsible to maintain an average load of classes and average grades throughout his program. IV. Client is responsible to file for any similar benefits which might help him pay for his program. V. Client is responsible to furnish VR counselor with a resume and a list of potential employers to interview with during the first part of his senior year. VI. Client is responsible to notify counselor of any significant change of address, health, phones number of financial status.

6. CLIENT'S VIEW OF PROGRAM ___The client and I have discussed the services necessary to help him reach his vocational objective and we are in mutual agreement with his plan.

I HAVE PARTICIPATED IN THE DEVELOPMENT OF THIS PROGRAM AND I UNDERSTAND IT.
I UNDERSTAND AND ACCEPT THE STATEMENT OF UNDERSTANDING WHICH HAS BEEN EXPLAINED TO ME.

Roy Johnson   5-6-XX   *Susan Fields*   5/6/xx
Client's Signature   Date   Supervisor Signature   Date

© Cengage Learning 2006

Figure 7.1   Roy's service plan

relating actions to outcomes is not. For accountability reasons, service provision is tied to outcomes. This makes writing goals a critical step in plan development. Remember that these broad statements of intent can be achieved only to the degree that their meaning is understood, so well-stated, reasonable goals are essential to problem resolution.

At one intensive case management program in Los Angeles, California, clients decide on their goals. "When we create a record of the plan, we use the client's own words. We work hard with the client to list problems, issues, strengths, and barriers to reaching the goal. In other words, we ask the client, 'What will get in your way to accomplishing this goal?' Then we talk about next steps. We break down the steps into little pieces. I have some clients who can not even make one goal, no matter how small. Sometimes I have to write it down for them and list steps. The next step is to decide who does what: What do I do? What does the client do? What does the client's family do?"

How does one write goals that are well stated and reasonable? Three criteria help us achieve this. First, the goal should be expressed in language that is clear and concise; second, the goal statement should be unambiguous; and third, the goal must be realistic and achievable. These criteria are illustrated in the following goals, which were established for a seventy-four-year-old woman who will attend the Daily Living Program at the Oakes Senior Citizens Center.

Draft 1 is a *goal statement* for Ms. Merriweather; Draft 2 improves the statement by making it more clear and concise.

> *Draft 1:* Ms. Merriweather will participate often in many of the Oakes programs that relate to sports, games, music, communication, exploring other cultures, and other educational programs as they are developed by the creative staff in the activities area.
>
> *Draft 2:* Ms. Merriweather will increase her social opportunities by participating in center activities.

A description of the *plan* is presented in Draft 1. In Draft 2, it is restated less ambiguously by defining who will help with medications and what the help entails.

> *Draft 1:* They will work with Ms. Merriweather and her numerous family members to help with medications.
>
> *Draft 2:* Nursing staff will develop a plan to administer Ms. Merriweather's medication.

The goal in Draft 1, below, is to establish general physical goals for Ms. Merriweather. Draft 2 restates these goals in realistic and achievable terms.

> *Draft 1:* Ms. Merriweather will increase her range of motion, physical strength, and stamina.
>
> *Draft 2:* Ms. Merriweather will participate four times a day in an exercise program that includes walking, weight lifting, and stretching.

Thus, goals are an important part of the service plan. They increase the chance of solving the problem by providing direction and focusing attention on well-expressed, reasonable statements. Because formulating goals requires collaboration between the client and the case manager, writing them also highlights the shared responsibility for the case.

Once the client and the case manager have agreed on a broad statement of intent, it is time to identify the activities that will lead to the desired outcomes.

This process continues as a cooperative effort between the client and the case manager. Activities are identified as objectives.

An **objective** is an intended result of service provision rather than the service itself. It tells us about the nuts and bolts of the plan—what the person will be able to do, under what conditions the action will occur, and the criteria for acceptable performance—so that we can know whether the objective has been accomplished. Objectives are useful for several reasons. First, they tell us where we are going. Second, they give the client guidance in organizing his or her efforts by stating the intervention or action steps. Third, they state the criteria for acceptable performance or outcome measures, thereby making evaluation possible. Objectives are all-important for the case manager since they provide the standards by which progress is monitored. As progress is made, the case manager adjusts the plan as needed.

Writing clearly defined objectives benefits the client, the case manager, and the agency. The Department of Health and Human Services, Centers for Disease and Control (2009) provide the following five guidelines for writing and evaluating service objectives. We have added an additional six guidelines.

1. *Specific:*

- Objectives should provide the "who" and "what" of program activities.
- Use only one action verb, since objectives with more than one verb imply that more than one activity or behavior is being measured.
- Avoid using verbs that may have vague meanings to describe intended outcomes (e.g., "understand" or "know"), since it may prove difficult to measure them. Instead, use verbs that document action (e.g., "At the end of the session, the students will list three concerns . . . ").
- Remember, the greater the specificity, the greater the measurability.

2. *Measurable:*

- Objectives are the basis for monitoring to determine whether objectives have beenmet, unless they can be measured.
- The objective provides a reference point from which a change in the target population can clearly be measured.

3. *Achievable:*

- Objectives should be attainable within an implementation of your strategies and progress toward achieving your program goals. Objectives also help set targets for accountability and are a source for program evaluation questions.

4. *Realistic:*

- Objectives are most useful when they accurately address the scope of the problem and programmatic steps that can be implemented within a specific time frame.
- Objectives that do not directly relate to the program goal will not help achieve the goal.

5. *Time-phased:*

- Objectives should provide a time frame indicating when the objective will be measured or a time by which the objective should be achieved.
- Including a time frame in the objectives helps in planning and evaluating the program. (Department of Health and Human Services, Centers for Disease and Control, 2009). Evaluation Briefs, SMART Objectives. Retrieved from http://www.cdc.gov/healthyyouth/evaluation/pdf/brief3b.pdf

6. *Written:*

Each of us, whether consciously or unconsciously, has a convenient memory: We tend to remember the things that turn out the way we want them to and either forget or modify those things that do not go as we wish. If we did not put objectives in writing, it would be relatively easy to look on accomplishments as if they were in fact planned objectives. On the other side of the coin, one of the sharpest areas of conflict among case manager, client, and supervisor is illustrated by such phrases as "I thought you were working on something else!" or "That's not what we agreed to do" or "You didn't tell me that's what you expected." Having objectives in writing will not eliminate all these problems, but it will provide something more tangible for comparison. Furthermore, written objectives serve as a constant reminder and an effective tracking device by which the case manager, the client, and the supervisor can measure progress.

7. *Consistent with resources:*

A statement of objective must be consistent with the available or anticipated resources.

8. *Individually accountable:*

Ideally, an objective should avoid or minimize dual accountability for achievement when joint effort is required. If dual accountability is needed, then assign specific tasks to each individual.

9. *Consistent with rules:*

Objectives must be consistent with basic agency policies and practices.

10. *Voluntary:*

The client must willingly agree to the objectives without undue pressure or coercion.

11. *Communicated personally:*

The setting of an objective must be communicated not only in writing but also in face-to-face discussions with the client and the resource persons or agencies contributing to its attainment.

The following case example illustrates the development of goals and objectives (including intervention and outcome measures) with a client who is elderly and needs assistance.

Mary Sue Davis is an eighty-six-year-old white married female. She recently placed her husband in a nursing home facility so that he can receive full-time care. Mrs. Davis has a severe heart condition and has been ordered by her physician to rest every two hours and not to travel by herself because of dizzy spells. The nursing home is now receiving her husband's Social Security income. Mrs. Davis lives in a two-bedroom apartment. They have one son who lives an hour away and also has a heart condition. Mrs. Davis is requesting assistance with transportation in order to visit her husband on a more regular basis.

An interview with Mrs. Davis at her apartment revealed that her income consists solely of her Social Security checks. She does have Medicare to help with the costs of treatment for her heart condition. She currently uses public transportation (bus) to travel where she needs to go. During the interview, the service coordinator identified additional problems: the affordability of her current apartment, the availability of affordable housing, the need for an escort for travel, and possible grief issues regarding her husband's condition and placement in a nursing home.

Mrs. Davis agrees that she cannot afford her apartment and needs to seek more affordable housing. She is willing to apply for Community Action Committee (CAC) transportation that will pick her up at her door. She is very realistic regarding her husband's condition. Although she wishes he could come home, she has accepted that he will most likely remain at the nursing home. She realizes that she has to take care of her own health, but at the same time she has to get things done, and there is not always somebody around to help.

The service coordinator identified two main goals for Mrs. Davis: to find affordable housing and to secure transportation that is appropriate. These are set forth in the Client Plan (Figure 7.2).

The first objective toward the housing goal was to complete an application for a rent-controlled apartment with the city housing authority. Due to long waiting lists, this needed to be done within the week. The next step was to determine where she preferred to live (probably close to the nursing home). After the application was completed, the service coordinator arranged for a volunteer to take Mrs. Davis to look at several apartments and to meet with apartment managers to find out about waiting lists. (Mrs. Davis couldn't afford to wait for long.) The service coordinator found a volunteer to help with this. Once Mrs. Davis decided on an apartment, other volunteers assisted with the move. Her son could afford

COUNCIL ON AGING
CLIENT PLAN

CLIENT _Mary Sue Davis_____

DATE _11/6/XX_____

GOAL 1 _Locate affordable housing_____

    Objective 1: _To contact the city housing authority this week for applica-_

        _tion for rent-controlled apt.—service coordinator._____

    Objective 2: _To review list of apts., decide which ones to see, and select_

        _one (1 month)—Mrs. Davis._____

    Objective 3: _To request volunteer assistance to escort Mrs. Davis on_

        apartment visits and to help with the move (1 week)—

        _service coordinator._____

GOAL 2 _Provide transportation_____

    Objective 1: _To complete application for K-Trans lift and CAC vans (this_

        _week)—Mrs. Davis._____

    Objective 2: _To determine eligibility for medical escort service_

        _(2 weeks)—service coordinator._____

    Objective 3: _____

        _____

(continues)

Figure 7.2   Client plan for Mrs. Davis

to rent a moving truck and to drive the truck, although he couldn't lift or carry due to medical problems. The time allotted for these objectives was workable, and the objectives were met within a month.

    The objectives for the goal of transportation were to apply for the K-Trans lift along with CAC vans. Obtaining an assessment from the state office on aging was also an objective; that agency provides escorted transportation for medical appointments and necessary errands for people over sixty. This service would be available until Mrs. Davis was accepted by another agency that provides transportation.

GOAL 3 _____

   Objective 1: _____

             _____

             _____

   Objective 2: _____

             _____

             _____

   Objective 3: _____

             _____

             _____

GOAL 4 _____

   Objective 1: _____

             _____

             _____

   Objective 2: _____

             _____

             _____

   Objective 3: _____

             _____

             _____

Service Coordinator _____ Client _____
                   SIGNATURE   SIGNATURE

Figure 7.2 (*Continued*)

    In this case, the plan identified services and then guided the delivery of those services. The goals and objectives in the plan were developed using the guidelines suggested previously. Note that each objective clearly defined the intervention or action steps, stated who would provide the service, and stated a time frame for service delivery. The outcome measures were clear and the plan was implemented successfully.

    Often planning is not quite so easy. Suppose Mrs. Davis refuses to rest as prescribed or is insistent that she will continue to ride the bus. Or perhaps there are no transportation services in her community, or agency rules limit services to those who have no other family. As you can see, a number of challenges may appear during plan development. Sources of these challenges include but are certainly not limited to clients themselves, family members, funding restrictions, agency policies and procedures, eligibility requirements, or lack of community resources. Barriers can also be more intangible: client values, the denial of problems, cultural prohibitions, reluctance, or lack of motivation. All of these possibilities present opportunities for the case manager's resourcefulness and creativity; for example, working with a client to develop a plan that is congruent with client values and desires, understanding cultural norms, mobilizing resources, consulting with colleagues, and networking with other agencies. Many of these challenges must be resolved in order to move forward with identifying services.

 # Identifying Services

Once the plan is complete and has been agreed on by the client and the case manager, it is time to begin thinking about the delivery of services. A well-developed plan provides information about what the service is, who will provide it, what the time frame is, and who has overall responsibility for service delivery. It is the case manager's responsibility to implement the plan. What are these responsibilities? How does one begin implementation? These questions are explored next.

Identifying services has been compared to the brokering role. In both situations, the case manager is involved in the legwork and planning that is necessary for implementation. As a broker, the case manager helps clients access existing services and helps other service providers relate better to clients. This linking of clients and services also occurs as the case manager arranges for service delivery. The steps are similar.

Information and Referral Systems

One of the most helpful tools for a case manager is knowledge of the human service delivery system in the community. Who do you know? What services are available? How does one access the services? Is there a waiting list? One of the challenges facing new case managers is to establish an **information and referral system.** For case managers with experience, the challenge consists of continually developing and updating their systems. Knowing what an information and referral system is, how to set one up, and how to use it are valuable skills in case management.

Human service employers believe that the people their agencies will serve in the future will be multiproblem clients, such as people with dual diagnoses, diverse problems, and problems of long standing (Woodside & McClam, 2012). The needs of clients such as these rarely match the services available from a single agency. In these cases, the case manager finds it invaluable to have information about other available services. Many helping professionals have personal service directories to supplement existing community or agency directories.

There are three components to information and referral. One component is the **social service directory**, which usually lists the kinds of problems handled and the services delivered by other agencies. In some communities, these are published by a social service agency, by a funding source such as the United Way, or (as a community service) by a business or organization. Almost always these services are located on the World Wide Web.

Another component is the **feedback log.** Such logs provide feedback to the agencies that deliver services to help ensure quality information and referral services. Some agencies accomplish this through referral forms that record referrals, give information on the services needed, and provide the referral agency with information on the services received. If the client takes the form to the agency providing services, it may also serve to remind the client of the appointment.

A third component of information and referral systems is staff training and networking. In the training sessions, the helper may be introduced to the services of the employing agency as well as those of other agencies. Other information and referral data that are shared during staff training may include reviewing and updating referral procedures, announcing new services or ones that no longer exist, and discussing the effectiveness and efficiency of service delivery.

As new case managers build their list of professionals with whom they can consult, they expand the number of peers who can help them find services for clients.

Usually social service directories have two indices: one that is an alphabetical listing of agencies and one that is a categorical listing of services. Each entry in the directory lists the agency's name, address, phone number, e-mail addresses, URLs, and services. Also listed may be fees, hours of service, eligibility criteria, and sources of agency support. Many agencies also have discussion boards, Facebook pages, and accounts on Twitter and other social media. Here is an example of an entry. The name of the agency, location, a site map, and a home page with a general description of the agency is illustrative.

RUNAWAY SHELTER

CENTER FOR CHILDREN, EL PASO, TEXAS

Our Site Map:
Home
About the Center
Therapeutic Homes Program
Runaway Shelter & STAR
Healthy Families Initiative
Contact
Our Logo
Donations
Directions
Employment Opportunities

Runaway Shelter & Star

(Services to At-Risk Youth)

The Center's Runaway Shelter and STAR program serve ten- to seventeen-year-old runaway and homeless youths and their families, as well as youth and families that are at risk of abuse, delinquency, or family separation. The program's purpose is to protect the safety and promote the healthy development of at-risk youth by providing each participant with an individualized program of shelter, counseling, skills training, and family reunification and support services.

The STAR program has served El Paso County since 1985 and has received continuous federal financial support. The program is grounded in the core values and principles of Positive Youth Development, and meets or exceeds federal Program Performance Standards. Major components of the project include Project Safe Place, the Runaway Shelter (a licensed, six-bed emergency shelter), and an extensive nonresidential counseling and outreach program. Annually, the program provides more than 4,000 hours of counseling and skills training to 800 youths and 1,300 days of emergency shelter care to 200 youths.

This program receives annual financial support from the United Way of El Paso County, the City of El Paso, and the U.S. Department of Health and Human Services. (Retrieved from http://www.epccinc.org/newlogo1.htm)

Existing directories are helpful to the case manager, but sometimes establishing one's own system is useful for filling in the gaps in published directories or for recording detailed information that may be of special interest to the individual helper.

Want More Information: Social Service Directories

The Internet expands our knowledge of the services in our community. Test your ability to use the Internet to find services in any community of your choosing.

- Use search terms such as "social service directory" and "human service directory." For services for specific populations or problems, search on terms like "homeless services" or "veterans services." NOTE: Placing quotation marks around a phrase directs the search engine to look for the entire phrase rather than the individual words in it.
- Narrow your search to a specific location or a specific age group. Report what you find to your classmates.
- Begin to build an electronic file of services in your own community. This will help you when you begin your field experience or internship as a case manager.

Setting Up a System

The first step in establishing one's own information and referral system is to identify all agencies and available services. This includes listing agencies previously contacted, browsing the Internet, and talking with other professionals. Each agency and service becomes part of a card file, a computer file, a spreadsheet, or an online directory that is easy to update. The file can be uploaded to a mobile phone, tablet, or "cloud" repository for easy access. The file can also be expanded by talking with clients (particularly those who have been in the human service system for some time), meeting other professionals at meetings and workshops, and attending community meetings.

Whether recorded on cards or electronic files, this information is easy to use when identifying the client problem and matching it with a service. However, since a client rarely has only one problem, using the file may not be so simple. First, the client and the case manager must prioritize the problems. Once this has been completed, the case manager identifies which problems the agency will address and which ones need referral. These additional services can be found by checking the file. If there is more than one resource to serve the client's particular need, the case manager works to identify the agency that can meet the client's needs in a manner responsive to the client's values and concerns.

Deborah Caudill is an eighteen-year-old client who needs long-term counseling to work on the anger she feels toward her father for deserting the family when she was eleven. Lou Levine, her case manager, knows that the counseling Deborah needs is beyond the scope of the services provided by the agency where Lou works. Two other agencies in their community offer long-term counseling for adolescents. Because Deborah and Lou agree that counseling would be beneficial, they discuss these two agencies. Deborah has questions about their locations, who provides the counseling, whether it is group or individual, and how much it will cost. Lou consults her file for the answers to these questions and provides Deborah with the information, and then they discuss the pros and cons of each option. The case manager's file indicates that one center provides counseling services and is well known for its work with adolescents. In addition, the latest entry in the file indicates that Jane Barkley, a previous client, had a positive experience there.

Establishing and using an information and referral system requires certain skills of the case manager. Being able to identify the client's problem, the community resources available to solve it, and the viable alternatives are all critical to the success of the system. Choosing a resource or a service requires the client's participation. The client may have the final say in the selection of the agency or service, so the more accurate and complete the information about the agency, the better the decision will be. Finally, good research skills are helpful, because the case manager continually works to locate potential community resource alternatives and to update data on existing agencies and services. These skills include computer searches, building networks, and experience.

Part of the development of a plan is identifying services to meet the client's needs. The development of an information and referral system is useful here. Throughout plan development, data gathering continues to take place.

Gathering Additional Information

Gathering additional information may be part of the planning process or part of the plan itself. To decide whether additional information is necessary, there must be a review of available information from other agencies, the referral source, employers, and others. The key to determining what is needed is relevance. Is the needed information relevant to the client and to service provision? Will it contribute to a complete array of social, medical, psychological, vocational, and educational information about the client? Once it is determined that additional information is necessary, the case manager decides how the information will be obtained. In some cases, the case manager can personally acquire the information, but it may also be necessary to consult family members, a significant other, or professionals such as psychologists, physicians, and social workers. The client also continues to be a primary source of information and is part of the decision-making process regarding the additional information needed and who can provide it. Next, we introduce two data collection methods that case managers use. In addition, Chapter 9 explores what data are available from other professionals.

Data Collection Methods for the Case Manager

Two primary tools are available to the case manager for data collection: interviewing and testing. They are similar in several ways. The information is used to describe, to make predictions, or both. Each may occur in an individual or group situation in which some type of interaction occurs. The group situation may be an interview with a family or a test administered to more than one examinee. Both interviews and testing have a definite purpose, and the case manager assumes responsibility for conducting the interview or administering the test.

Interviewing

There are different types of interviews (Kaplan & Saccuzzo, 2009). The **assessment interview** is an interaction that provides information for the evaluation of an individual. The interview may be structured or unstructured; it uses both open-ended and closed questions. The intake interview is an example of an assessment interview in which the applicant provides information that helps the case manager evaluate him or her and the problem in relation to the mission, resources, and eligibility criteria of the agency.

A **structured clinical interview** consists of specific questions, asked in a designated order. This type of interview is structured by guidelines to ensure that all clients are handled in the same way. The structure also makes it possible to score the responses. One advantage of this type of interview is its reliability or consistency. Flexibility is limited. Although it is a valuable source of information, the interview results should be interpreted with caution. The major limitation is that it requires the respondent to be an honest and capable interviewee who has skills for self-observation and insight.

A more comprehensive interview is the **case history interview.** This interaction includes both open-ended questions and specific questions. Topics may include a chronology of major events, the family history, work history, and medical history. Usually an interview of this type begins with an open-ended question or statement: *"What was school like for you?" "Tell me about your work history." "What do you remember as the happiest times when you were growing up?" "Describe your relationship with your parents."* These probes may be followed by specific questions, which may or may not be dictated by agency forms or guidelines. *"When did you quit your last job?" "What grade did you complete in school?" "Are you the oldest child?"* These questions help the case manager understand the client's background and uncover any pertinent information.

Technology is also an influence on interviewing. Many times a computerized interview takes place via computer rather than face-to-face. Questions are presented and followed by a choice of responses.

Are you married?   Yes   No

If the answer is yes, then another question related to marriage may follow.

Is this your first marriage?   Yes   No

If the answer to the first question is no, then another question appears.

Did you complete high school?   Yes   No

The computerized interview is a good way to collect facts about a person. The limitations are that there is no nonverbal communication, and the feelings of the client are not shared. Important information may be lost as a result of these limitations.

The *mental status examination,* described in more detail in Chapter 8, is a special type of interview used to diagnose psychosis, brain damage, and other major mental health problems. Its purpose is to evaluate a person thought to have problems in terms of factors related to these problems. This type of interview requires the case manager to have some expertise on major mental disorders and the various forms of brain damage.

The skillful interviewer also needs to know about **sources of error** in the interview. Awareness of sources of potential bias in the instrument itself or in the interviewer enables the case manager to compensate for any resulting distortions. A look at interview validity and reliability will help us identify potential sources of error.

For a number of reasons it is often difficult to make accurate, logical observations and judgments. One is the **halo effect** (Whiston, 2012), which occurs in an interview situation when the interviewer forms a favorable or unfavorable early impression of the other person, which then biases the remainder of the judgment process. For example, an unfavorable initial impression can make it difficult to see positive aspects of a client or a case. If a home visit to a single parent who lives in a housing project reveals an unkempt, dirty, and very sparsely furnished living area, the case manager may find the visit unpleasant. The resulting interview with the parent is likely to be rushed and cursory, with little chance of gaining insight into any problems. The case manager may also find it difficult to maintain eye contact with the parent, thereby missing important nonverbal cues. Other contacts with this parent may be influenced by the memory of the physical setting.

The case manager may visit families from different cultures. In many cultures several generations live together in a small space, sharing kitchen and bedrooms, with little individual space allotted. Interactions may vary; in some cultures, there are few individual matters, and the case manager sees the entire family rather than just one or two members of the family (Gulanti, 2008).

A second cause of invalidity in an interview is "general standoutishness" (Hollingsworth, 1922). This is the tendency to judge on the basis of one outstanding characteristic, such as personal appearance. A more attractive, well-groomed individual might be rated more intelligent than a less attractive, unkempt individual. Consider a case manager who makes a home visit to investigate a child abuse report. The address is in an affluent suburb and the house is a stately two-story brick house with elaborate landscaping. The initial impression of neatness, money, and social standing may influence the investigator's interaction with the parents and the subsequent course of the investigation.

Cultural differences can also contribute to error. To take an extreme example, a case manager has been asked to visit a family that recently emigrated from India and has just moved into a rent-controlled apartment in the city. It is her last stop of the day, and she finds that she has interrupted a ceremony of *puja* (prayers of thanks for their new home). She finds family members seated on the floor around a fire. Appalled that they have started a fire on the floor, she stamps it out and begins lecturing the family on fire safety. When she finally begins to talk about the services that are available, the family does not respond.

As you can see, sources of error can prejudice interview validity. Error reduces the objectivity of the interviewer, often leading to inaccurate judgments. The more structured the interview is (see Chapter 5), the less error there will be. Because an interview does provide important information, the case manager can consider the information tentative and seek confirmation from other sources, such as more standardized procedures. Similarly, test results are more meaningful if placed in the context of a case or social history or other interview data. The two can complement each other.

The reliability of an interview is its consistency of results. In interviewing, this means that there is agreement between two or more interviewers in their conduct of the interview, the questions they ask, and the responses they make. As you might imagine, reliability varies widely. The reliability of structured interviews is higher because they have more stringent guidelines concerning the questions and even the order of the questions. (The downside is that this structure limits the information the interviewer obtains.) In general, interview data have limited reliability, because interviewers look for different things, have various interviewing styles, and ask different questions. It is important for the case manager to verify information with other sources over time.

Voices from the Field: Culturally Sensitive Interview Questions

As discussed earlier in this chapter and in Chapter 6, culture plays a critical role in the case manager's interaction and success with the client. Interviewing clients with sensitivity requires a desire to understand the client and cultural competence and skills. Gulanti (2008) suggests using the 4 C's of Questioning to explore case manager and client awareness of culture. The four questions relate to calling (i.e., naming or labeling), causation, coping, and concerns. Each question is designed to help case managers understand the client's perspective.

Question 1:

What do you call your problem (issue, strength, barrier)?

This question allows the case manager to enter the world of the client. How we name aspects of our lives provides us information about culture, values, beliefs, and personalities. As clients answer this question, we communicate to the client that we are interested in them and their experiences; this begins building the foundation upon which trust develops. Responses also help case managers understand how clients conceptualize their situations.

What do you think caused your problem (issues, strengths, barriers)?

Answers to this question allow case managers to understand the power structure in the client's world. Causes of or relationships to the issues at hand might be scientific, social, emotional, educational, financial, or spiritual, and may reflect an internal or external locus of control. During the assessment and planning phase of case management, the view of the client must be considered. Then the intervention, from the client perspective, is related to perceived cause.

How do you cope with this problem (issues, strengths, barriers)?

When the case manager asks this question, he or she learns much about the culture of the client and what interventions are important to the individual, as well as what is available within the cultural context. Sometimes clients have tried nontraditional or non-Western approaches and are reluctant to share this knowledge. They fear ridicule. Asking this question and validating the approaches the client has taken helps build client strength and empowerment.

What concerns do you have regarding your problem (issues, strengths, barriers)?

Hopefully, as the client answers this question, the case manager learns about the client's worries, fears, and sense of seriousness of the problem, and possibly his or her concerns about working with the case manager. Addressing these concerns demonstrates respect. It also helps the case manager understand any potential barriers during the case management process.

Testing

In the previous section, testing was recommended as one way to verify the information gathered in an interview. Most people encounter tests shortly after beginning school. How we perform on tests affects our lives, and test scores have become key factors in many decisions. They influence placement in special academic classes; the assignment of labels such as *high achiever*, *mentally challenged*, and *average*; admission to schools and colleges; and job selection. In fact, test scores are more important today than ever before.

Case managers encounter tests in various contexts; for example, they receive test reports from other professionals. In some cases, the information consists of test scores and nothing more. Figure 7.3 shows one example of how test results may be communicated.

TEST DATA

Name Joe Brown   Date X/X/XX

Date of Birth X/X/XX   Age 26   Sex: (M) F   Race: W B   Other _____

Marital Status: M S D W   Education 11th grade

Wide Range Achievement Test (WRAT)

Reading grade	13.2	SS	116	%ile	82
Spelling grade	11.2	SS	106	%ile	66
Arithmetic grade	8.5	SS	93	%ile	32

Wechsler Adult Intelligence Scale

Verbal IQ 119   Performance IQ 106   Full Scale IQ 114

General Clerical Test

Clerical subscore	26	%ile	10
Numerical subscore	22	%ile	35
Verbal subscore	56	%ile	40
Total score	104	%ile	25

Other tests:

Figure 7.3   Test data

BOX 7.1	**Test Administered and Results**

Wide Range Achievement Test: A measuring device to estimate grade levels in three academic areas. Results from administering the WRAT show that the client is functioning at the 17.4 grade level in reading, 15.0 in spelling, and 7.1 in arithmetic. These scores are within the very superior, superior, and average classifications, respectively, when compared with the appropriate normative age group.

In other cases, test scores are part of a written report that also gives some explanation of the scores. Box 7.1 is an excerpt from a report on a thirty-seven-year-old white male who was hospitalized for depression. He has completed two years of college and has been a personnel interviewer for ten years. To use this information, the reader of the report must have knowledge of tests and an understanding of test data.

A case manager may also encounter testing as a service offered by an agency. For example, a statewide evaluation facility located on the campus of a school offers services that include achievement testing for placement at the school and vocational testing for career development. Workers at the evaluation facility administer these tests to each client who is referred to the facility. Scores are interpreted and included in their evaluation reports, which are sent to the referring counselor. There may be other situations in which knowledge of testing is important. For example, a case manager may be asked to select tests to be administered as part of the services required in a plan. This task requires knowing the sources of information about tests, the criteria for selecting a test, and eligibility for purchase and use. Such knowledge is also important when the case manager encounters a situation like the following.

A family on my caseload had trouble understanding the results of a recent assessment test that was administered at their son's elementary school. The school counselor who originally explained the results of the test used terms unfamiliar to the parents and did not answer the questions they asked. The parents feel that if they understood the results of the test, they could help their son in the areas where he was weakest. The parents have asked me to look at the test results and explain them again.

The case manager needs an appropriate level of testing knowledge in order to use tests as a resource. Because tests have assumed such importance today, particularly in decision making, case managers must think carefully about the role of testing in their work with clients. To make proper use of test results, one must understand the test being used, including the purpose of the test, its development, its reliability and validity, administration and scoring procedures, the characteristics of the norm groups, the cultural dimensions, and its limitations and strengths. This section presents an overview of these areas.

What is a test?   A **test** is a measurement device. A **psychological test** is a device for measuring characteristics that pertain to behavior. It is a way to evaluate individual differences by measuring present and past behavior. For example, the test your instructor will give you to measure your mastery of this material will provide an indication of what you know now. Tests also attempt to predict future behavior. You probably took the Scholastic Assessment Test (SAT) as part of the admission requirements for college. SAT scores are usually required by higher education institutions as a predictor of success in college.

One important caution needs to be noted here: A test measures only a sample of behavior. Tests are not perfect measures of behavior; they only provide an indication. It is therefore important that case managers not make decisions based solely on test scores.

Types of tests   Thousands of tests are in use today. One way to make sense out of all the tests that are available is to know how they are categorized. One classification is by type of behavior measured. Two categories are identified in this system. *Maximum performance tests* measure ability, and *typical performance tests* give an idea of what an examinee is like. These and other helpful categories are discussed in the pages that follow.

Maximum performance tests include achievement tests, aptitude tests, and intelligence tests. On these tests, examinees are asked to do their best. **Achievement tests** are used to evaluate an individual's present level of functioning or what has previously been learned. Achievement tests that a case manager will often encounter include the Test of Adult Basic Education (TABE) and the Wide Range Achievement Test (WRAT).

Aptitude tests provide an indication of an individual's potential for learning or acquiring a skill. Because aptitude tests imply prediction, they are useful in selecting people for jobs, scholarships, and admission to schools and colleges. The SAT is an aptitude test. In your work with clients, you will likely read about aptitude tests such as the General Aptitude Test Battery (GATB), the Differential Aptitude Test (DAT), and the Minnesota Clerical Test.

When we think about how smart someone is, usually we mean intelligence. Tests such as the Wechsler Adult Intelligence Scale (WAIS), the Wechsler Intelligence Scale for Children (WISC), the Revised Beta Examination (Beta IQ), and the Peabody Picture Vocabulary Test are **intelligence tests.** Careful consideration should be given to these tests and test scores, because intelligence can be defined in a number of ways. Some tests measure verbal intelligence, some measure nonverbal intelligence, and others measure problem-solving ability. The WAIS, for example, yields a Verbal IQ, a Performance IQ, and a Full-Scale IQ. On the other hand, the Revised Beta Examination yields only a performance IQ score called a Beta IQ.

The other major category is the typical performance test. Such tests provide some idea of what the examinee is like—his or her typical behavior. In this category are interest inventories (California Picture Interest Inventory, Strong Interest Inventory Test, Kuder Preference Record); personality inventories (Edwards Personal Preference Schedule, Minnesota Multiphasic Personality Inventory, Sixteen Personality Factor Questionnaire); and projective techniques (Rorschach Inkblot Test, Thematic Apperception Test, Rotter Incomplete Sentences Blank). Other well-known performance tests are the Bender–Gestalt Test, the Vineland Social Maturity Scale, and the McDonald Vocational Capacity Scale.

There are a number of other categorization schemes for tests. Individual tests, such as the WAIS or the projective techniques, are administered to one person at a time. Group tests are administered to two or more examinees at a time. The Revised Beta Examination and the Otis Lennon School Ability Test are group tests, although they can also be administered on an individual basis. Tests can also be classified as standardized or informal. Standardized tests are those that have content, administration and scoring procedures, and norms all set before administration. Informal tests are developed for local use. An example is the test your instructor will give to you to measure your mastery of this course material. Verbal tests use words, whereas nonverbal tests consist of pictures and require no reading skills. Tests in which working quickly plays a part in determining the score are speed tests. Tests such as the Revised Beta Examination are closely timed. In contrast, power tests have no time limit, or one that is so generous that it plays no part in the score.

As you begin to explore the testing literature, you will discover that testing has a language of its own. Recognizing the categories and knowing their meanings will help you develop the vocabulary to understand testing concepts and the advantages and limitations of tests. Selecting tests requires an understanding of the terms in the following list, as well as others.

Edition: the number of times a test has been published or revised
Forms: equivalent versions of a test
Level: the group for which the test is intended (e.g., K–3 is kindergarten through third grade)
Norm: the average score for some particular group
Norms table: a table with raw scores, corresponding derived scores, and a description of the group on which these scores are based
Percentile rank: the proportion of scores that fall below a particular score
Reliability: the extent to which test scores or measures are consistent or dependable—that is, free of measurement errors
Test: a measurement device
Test administrator: the person giving a test
Test profile: a graph that shows test results
Validity: the extent to which a test measures what it claims to measure
Test selection: where to find out about available tests, and how to select a test

There are many sources of information about tests, ranging from test publishers to reference books available in the library. These sources refer to journal articles that provide more detailed information about a test. The general information about a test can help you narrow the choices to those of interest, and it is for these tests that more specific information is gathered. Let's begin with the more general information.

Thousands of commercially available tests in English are described and critically reviewed in the many editions of the *Mental Measurements Yearbook*, or *MMY*, published by the Buros Institute of Mental Measurements at the University of Nebraska–Lincoln. Begun in 1938 by Oscar K. Buros, the *MMY* provides comprehensive reviews of tests by almost 500 notable psychologists and education specialists (Kaplan & Saccuzzo, 2009). For each test included in the *MMY*, there is a detailed description and price, followed by references to articles and books about the test, along with original reviews prepared by experts. The *MMY* contains no actual tests.

Another reference that summarizes information on tests is *Tests in Print*. This volume is helpful as an index to tests, test reviews, and the literature on specific tests. Entries include the title and acronym of the test, who it was designed for, when it was developed, its subtests, the cultural dimensions, the authors and publishers of the test, and cross-references to *MMY*. Other references may be helpful to you as you narrow your selection, but *MMY* and *Tests in Print* provide the most comprehensive overviews of published tests now available.

Once you have narrowed your choice, specimen sets of tests are available for purchase from test publishers. Although there are approximately 400 companies in the test industry, the top 10% are responsible for 90% of the tests used in the United States. Test publishers have catalogs that provide lists of tests and test-related items sold by that company. Companies usually offer specimen sets for sale: the test manual, a copy of the test, answer sheets, profiles, and any other appropriate material related to a particular test. The test manual—the best source of information about a particular test—provides statements about the purposes of the test, a description of the test and its development, standardization procedures, directions for administration and scoring, reliability and validity information, norms, profiles, cultural dimensions, and a bibliography. This specific information can help the case manager decide whether to use the test.

Once available information about a particular test is gathered, the case manager decides whether to select it. Then the second question surfaces: What are the criteria for selection? One helpful source of information is *Standards for Educational and Psychological Testing*, published by the American Psychological Association. The *Standards* is a technical guide that provides the criteria for the evaluation of tests. Among the standards discussed are validity, reliability, cultural dimensions, test administration, and standards for test use. Any case manager who is involved in testing should carefully review the complete standards.

Criteria for selection   Many considerations go into a decision to use a test. Three primary considerations are validity, reliability, and usability. This section provides a brief overview of each consideration by defining each, identifying the types, and discussing how to evaluate, including the issue of bias. For more information, consult the *Standards for Educational and Psychological Testing* or testing textbooks that present this material in more detail.

The issue of bias is important to consider when considering testing and determining what particular tests to use. Bias refers to both technical and social aspects of the test. For example, if a test if technically biased, this means that the test has different validity for a specific population (Whiting & Ford, 2009). Social bias means that the test, while valid, still might be unfair. We explore both technical and social bias below.

Validity, the most important of the three considerations, is the extent to which a test measures what we actually wish to measure. Think about a time when you have had to move furniture. A yardstick was probably helpful to you as you measured the available space and the piece of furniture for a fit. Long experience has confirmed the validity of a yardstick as a tool for measuring length. Validity works the same way for a test, which is a tool for measuring behavior. Among the questions related to validity are the following: How well does a given math test measure math achievement? Could we make predictions based on those scores? Does this test measure other qualities as well? These questions are answered as content-, construct-, and criterion-related validity are established for a test. (These terms are likely to appear in the test manual's section on validity.)

Validity must be evaluated in light of the intended use. There are a number of factors to consider. One is *evidence* of several types of validity; the more sources of evidence, the better. The quality of the evidence is also important—more important, in fact, than the quantity, should the evidence be questionable. Included in the reported validity data should be a description of the subjects, the testing situation, and how the test was used. Finally, the evidence should relate to the purpose of the test. For example, evidence for the validity of an achievement test should be relevant to content, whereas criterion-related validity is important for tests that predict outcomes. In terms of validity, bias can be against a specific race, ethnicity, or linguistically different group. For example, when discussing scores on a test, bias means that the scores have different meaning for one cultural group than another. Since validity answers the question "Does the test measure what it intends to measure?", the case manager should ask, "Does the test measure what it intends to measure for the population with which I work?" This question is at the heart of the issue of bias and validity.

There are ways in which a test might not be valid for a specific race, culture, or linguistic group (Whiting & Ford, 2009). First, the test takers may lack the situation, education, or opportunity to gain knowledge that helps them answer the question. For example, a rural student might not understand questions that use city references. A second example occurs when an individual provides a "correct answer" from his or her own culture. Unfortunately, the answer is scored "wrong"; the right answer represents a different culture. A third example of bias occurs when the structure of the question or its wording does not make sense to the test-taker. The format is unfamiliar.

Reliability, the second consideration, is the degree of consistency with which a test measures whatever it is measuring—the degree to which test scores are free from errors of measurement. Such errors result in inconsistent scores from one test form to another or from one testing time to another, changes that are not attributable to such factors as the examinee's motivation, fatigue, anxiety, or guessing.

As with validity, a review of the reliability evidence of a test takes into account any errors of measurement in light of the test's intended use. For example, split-half reliability, which is often used with longer tests, yields an inflated estimate of reliability when applied to speed tests. Test–retest reliability (which involves testing the same group with the same test on two occasions) would not be appropriate for tests in which memory plays a role from one testing situation to another. When reviewing reliability evidence, it is also important to read the descriptions of the number and characteristics of the sample, the testing situations (including the interval between administrations), and the reported reliability estimates. This information should be evaluated in relation to the test's intended use.

The final consideration is test *usability*, which includes all the practical factors that are part of a decision to use a particular test. The factors include economy, test administration, social bias, and interpretation and use of scores. The following questions relate to usability.

ECONOMY
Can we reuse the test booklets?
Are separate answer sheets available?
How easy is it to score the test?

TEST ADMINISTRATION
Are the instructions clear and complete?
Is close timing critical?
What is the layout of test items on the page (print size, clarity of pictures, and so on)?
Can the test be administered via computer?

SOCIAL BIAS
Are there specific populations (races, ethnic groups, or linguistic groups) that do not score "well" on a test?
Are there negative consequences relating to, for example, school placement or job placement? (Whiston, 2012; Whiting & Ford, 2009).
Is the test fair to those taking it?

INTERPRETATION AND USE OF SCORES
Are scoring keys and instructions provided?
Are norms included for appropriate reference groups?
Is there evidence that test scores are related to other variables?
Is there a clear statement about the function of the test and its development?
Are there multicultural considerations?

These questions, as well as the validity and reliability evidence, are all factors to be considered in test selection. Once a test has been selected, one purchases the test, answer sheets, scoring keys, interpretive manuals, and so on from the test publisher.

Administration and interpretation   The test manual is an important source of information about the administration of a test and the interpretation of test scores. It should identify any special qualifications in training, certification, and experience that are necessary for the people who will administer and interpret the test. Needless to say, only individuals who have the training and experience for using tests should be entrusted with them. The test manual also describes the recording and scoring of answers. Can answers be recorded in the test booklet? Are separate answer sheets available? If the answer to both questions is yes, are the results interchangeable? Also described in the test manual are procedures to facilitate the use of the test and reduce bias. Information about the interpretation of test scores will be included; in some cases, handouts are provided. Note that it is the responsibility of the test administrator to read the manual carefully and be knowledgeable about the test and its uses.

Standardized test administration procedures are important. Many sources of error can influence test scores, and some relate directly to test administration. The testing situation itself is influenced by various factors, including the lighting, ventilation, room temperature, crowding, adequate work space, and noise level.

The test administrator's behavior can also be a source of error. Giving extra help, pointing out errors after the test begins, establishing varying degrees of rapport with different examinees, and allowing additional time all influence test scores. The administrator of a test should eliminate as many of these conditions as possible before beginning the test and avoid deviating from the standardized procedures.

Today, computer-assisted test administration is a reality. In this testing situation, test items are presented via a computer terminal or a personal computer, and an automatic recording of test responses occurs. There are two important advantages to testing by computer: (1) standardization of administration and scoring and (2) bias control. If a test is administered via computer, it is important to ensure that the items are legible, the screen is free from glare, and the terminals are properly positioned.

What about the client who can't read, or who reads below fifth- or sixth-grade level? Or one who has impaired vision or hearing? Or a person with epilepsy who is heavily sedated, causing his or her motor processes to slow down? An initial strategy is to find individual psychological tests that can accommodate the client's problems and still measure interests, intelligence, achievement, aptitudes, or personality. The limitations of each disability need to be carefully considered in relation to the person's ability to perform on a test. For example, a client who writes slowly because of a disability should not be given a test that places a premium on speed. In some testing situations, however, special consideration cannot and should not be given to a person with a disability. If a client has an amputation below the knee and is interested in attending college, he or she needs no special considerations when taking the SAT. The limitations imposed by the disability do not interfere with the traits measured by the SAT. When adjustments are made for the person with a disability, avoid deviating from standardized procedures as much as possible. The reason for using tests is to obtain an objective comparison of the individual with a standardized group. If the test is administered in a nonstandardized way, the norm-based scores obtained are not accurate.

When the test is given to people who are different from the groups the test was designed for and normed on, it is more difficult to get accurate results. The instructions, item content and format, methods of answering items, and many other aspects of the test have been designed to make it useful for a specific population. Knowledge of disabilities can help you decide whether modifications are needed. If deviation from established procedures or content is necessary, it is best to consult with a colleague who has psychometric expertise.

The person who takes a test has the right to receive a correct interpretation of the score. It is the responsibility of the test user to interpret scores accurately and meaningfully. Test scores are usually reported as raw scores, which have very little meaning. The raw scores can be converted and reported as standard scores—percentiles, stanines, or T scores—which are easier to understand and interpret. The conversion is based on the norm group or standardization group.

Once the standard scores are available, it is time to interpret the test. There are two essential steps in test interpretation: understanding the results and communicating them to another person, orally or in writing. The following suggestions will guide your preparation for test interpretation.

- Know the test—its purpose, development, content, administration and scoring procedures, validity and reliability, cultural perspectives, and advantages and limitations.
- Avoid technical discussions of tests. Use short, clear explanations of what you are trying to communicate.
- Use the test profile as a graphic presentation of the test results. The examinee may find this easier to follow as you explain the score are explained.
- Explain what the score means in terms of behavior.
- Go slowly. Give the examinee time to process the information and react.

Tests are helpful tools for measuring traits common to many people. A score helps show where a person stands in a distribution of scores of peers. How high or low a score is does not measure an individual's worth or value to family, friends, or society. A guiding principle for professionals who use tests is to consider scores as clues. They do mean something, but in order to know *what*, we must consider each examinee as an individual, combining test evidence with everything else we know about the person. It is unsound practice for case managers to base important decisions on test scores alone. It is important to remember this in test selection, administration, and interpretation.

 # Specific Cultural Guidelines

Whiting and Ford (2009) suggested the following guidelines for considering bias in testing. Although some of the information is relevant for all testing, case managers should follow these guidelines to advocate for their clients. We believe these guidelines are appropriate to consider for all clients involved in the case management process.

- Translate tests into the language of the examinee.
- Use interpreters to translate test items for examinees.
- Examine all test items/tasks to see if groups perform differently and eliminate those items/tasks.
- Eliminate items that are offensive to examinees.
- When interpreting test scores, always consider the examinee's background experience.
- Do not support the assumption of homogeneous experience or equal opportunity to learn; groups have different backgrounds and experiences that affect their test performance.
- Never base decisions on one test and/or one score. One piece of information or lone score cannot possibly be useful in making effective and appropriate decisions.
- Do not interpret test scores in isolation; collect multiple data and use this comprehensive method to make decisions.
- When an individual or group scores low, consider that the test may be the problem; it may be inappropriate and, if so, should be eliminated.
- If a group consistently performs poorly on an intelligence test, explore contributing factors and the extent to which it is useful/helpful for that group.
- Always consider the technical and social merits of tests. A test can be technically unbiased but still be unfair (i.e., have a disparate impact).
- Review norming data and sample sizes; while diverse groups can be proportionately represented in the standardization sample, their actual numbers may be too small to be representative, which hinders generalizability.
- Include culture-fair or culture-reduced tests in the assessment or decision-making process; these tests are designed to minimize irrelevant influences of cultural learning and social climate and, thereby, produce a clearer separation of ability or performance from learning opportunities. Nonverbal intelligence tests, with their reduced cultural and linguistic content, fall into this category.
- Always use and interpret test scores with testing principles and standards in mind.

Summary of testing   Test misuse can easily occur. Let's review some guidelines for the selection, administration, consideration of bias, and interpretation of tests.

First, case managers should select tests that they have carefully reviewed. The validity, reliability, cultural sensitivity, and usability of a test; its statement of purpose, content, norm groups, administration, and scoring procedures; and its interpretation guidelines should all be evaluated in light of the intended use. The case manager should check any reviews by experts to add to his or her knowledge of the test. It is also a good idea for the case manager to take the test.

Second, case managers should use only tests they are qualified to administer and interpret. This often depends on one's ability to read the manual. Other tests require advanced coursework and supervision or practicum experiences for proper administration and interpretation. Test catalogs usually indicate how much expertise is required for the tests listed. Another helpful source of information is the *Standards for Education and Psychological Testing*.

Third, case managers who administer tests have an obligation to provide an interpretation of the test results. An understanding of raw scores and their conversion to standard scores, coupled with the ability to communicate the meaning of the scores, is necessary to do this right. In addition, it is essential to be aware of the norm groups and their applicability to the examinee. Some groups, such as Latinos, African Americans, and rural populations, may be underrepresented in the establishment of norms. Follow guidelines to insure that you consider technical and social bias when performing testing and assessment.

Deepening Your Knowledge: Case Study

Scarlett is a case manager at a state-supported boarding school for the blind in Denver, Colorado. She holds seven years of experience on the job, working with recent graduates of the school as they prepare for the transition from the boarding school to independent living and work. All of the students are eligible for transitional services but have to agree to the services before they are provided. During the assessment phase, Scarlett focuses on strengths, needs, barriers, cultural issues, and vocational interests for each client. After completing pretransitional assessments for all clients, Scarlett reviews these assessments and then develops individual plans for service delivery.

One of her newest clients, Edward, is an only child who moved to Denver from rural Colorado to attend the school at age six. He is seventeen, classified as legally blind, cannot hold a driver's license, and can see well at distances of two feet or less. At the age of fourteen, his parents were killed in an automobile accident and he has no known living relatives. Edward received counseling for his grief and excelled in his studies and social behaviors despite his challenges. While at school, Edward developed an interest in mathematics and computers, benefitting from the excellent vision-impaired technology at the school. He would like to pursue his academic interests at the Denver Institute for Technology, a four-year school where he has taken one class and met briefly with an admissions counselor. He plans to obtain a B.S. degree in computer engineering. Edward is open to help and has told Scarlett that he knows he needs assistance in applying to college, finding somewhere to live, developing a support network, and securing transportation.

Scarlett works with Edward to develop a written plan for service delivery, carefully documenting agreed-upon goals, objectives, and specific interventions. Due to her experience at the school, Scarlett has an extensive referral information system with services for the visually impaired. This disability group has a well-organized collaborative of service providers and advocates. This task becomes easier in the case of Edward, since he wants to live and attend school locally. Scarlett, however, must set up many important aspects of service delivery so Edward can experience success with his goals.

Since Scarlett already has an assessment interview and case history information on file for Edward, she proceeds to planning specific interview and testing needs based on Edward's disability, social position, and academic plans. Edward agrees to meet with Scarlett twice each month during his senior year at the school to begin working on his goals, objectives, and service delivery plan. At this time, Scarlett focuses her work with Edward on identifying testing goals and service providers for his unique needs.

Morgan, C. (2012). Unpublished manuscript, Knoxville, Tennessee. Used with permission.

Case Study Discussion Questions

1. What types of testing needs should Scarlett consider for her work with Edward? How can she work to help Edward understand the purpose of the tests and interpret the results?
2. What are some of the goals, objectives, and services that may fit Edward's case? Why is an information referral system especially important in this instance?
3. What are some ways that Scarlett can empower Edward to begin working towards his short- and long-term goals? Which areas might require special attention to advocacy?

Chapter Summary

This chapter has introduced the planning phase of case management, which includes a number of stages and strategies. Planning begins with a review and continuing assessment of the client's problem. Two areas of concern during this time are a consideration of the problem and the available information about the problem: sources, previous attempts to resolve the problems, and barriers.

The development of a plan guides service provision. Planning is a process of setting goals and objectives and determining intervention. A goal is a brief statement of intent concerning where the client wants to be at the end of the process. Objectives provide the standards by which progress is monitored, the name of the person who is responsible for what actions, when they must perform the actions, and the criteria for acceptable performance.

Identifying services is a critical part of a well-developed plan. Linking clients and services is facilitated by information and referral systems, which include social service directories, feedback logs, staff training, and updated directories. Some professionals prefer setting up their own information and referral systems.

Part of the planning process or the development of the service plan may be gathering additional information. Two data collection methods used by case managers are interviewing and testing. Assessment interviews, structured clinical interviews, case history interviews, and mental status examinations are the four types of interviews. Potential sources of error include the halo effect and general standoutishness. Cultural consideration of interviewing is critical to this process.

Testing, a second data collection method, is encountered by the case manager in reports, in case files, and as a service offered by an agency. Understanding test language, concepts, sources of test information, and the factors included in selecting, administering, and interpreting tests facilitates the meaningful use of test scores with an effort to eliminate bias.

Chapter Review

Key Terms

Plan

Goals

Information and referral system

Feedback logs

Structured clinical interview

Sources of error

Test

Maximum performance test

Aptitude test

Plan development

Objectives

Social service directory

Assessment interview

Case history interview

Halo effect

Psychological test

Achievement test

Intelligence test

Reviewing the Chapter

1. Describe the two areas of concern addressed by revisiting the assessment phase.
2. What sources help a case manager review a client problem?
3. What role does documentation play in the review of the problem?
4. Define *plan*.
5. What activities occur before development of the plan?
6. List the characteristics of a service plan.
7. What are the benefits of establishing goals?
8. List the criteria for well-stated and reasonable goals.
9. Distinguish between a goal and an objective.
10. Identify a problem you would like to address, and develop a plan with goals and objectives.
11. What are some of the challenges of plan development?
12. List the three components of information and referral and give an example of each.
13. Discuss the similarities between interviewing and testing.
14. Compare the four types of interviews and their roles in case management.
15. Describe the 4 C's of questioning that support culturally sensitive interviewing.
16. Illustrate how sources of error may affect an interview.
17. How do case managers use tests?
18. Describe the different ways to categorize a test.
19. Describe how to select a test to measure vocational interests.
20. How do you consider testing in light of racially, ethnically, or linguistically different populations?
21. Define *validity* and *reliability*, and describe their roles in test selection. Include technical and social bias in your definitions.
22. What makes the test manual the best source of information about a test?
23. Identify some sources of error in testing.
24. Under what conditions should special considerations be made for a person with a disability?

25. Identify the two essential steps in test interpretation.
26. Describe the guidelines you would use to insure cultural sensitivity in the testing process. Explain for each the possible insensitivity you are trying to address.

Questions for Discussion ....

1. Why do you think developing a plan is important?
2. If you were a new case manager, how would you begin to develop a network of available services?
3. What kinds of criteria would you use to determine whether you need to conduct a structured interview with an eight-year-old?
4. Do you believe that you will be able to determine what errors exist in the information that you gather? What problems do you expect to encounter in finding errors?
5. Describe the skills you need to be a culturally sensitive interviewer and test administrator.

References ....

Department of Health and Human Services, Centers for Disease Control. (2009). *Evaluation Briefs: SMART Objectives*. Retrieved from http://www.cdc.gov/healthyyouth/evaluation/pdf/brief3b.pdf

Gulanti, G. (2008). *Caring for patients from different cultures*. Philadelphia: University of Pennsylvania Press.

Hollingsworth, H. L. (1922). *Judging human character*. New York: Appleton-Century-Crofts.

Kaplan, R. M., & Saccuzzo, D. P. (2009). *Psychological testing: Principles, applications, issues*. Pacific Grove, CA: Brooks/Cole.

McClam, T. (1992, September–October). Employer feedback: Input for curriculum development. *Assessment Update, 4*(5), 9–10.

McClam, T., & Woodside, M. R. (2012). *Introduction to human services*. Pacific Grove, CA: Brooks/Cole/Cengage.

Whiting, G., & Ford, D. (2009). *Cultural bias in testing*. The Gale Group. Retrieved from http://www.education.com/reference/article/cultural-bias-in-testing/

Whiston, S. (2005). *Principles and applications of assessment in counseling* (2nd ed.). Belmont, CA: Thomson Brooks/Cole.

. . . .

CHAPTER EIGHT

Building a Case File

Thriving and Surviving as a Case Manager

Susan

The case file ends up being really important to our work. Since we are a government agency, it provides the basis for our applications for funding for services for the client. The case file also provides the information for us; it is the basis of the treatment plan once the client is accepted for services. But a lot has to happen before the client receives services. A majority of what occurs reflects the building of the case file. Since we serve adults who apply for vocational rehabilitation, we ask for evaluations in various areas—physical, psychosocial, vocational, skills, independent living, educational, financial—and we also require a detailed intake and social history for the individuals and for their families. We also conduct a strengths and needs assessment. No one person can make all of these assessments. We used to have various professionals working for our agency and had a one-stop shopping service. We could conduct psychological assessments, medical assessment, vocational assessments, and others. Now we refer the client to professionals who contract with our agency. The case file and the compiling of information are really complicated. And if we can't get information from one of the sources, or if we can't schedule an assessment in a timely manner, or if the client or the client's family can't travel for the assessment, then the time between initial intake, evaluation of application for services, and beginning services is delayed. The process also is cumbersome and complicated, and many clients get discouraged. They see us asking them to do one more thing—then another thing—before we can even consider them. The good news is that once we compile this information, we are able to understand and help our clients.

> —Permission granted from Susan Grant (pseudonym), 2012, text from
> unpublished interview

When I am at work it is important for me to remember that I need to know about medical assessments, especially medical terms and how they are used. I use my computer to search terms I don't know. I also have a hard copy of the Physician's Desk Reference, *although, quite frankly, I primarily use the Internet. I try to have a basic knowledge of special therapies like speech therapy or occupational therapy.*

—Case manager, services for children and families, New York, New York

Our shelter provides mid-term shelter. A few shelters only provide lodging and services for four nights. We allow our clients to remain with us for up to thirty days. Other shelters' clients have to commit to a year of residence. For our month, we provide clients with a wide range of services. We try to meet all of their needs. Comprehensive assessment is key, and we provide an on-staff psychologist, group counseling, and individual and family work.

—Director, emergency shelter, St. Louis, Missouri

When a child is referred to our agency, we immediately seek more information. For example, the first thing a case manager does is contact the school and ask for records. Of course we have a release from the parents to get the records. . . . We need the school records even if the child is referred for a medical issue.

—Care manager, high school, Los Angeles, California

Information from other professionals comes to the case manager in two ways. When he or she receives a case file on a client from another agency or professional, it may contain reports or evaluations from other professionals. In other situations, the plan developed by the case manager and the client may include referrals to other professionals for evaluations. In both situations, the case manager must be able to understand the information provided and (if asking for help from other professionals) know what to request.

The chapter-opening quotations illustrate the kinds of information that a case manager may need to get from other professionals in order to develop a plan or to provide services. The medical information, histories, or exams these three helpers mention are part of the case files of clients who have medical problems. The case manager providing services for children and families speaks of the advantages of being familiar with medical terms and medical references when trying to decipher medical reports. Physical assessments and psychological assessments offer important information to the emergency shelter staff as they work with homeless and runaway female teens. Professional staff at the high school in Los Angeles gather much information about the student from other schools.

This chapter examines the types of information that may be found in a case file or that must be gathered to complete one. Exactly which information is needed depends on the individual's case and the agency's goals, but many cases involve medical, psychological, social, educational, and vocational information. We introduce each type of information, give a rationale for gathering it, describe the kinds of data likely to be provided, and discuss what the case manager needs to know in order to make the best use of the report. For each section of the chapter, you should be able to accomplish the following objectives.

MEDICAL INFORMATION
- Tell how medical information contributes to a case.
- Decode medical terms.

PSYCHOLOGICAL EVALUATION
- List the reasons for a psychological evaluation.
- Make an appropriate referral.
- Identify the components of a psychological report.
- Describe the type of information provided by the *DSM–IV–TR*.

SOCIAL HISTORY
- State the advantages and limitations of a social history.
- Name the topics included in a social history.
- List the ways social information may appear in the case file.

OTHER TYPES OF INFORMATION
- List the types of educational information that may be gathered.
- Define a vocational evaluation.

 # Medical Evaluation

Knowledge of medical terminology, conditions, treatments, and limitations is important in understanding a case. Medical information may be provided on a form or in a written report. The exam and report may have been done by a general practitioner or by a specialist in a field such as neurology, orthopedics, or ophthalmology. In some cases, the case manager can interact with the medical service provider and thus be able to ask questions, request specific assistance, or offer observations. Often, however, he or she does not have this opportunity and must rely on the written report. Then the resources mentioned at the beginning of the chapter may prove particularly helpful. Many agencies have a copy of the *Physician's Desk Reference* (PDR) or other medical guide or access the PDR on mobile devices. Some also have a physician serving as a consultant who is available to answer questions. This section introduces basic medical information to help you understand medical terminology.

Agencies approach medical information in different ways. Some require documentation of a mental or physical disability or condition when determining eligibility for services. Others use a medical examination as part of their assessment procedures. In certain situations, medical information is not gathered unless there is some indication or symptom of a disease, condition, or poor health that would affect service delivery.

Medical knowledge is particularly crucial when working with people who have disabilities. A general medical examination and specialists' reports help determine the person's functional limitations and potential for rehabilitation. It is important to set objectives that are realistic in light of the client's physical, intellectual, and emotional capacities. When a medical report covers a disability in functional terms,

> [I]t addresses the following factors [and] the description can read like the following: strength, climbing, balancing, stooping, kneeling, crouching, crawling, reaching, handling, fingering, feeling, talking, hearing, tasting, and smelling, near acuity, far acuity, depth perception, visual accommodation, color vision, and field of vision. (Debates, Rondinelli, & Cook, 2000, p. 81)

Each medical evaluation includes recommendations relating to the individual's physical, emotional, and intellectual capacities. What follows is a sample medical recommendation.

The individual has a diagnosis of obsessive-compulsive disorder and has limited strength, balancing, hearing, and near-acuity functionality. This person needs work with supervision, few stressors, limited lifting, and limited need for close work.

Often, however, the form for a general medical examination allows only a small space for the diagnosis, so the case manager reads a phrase such as "chronic back pain," "normal exam," or "emotional problems." Not very helpful, is it? Remember that the client is an important source of information; he or she can tell you about any problems. You may then need to decide whether or not a specialist's evaluation would be helpful.

It is important when referring a client for a medical exam the case manager prepare the client for that experience. This is especially critical from the multicultural perspective. For many individuals the medical establishment represents a place where they have little knowledge, no power or authority, or have had previous difficult experiences. Many case managers find that the best resource for culturally sensitive physicians is the client him or herself. When you follow up on a medical referral, you can ask clients about their experiences.

Voices from the Field: Conducting Culturally Sensitive Medical Exams

An approach to medicine that is client-centered is important. The University of Washington (2009) uses guidelines to educate and train their medical students on how to conduct an interview. The guidelines that follow provide case managers with specific ways they may assess the sensitivity of the physicians conducting the medical exam.

Cultural sensitivity in this setting means "Appreciating the ethno-cultural, spiritual, and religious perspectives of patients, families, and communities.... The term *cultural humility*, coined by Tervalon and Murray-Garcia (1998), expands this to include the recognition of power dynamics in health care and the community at large and encourages physician advocacy to address imbalances" (p. 77).

Goals of culturally sensitive medical interviewing and their responsibilities related to each follow.

- Demonstrate *contextual sensitivity* and use *cultural sensitivity*. Be aware of family, cultural, and religious values, and the influences of gender, age, socioeconomic status, and education level.
- Gather information regarding patient and family perspectives on, and use of, traditional and/or complementary healing strategies.
- Exploring and understanding the approaches patients have used in treating their illness is very important.
- Eliciting this type of information may be challenging. Historically, some patients and family members have been misjudged and even chastised by health care professionals for admitting use of alternative or traditional remedies.
  - Be cautious and sensitive when trying to elicit this important information.
  - Be attentive to any verbal or nonverbal cues that the patient may be uncomfortable discussing alternative healthcare practices (i.e., silence, eye deviation, a shift in their seated position, crossing their arms, etc.).
  - Questions such as the following may be helpful to ask:
    - Have you seen anyone else about this problem besides a physician?
    - Who do you think gives you good health advice?
    - Who else do you trust?
    - Have you participated in any healing practices or ceremonies to treat your problem?
    - Have you used nonmedical remedies or alternative or traditional treatments for your problem?
    - What role do they serve in your care?
    - Who in your family or community advises you about this condition?
    - How common is this condition in your family and/or community?
    - What is done commonly to heal this illness?

The University of Washington suggests a model to guide the cultural sensitivity of the physician during the medical exam.

Beliefs about health (What caused your illness/problem?)
Explanation (Why did it happen at this time?)
Learn (Help me to understand your belief/opinion.)
Impact (How is this illness/problem impacting your life?)
Empathy (This must be very difficult for you.)
Feelings (How are you feeling about it?) (p. 80)

Medical Exams

Generally, medical information contributes to a case in two ways. **Medical diagnosis** appraises the general health status of the individual and establishes whether a physical or mental impairment is present. For example, ten-year-old Javier Muldowny comes into state custody, abandoned by his parents. A case manager at the Department of Children's Services assigns an assessment, care, and coordination team to provide support to Javier. One member of the team takes him to the agency's health department for an examination. The examination results in a diagnosis of otitis media.

Diagnostic medical services include general medical examinations, psychiatric evaluations, dental examinations, examinations by medical specialists, and laboratory tests. A medical diagnosis is helpful when the client has a medical problem or is currently receiving treatment from a physician, who may provide important information about social and psychological aspects of the case in addition to the medical aspects. When making a referral for a medical diagnosis, the case manager should help the client understand why the referral is necessary, the amount of time it will require, what the client can expect to learn, and what use the agency will make of the report.

Medical consultation is used in several ways. First, the consulting physician can provide an interpretation of medical terms and information. For example, Javier Muldowny was diagnosed with otitis media. The case manager received this report, asked a colleague what the diagnosis meant, and learned that it was an ear infection. A consultation with a physician would reveal that otitis media is a severe ear infection that sometimes results when the eustachian tubes are not properly angled. The consultation might also explain the report further and clarify possible treatments. In Javier's case, the case manager may need further information about the advantages and disadvantages of two possible treatments: insertion of tubes in the ears or a regimen of antibiotics. A consultation with an otorhinolaryngologist (ear, nose, and throat specialist) could shed light on the medical prognosis and the extent of any hearing disability that might be expected.

The role of a medical consultant is to interpret the available medical data, determine any implications for health and employment, and recommend further medical care if needed. The case manager can make the best use of a consultant by being prepared for the meeting, perhaps specifying in writing what is needed from the consultant. This usually involves identifying problems that need to be resolved and setting forth the significant facts of the case. The case manager needs to understand medical terminology, the skills of specialists in diagnostic study and treatment programs, and the effects of disability on a client.

The medical service used most often in human services is the **physical examination**, in which a physician obtains information concerning a client's medical history and states his or her findings. The exam data are entered into the medical record. Here we give an overview of the physical examination, including the kinds of information obtained and what the case manager needs to know to make such a referral and to understand the physician's report.

Diagnosis involves obtaining a complete medical history and conducting a comprehensive physical exam (also called a *physical*, a *health exam*, or a *medical exam*). The results of the exam may be reported on a form provided by the referral source. Sometimes physicians use preprinted schematic drawings of various body parts or organ systems to enhance or clarify the written report. However the information is transmitted, the quality of the reporting depends on the relationship between the physician and the patient. In some cases, the patient may have mixed feelings about the referral for a physical exam. He or she may need an explanation of why the referral is necessary, the amount of time the exam will take, what outcome is expected, and how the information will be used. Keep in mind that the client's socioeconomic status, language skill limitations, or cultural background may also influence how he or she feels about the referral. If the request is communicated with sensitivity, and if a good relationship with the physician is established, the client can overcome any barriers of anxiety, depression, fear, or guilt.

The general medical exam is done by a physician who takes an overall look at the person's medical state. Its purpose is to evaluate the person's current state of health, focusing on two areas. First, a complete medical history records all the factual material, including what the client states and the physician's inferences from what is not said. A typical starting point is the chief complaint, as expressed by the individual. (**See Figure 8.1.**) If there is an illness present, it is described in terms of onset and symptoms (including location, duration, and intensity). A family history relates significant medical events in the lives of relatives, particularly parents, grandparents, siblings, spouse, and children. Extensive information about the individual's past medical history is also collected. This may include childhood diseases, serious adult illnesses, injuries, and surgeries. A review of symptoms focuses on information about present and past disorders, which the physician elicits through questions about organs and body systems. After completing the physical exam, the physician records a diagnostic impression. The actual diagnosis is made once there is conclusive evidence, which may mean getting further studies or referring the client to a specialist for consultation.

What exactly makes up a medical exam? Techniques used during a physical exam are inspection, palpation (feeling), percussion (sounding out), and auscultation (listening). Usually, the examining physician works from the skin inward to the body, through various orifices, and from the top of the head to the toes (Felton, 1992). Special instruments are used to look, feel, and listen. More time is spent in particular areas to ascertain whether a certain finding truly represents a change in an organ or tissue. Some parts of the exam are carried out quickly, and others require more time. More important areas may receive a second, more thorough examination. The physician records the findings as soon as possible after completing the exam and shares the results with the client.

MEDICAL REPORT

CHIEF COMPLAINT: Weakness and malaise

PRESENT ILLNESS: Three weeks ago, this 40-year-old single African-American male had a cold, with associated mild cough and temperature elevation, which lasted two days. At that time there was a loss of appetite and a decrease in food intake.

After the cough and temperature elevation subsided, the patient noted increased weakness and general malaise. The patient reports he tires easily and is unable to sustain exercise, which was tolerated well before the onset of the cold.

FAMILY HISTORY: This patient's mother, 75, has had breast cancer and a radical mastectomy and was diagnosed with lupus ten years ago. His father, 80, had quadruple bypass surgery five years ago. There is extensive evidence of heart disease in father's family. There is a history of diabetes and tuberculosis in mother's family, although the mother has had neither.

PAST HISTORY: Patient had chicken pox at age 7. Sustained multiple fractures in a motor vehicle accident at age 16, but he is without permanent motor or neurological damage. Frequent lower respiratory infections characterized by productive cough and yellow-mucus-producing hacking cough. The patient has been exposed to tuberculosis, showing a positive PPD test but negative chest X-ray five years ago, and was treated with usual course of preventative medications. Since that time patient has been free of other persisting symptoms.

SOCIAL HISTORY: Unmarried, owns home, smokes one and a half packs of cigarettes a day, drinks occasionally.

OCCUPATIONAL HISTORY: Worked construction jobs following high school. Has experience with foundation work, masonry, and plumbing. Last spring he graduated from a local college with degree in accounting. Plans to start his own construction business next year.

(continues)

Figure 8.1   Medical report

REVIEW OF SYSTEMS:

HEAD, EYES, EARS, NOSE, THROAT: No frequent or severe headaches or head injuries. Wears corrective lenses for nearsightedness. Had frequent ear infections as child, but has suffered no hearing loss. Has two or three colds a year, but is without sinus pain. No problems with throat.

NECK: No significant abnormality.

RESPIRATORY TRACT: No expectoration of blood. Wheezing and shortness of breath with exertion—walking up stairs. When colds "go to chest," takes over-the-counter cough medications. Early morning nonproductive cough present. States he is not trying to quit smoking.

CARDIOVASCULAR SYSTEM: No chest pain or palpitations. No history of murmur, coronary artery disease, or hypertension. Fatigue has been increasing over the last month, making him unable to complete his morning walk, which is normally 2 miles.

GASTROINTESTINAL SYSTEM: Appetite has decreased since this cold. No indigestion. States he is on no special diet.

GENITOURINARY SYSTEM: Denies venereal disease.

NERVOUS SYSTEM: No significant abnormality.

MUSCULOSKELETAL SYSTEM: Negative.

ENDOCRINE SYSTEM: Negative.

© Cengage Learning 2006

Figure 8.1   (*Continued*)

For some clients, one of the first things that occurs in the case management process is a referral to a physician for a general medical exam. As the physician conducts the exam, he or she completes a form like the one shown in **Figure 8.2**, which is then sent to the referring counselor. It becomes part of the client record.

GENERAL BASIC MEDICAL EXAMINATION RECORD

This record is CONFIDENTIAL

Section I. - (To be filled out by rehabilitation agency)

Client No. _____

Johnson, Ray R.   7/16/60   W   M   S ✓ M__ D__ Sep_
(Last Name)   (First Name)   (Middle Name)   (Date of birth)   (Race)   (Sex)   (Marital Status)

Rt. 1   Box 68   Centerville   TN
(Home address: Street and number or R.F.D.)   (City or town)   (County)   (State)

Usual occupation **Plumber** _____ Description of last job **Same**

Last time hospitalized __**5-XX**__ __**Surgery**__ _____
(Date)   (Reason)   (Name and location of hospital)

Last visit to physician _**18 mos.**_ __**Spinal cord injury**__ __**Dr. Alderman**__
(Date)   (Reason)   (Name and address of physician)

Is patient now under care of physician? __**Yes**__ _____ __**Dr. Brown**__
(Yes or no)   (If answer is "Yes," give name and address of physician)

Patient's statement of disabilities __**Spinal cord injury**__

Signature of rehabilitation counselor __**Tom Chapman**__ _____ Date **7-20-XX**

Section II. PERTINENT HISTORY (To be filled out by physician.)

Back surgery; surgery twice continued pain

Section III. PHYSICIAN EXAMINATION. (To be filled out by physician. Items checked ✓ were examined and found
normal. Deviations from normal are noted. If items require additional description, please record on extra sheet.)

Height (without shoes): __**5**__ ft. __**10**__ in. ☐ Weight (without clothing) **220** pounds ☐ Temperature **98⁴**

Eyes - Right _____✓_____ LEFT ____✓____
(Discharge; corneal scars; strabismus; pterygium; ptosis; trachoma; fundi; cataract; intraocular tension)

Distant vision: Without glasses: R. 20/**30** L. 20/**70**. With glasses: R. 20/**20** L. 20/**20**
(If vision is too low to be recorded at 20 feet, indicate by recording as "less than 20/200"

Ears - Hearing: Right **20** Left **20** ☐ Other findings: R ____✓____ L ____✓____
20 feet   20 feet   (Evidence of middle ear or mastoid disease. Drums: Normal, absent
(Consider denominators here indicated as normal   perforated, dull, retracted, discharge).
Record as numerators greatest distance heard)

Nose _____✓_____ ☐ Throat ____✓____
(Obstruction, evidence of chronic sinus infection, polypi   (Tonsils: Normal, enlarged, removed, etc.)
perforated septum, etc.)

Mouth _____✓_____ ☐ Neck ____✓____
(Missing teeth, pyorrhea; abnormality of tongue or palate)   (Thyroid enlargement, nodules, etc.)

Lymphatic system _____✓_____ ☐ Breasts ____✓____
(Especially cervical, epitrochlear, inguinal)   (Abnormal discharge, nodules, tenderness, hypoplasia)

Lungs: Right _____✓_____ ☐ Left ____✓____
(If history or physical findings indicate active or arrested tuberculosis, recommend chest X-ray, sputum examination,
and consultation with chest specialist)

(continues)

Figure 8.2   Medical examination form

Medical Terminology

Reports from health care providers often include **medical terminology**, which
may seem like a foreign language to a case manager who is unfamiliar with it,
because physicians rely on technical words and phrases for exactness. Medical

Circulatory System: Heart_____ ✓
_____(Enlargement, thrill, murmurs, rhythm)

Blood pressure { Systolic } 118/80 ___ Pulse rate _76_ Dyspnoea _O_ Cyanosis _O_ Edema_O_
_____Diastolic

Evidence of arteriosclerosis _____ None
_____(Type; degree; where found, as "cerebral," "brachial," etc.)

Abdomen _____ ✓
_____(Scars, masses, palpable liver, palpable spleen, etc.)

Hernia _____ ✓
_____(Type: Inguinal, ventral, femoral, etc. Right, left, bilateral)

Genito-urinary _____ ✓
_____(Urethral discharge, varicocele, hydrocele, scars, epididymitis, enlarged or atrophic testicle)

and
Gynecological _____
_____(Prolapse, cystocele, rectocele, Cervix)

Ano-rectal _____ ✓
_____(Hemorrhoids, prolapse, fissures, fistula. Prostate)

Nervous system _____ ✓
_____(Paralysis, Sensation, Speech, Gait, Reflexes: Pupillary, Knee, Babinski, Romberg)

_____(Memory, Peculiar ideas or behavior. Spirits; Elated, depressed, normal)
_____ ✓
_____(Neurological or psychiatric abnormalities should be described on separate sheet)

Skin__ ✓ _____ ☐ Feet _____ ☐ Varicose veins __ ✓ __
_(Moist, dry, clear)_____(Weak feet, Congenial or traumatic defects)_____(Site)

Orthopedic Impairments: (Describe) _Low back pain_

Laboratory: Urinalysis: Date _8-7-XX_ ☐ Specific gravity _1.020_ ☐ Reaction _7_
_____☐ Albumen _neg_ ☐ Sugar _neg_

DIAGNOSIS: (Indicate major and minor) _Chronic lower back pain after two spinal_
___ops for herniated disc_

STATUS: of major disability: (Check appropriate terms) Permanent _✓_ Temporary _____ Stable _____

Slowly progressive _____ Rapidly progressive _____ Improving _____

PROGNOSIS: Can the major disability be removed by treatment? ☐ ☑ Substantially reduced by treatment? ☑ ☐
_____(Yes) (No)_____(Yes) (No)

Physical Capacities: (Under "Physical activities" and "Working conditions" use symbols as follows:
_____(✓) No limitation. (X) Limitation. (0) To be avoided.

Physical activities: Walking _X_ Standing _X_ Stooping _X_ Kneeling _X_ Lifting _X_ Reaching _X_ Pushing _X_
_____Pulling _X_ Other (Specify) _____

Working conditions: Outside _✓_ Inside _✓_ Humid _✓_ Dry _✓_ Duty _✓_ Sudden temperature changes _✓_
_____Other (Specify) _____

RECOMMENDATIONS:

☐ Is examination by specialist advisable? If so, specify which speciality _____

☐ Refraction ☐ X-ray of chest ☐ Other diagnostic procedures (Specify) _____
☐ Prosthetic appliances (Specify) _____
☐ Hospitalization (Specify reasons and approximately duration) _____
☐ Treatment (Specify type and approximately duration) _____

Remarks: Please use additional sheet for remarks and expansion of any observations.

Date _8-7-XX_ _____ _____ M.D.
_____(Physician)
_____ Suite 201 Physicans Office Bldg.
_____(Address)

Figure 8.2   (Continued)

specialties also have special terminologies. Other professionals who may write reports using medical terminology are dentists, podiatrists, veterinarians, pharmacists, nurses, physical therapists, and occupational therapists. It can be a challenge for the case manager to make sense of these reports; to do so, he or she must have at least a rudimentary understanding of medical terminology.

Medical terminology follows simple rules. To analyze medical words, identify the four elements that are used to form such words: the word root, the combining form, the suffix, and the prefix. It may help to think of these elements as verbal building blocks. Let's examine each component.

Word roots   The main part or stem of a word is the **word root**. In medical terminology, the root usually derives from Greek or Latin and often indicates a body part. All medical words have one or more word roots.

GREEK WORD	MEANING	WORD ROOT
kardia	heart	cardi
gastro	stomach	gastr
nephros	kidney	nephr
osteon	bone	oste

Combining forms   A word root plus a vowel, usually an *o*, is the combining form, as in the following examples.

WORD ROOT		COMBINING VOWEL		COMBINING FORM	MEANING
cardi	+	o	=	cardio	heart
gastr	+	o	=	gastro	stomach
nephr	+	o	=	nephro	kidney
oste	+	o	=	osteo	bone

Suffixes   A *suffix* is a word ending. In medical terminology, the suffix usually denotes a procedure, condition, or disease, as in the instances listed here.

COMBINING FORM		SUFFIX		MEDICAL WORD	MEANING
arthr (joint)	+	-centesis (puncture)	=	arthrocentesis	puncture of a joint
thoraco (chest)	+	-tomy (incision)	=	thoracotomy	incision in the chest
gastro (stomach)	+	-megaly (enlargement)	=	gastromegaly	enlargement of the stomach

Suffixes also form adjectives, express relative size, indicate surgical procedures, and express conditions or changes related to pathological processes. Examples follow.

ADJECTIVES	EXAMPLE	MEANING
-al (means "pertaining to")	arterial	pertaining to an artery
-ible (indicates ability)	digestible	capable of being digested

RELATIVE SIZE	EXAMPLE	MEANING
-ole (means small)	arteriole	a small artery
-ule (means small)	granule	a small grain

SURGICAL PROCEDURE	EXAMPLE
-ectomy (means "removal of an organ or part")	appendectomy

PATHOLOGY	EXAMPLE
-mania (means "excessive excitement or obsessive preoccupation")	pyromania

Prefixes   The word element located at the beginning of a word is the *prefix*. It usually denotes number, time, position, direction, or negation.

PREFIX		WORD ROOT		SUFFIX		MEDICAL WORD	MEANING
hyper (exces- sive)	+	therm (heat)	+	ia (condi- tion)	=	hyper- thermia	condition of excessive heat
micro (small)	+	card (heart)	+	ia (condi- tion)	=	microcardia	condition of a small heart

Other common prefixes that modify word roots indicate position (e.g., *ab* means "away from," as in *abnormal*); quantitative information (e.g., *a* or *an* means "without," as in *anorexia*, or without appetite); qualitative information (e.g., *mal* means "bad," as in *malfunction*); and sameness or difference (e.g., *homo* or *hetero*). For other prefixes and suffixes that are common in medical terms, see **Table 8.1**.

There are three basic steps to working out the meaning of a medical term. First, identify the suffix and its meaning. Second, find the prefix, if any, and determine what it means. Third, identify the root words and their meanings. For

TABLE 8.1   COMMON PREFIXES AND SUFFIXES

Prefix	Meaning	Suffix	Meaning
dys-	bad, painful, difficult	-itis	inflammation
macro-	large	-algia	pain
hypo-	under, below	-toxin	poison
scler-	hard	-oma	tumor
tachy-	rapid	-pathy	disease
hyper-	over, above, excessive	-osis	abnormal condition, increase
eu-	normal	-glycemia	normal blood sugar

example, *thermometer* consists of a suffix (*meter*, meaning "instrument for measuring") and a word root (*thermo*, meaning "heat"). Thus, a thermometer is an instrument for measuring heat. Another example is *gastroenteritis*. The suffix is *itis* (inflammation), the prefix is *gastr* (stomach), and the word root is *enter* (intestine). Gastroenteritis is an inflammation of the stomach and intestine. Remember that the vowel *o* is a combining form, linking one word root to another to form a compound word. *Osteoarthritis* is another example. The suffix *itis* means "inflammation"; word roots are *oste*, which means "bone," and *arthr*, which means "joint." The *o* is the combining vowel. Osteoarthritis means inflammation of bone and joint. The following list contains some common medical terms that use suffixes, prefixes, and word roots introduced in this chapter. Can you fill in the columns with the meaning of each term? Other examples are shown in **Table 8.2**.

TERM	SUFFIX/PREFIX	WORD ROOT	MEANING
tachycardia			
dysfunction			
gastritis			
nephritis			
osteopathy			
hypodermic			

TABLE 8.2   SOME COMMON COMPONENTS OF MEDICAL TERMS

Component	Meaning	Example
-algia	pain	neuralgia
angio-	blood vessel	angiogram
arth-	joint	arthroscopy
contra-	opposed to	contraception
derm-	skin	dermatology
-emia	condition of the blood	polycythemia
enceph-	brain	encephalitis
glyco-	sugar	glycosuria
hepat-	liver	hepatitis
hyster-	uterus	hysterectomy
leuk-	white	leukocyte
lip-	fatty	hyperlipidemia
-oscopy	visual examination	laparoscopy
-ostomy	creation of an artificial opening	tracheostomy
-otomy	incision	craniotomy
-plasty	reparative or reconstructive surgery	rhinoplasty
pre-	before	precancerous
pyel-	pelvis	pyelogram
syn-	together	synarthrosis
tri-	three	triceps

It is a continuing challenge for case managers to keep current with terminology because of ambiguities, inconsistencies, and the changing course of medical knowledge. Although most word roots have Greek or Latin origins, some occur in both languages but have different meanings. The root *ped*, for example, means "child" in Greek (e.g., pediatrician), but in Latin *ped* means "foot" (e.g., pedicure). Many diseases are named for individuals, such as Alzheimer's disease, Parkinson's disease, and Hodgkin's disease. Some disorders are called *syndromes*: Cushing's syndrome, Horner's syndrome. Acronyms are formed from the initials of lengthy terms: MRI (magnetic resonance imaging) and ACTH (adrenocorticotropic hormone) are examples. In addition, medical terminology traditionally uses hundreds of abbreviations; some of the most common are listed in **Table 8.3**. However, one must be cautious about using abbreviations, since this often increases the likelihood of error. For example, "qid" mean four times a day, and "qd" means once a day. If "qd" is interpreted as "qid" and a drug is administered four times a day rather than once a day, serious complications could result. Keeping informed about trends in medicine increases one's understanding of the meanings of terms. For example, physicians increasingly prescribe generic drugs rather than brand names (e.g., the generic diazepam rather than Valium®). Keeping current with medical terminology entails awareness of chemicals, syndromes, and diseases that are newly named and sometimes given acronyms or abbreviations (e.g., AIDS for acquired immunodeficiency syndrome). It must also be remembered that words can have multiple meanings, and that several names may apply to a single entity.

TABLE 8.3   MEDICAL ABBREVIATIONS

Abbreviation	Meaning	Abbreviation	Meaning
a.c.	before meals	L-1, L-2, L-3	lumbar vertebrae (by number)
b.i.d.	twice daily	LLQ	left lower quadrant
B.P.	blood pressure	LMP	last menstrual period
C-1, C-2, C-3	cervical vertebrae (by number)	p.c.	after meals
CBC	complete blood count	p.r.n.	as needed
CNS	central nervous system	q.i.d.	four times daily
DX	diagnosis	RLQ	right lower quadrant
F.H.	family history	RX	treatment
GI	gastrointestinal	S-1, S-2, S-3	sacral vertebrae (by number)
GU	genitourinary	T-1, T-2, T-3	thoracic vertebrae (by number)
HDL	high-density lipoprotein	t.i.d.	three times daily
h.s.	at bedtime	WBC	white blood count
H & P	history and physical examination		

Psychological Evaluation

The objective of a **psychological evaluation** is to contribute to the understanding of the individual who is the subject. The report writer is a consultant who makes a psychological assessment that is practical, focused, and directed toward the solution of a problem. Thus the psychological report he or she prepares is more than a presentation of data. This section helps you determine when a psychological evaluation is needed, how to make the referral, and how to prepare the client. It also discusses the evaluation itself and the report.

Referral

Case managers may refer clients for psychological evaluations for a number of reasons. One reason is to establish a diagnosis in order to meet criteria of eligibility for services.

Nadine is a deeply depressed fifteen-year-old who is currently taking antidepressant medication. She is increasingly out of control. Yesterday, she slapped her grandmother, with whom she lives, and threatened to kill her. If she is to receive services in an inpatient treatment program, she must have a diagnosis confirming emotional disturbance.

Another reason for a psychological evaluation is to provide justification for a particular service.

Amal is a 28-year-old male whose divorce will be final in a month. As the court date approaches, Amal feels more and more depressed. He is having trouble getting up in the morning, showing up for work on time, and maintaining relationships with those who are close to him. His physician has suggested counseling, but Amal's insurance company insists that he have a psychological evaluation to determine whether or not he needs it.

Sometimes a psychological evaluation functions as a screening or routine evaluation to obtain information about a client's personality, aptitude, interests, intelligence, and achievement.

Greg is a thirty-five-year-old male who is the only child of elderly parents. He is developmentally disabled. His parents, concerned about who will care for Greg if something happens to them, have learned of a group home where the residents live under close supervision. One requirement for acceptance into the program is a recent psychological evaluation that assesses intelligence as well as ability to function independently.

A case manager may also order a psychological evaluation to resolve contradictions or ambiguities, or to add information that is missing.

Paloma is a ten-year-old who is enrolled in public school. Her teacher is concerned about her behavior. One day she is passive, rarely interacts with her classmates, and does not participate in class. The next day, she may be loud, talkative, and disruptive. Just yesterday, she started a fight with a classmate. This has prompted her teacher to request an evaluation from the school psychologist.

Finally, a psychological evaluation may be recommended to answer particular questions regarding the client. Is there brain damage? Why does the individual have trouble relating to others? How is this person adjusting to the recent amputation of her leg? Why is the client doing poorly in school?

In any of these situations, a referral for a psychological evaluation is appropriate. In each case, the case manager seeks help in order to provide the client with needed services. It is easiest to get what is needed if the consulting psychologist knows the general mission of the agency and understands the specific problem to be addressed. Having this information allows him or her to choose the most relevant and efficient approach to gathering the needed information. The referral for a psychological evaluation is usually made by a case manager, who specifies what is needed: a routine workup, testing, questions about the case, a diagnosis. Thus, the psychologist is charged with a mission. It is therefore critical that the referral be more than a general request, such as "psychological evaluation" or "for psychological testing." These terms communicate poorly; the referring professional has failed to express what prompted the referral. Two scenarios may result: The psychologist may ask the case manager for more specific information, or he or she may try to guess what is wanted or needed. When the reason for the referral is not clear, it is difficult for the psychologist to provide a useful report.

How does a case manager make a good psychological referral? First, it is important to be clear about the reason for referral. The case manager must clarify the need to document a condition or disability, obtain test scores, or explore behavioral inconsistencies. Specific questions also help the psychologist focus on the client's problems. The psychologist then makes recommendations to the case manager. The two professionals can discuss the case before the evaluation to clear up any questions or needs. Because many referrals are made by phone or direct personal contact, such a discussion can easily take place, but it may be even more important when the referral is made in writing.

Part of making a successful referral is preparing the client for the psychological evaluation. To do this, the case manager needs a clear understanding of the process and the ability to explain it to the client. Some clients may be suspicious of testing or may fear that the case manager considers them crazy. Demystifying the evaluation helps to dispel these attitudes.

The Process of Psychological Evaluation

The evaluation itself includes a study of past behavior, conclusions drawn from observations of current behavior, a diagnosis, and recommendations. This study requires the psychologist to assess which data are important to the client's presenting problems. In some cases, relevant information is in the client file; it is then helpful for the psychologist to have access to these documents in addition to the observations and questions from the referral source.

One of the primary ways that a psychologist observes current behavior is by testing. From the discussion of testing in the previous chapter, you know that testing gives samples of behavior. That discussion also introduced a number of tests that are useful in human services. Psychologists use many of them, notably the WAIS and projective tests (such as the *Rorschach* and *Thematic Apperception Test*). These tests are individually administered and scored, and psychologists are specially trained to use them. As a consultant, then, the psychologist decides what kinds of data must be gathered to carry out the assignment given by the referral source, which findings have relevance, and how these findings can be most effectively presented. We talked in Chapter 7 about the culturally sensitive approach to testing; this focus on bias in testing remains important in psychological testing.

The results of the psychological evaluation are communicated to the case manager in a written report. The **psychological report** is a written document that explains an individual's personal characteristics, mental status, and social history. This document provides information that helps determine what problems and challenges the client faces and what might be possible interventions. The report may appear in one of several forms, the most common of which is a narrative (illustrated by the report included in this section).

Results may also be communicated as a terse listing of problems and proposed solutions. Still another option is the computer-generated report, usually consisting of a sequence of statements or a profile of characteristics. Less frequently used are checklists of statements or adjectives, clinical notes, and oral reports relating impressions. Because the narrative is the form of psychological report that is most often used in human services, let's explore it further.

Usually, the content, sources, and format of narrative psychological reports follow a similar pattern. There are three components to the content of a report. One is the orienting data, which includes the reason for the referral and pertinent background information, such as age, marital status, social history, and educational record. Illustrative and analytical content is the second component; here one finds the interpretation of raw data, including test scores. The third component, the psychologist's conclusions, includes a diagnosis and recommendations, which are presented with supporting evidence. The sources of the information in all three components are the interview between the psychologist and the client; test data; behavior observed during the evaluation; any available medical reports and social histories; and any observations, case notes, or summaries written by other professionals involved with the case.

Among the headings that organize the report are "Reason for the Referral," "Identifying Data," and "Clinical Behavior." Under such headings one would find the reason for the assessment, identifying information, any social data, and the psychologist's observations of behavior during the evaluation. The subsequent headings—"Test Results," "Findings," "Test Interpretation," or Evaluation"—may be subdivided into intellectual aspects (e.g., an IQ score and what it means) and personality (e.g., psychopathology, attitudes, conflicts, anxiety, and significant relationships). The diagnosis section presents the main evaluative conclusions, usually expressed as a series of numbers followed by the name of a disorder or condition.

The classification system for diagnoses that will be used until May 2013 in the United States is the *Diagnostic and Statistical Manual of Mental Disorders, Text Revision*, Fourth Edition *(DSM–IV–TR)*, published by the American Psychological Association. From May 2013 forward, the Diagnostic and Statistical Manual of Mental Disorders-5 (known as "DSM-V") will be used. Many case managers may receive reports based upon a **DSM–IV–TR** diagnosis and, in the future, based upon the *DSM-V*. Understanding what the various diagnoses and scores mean will help the case manager understand the challenges the clients face. At times, professionals will submit a *DSM* diagnosis and a treatment plan. After consulting with the professional, the case manager may provide supportive services. Let's see what types of information a *DSM–V* diagnosis provides.

The new *DSM–V* (American Psychiatric Association, 2012) codes include a broad range of psychological disorder categories, such as anxiety disorders; trauma and stressor-related disorders; depressive disorders; bipolar and related disorders; and disorders usually first diagnosed in infancy, childhood, or adolescence, to name a few. The Web site DSMV (http://www.dsm5.org/) provides information related to changes in the organization and structure of the new manual, description of the disorders, a timeline for DSM-5 changes, frequently asked questions, and updated information about publications and presentations related to the DSM-5. During the revision of the DSM-5, specialized work groups were asked to consider the following aspects of diagnosis:

- Clarify boundaries between mental disorders to reduce confusion of disorders with each other and to help guide effective treatment.
- Consider "cross-cutting" symptoms (symptoms that commonly occur across different diagnoses).
- Demonstrate the strength of research for the recommendations on as many evidence levels as possible.
- Clarify the boundaries between specific mental disorders and normal psychological functioning (American Psychiatric Association, 2012).

As stated earlier, the *DSM–IV–TR* and the *DSM-V* describe criteria for specific disorders. The new *DSM-V* will continue to use a multi-diagnostic approach that helps assess clients using multiple factors. In addition, dimensional assessment can be used to document symptoms not tied to a specific diagnosis, but related to an individual condition. These assessments are a way of cross-cutting major diagnosis categories.

The diagnosis depends on various factors. Clients do not have to meet all the criteria to receive the diagnosis; this system allows for individual manifestations of the diagnosis. Because the *DSM–IV–TR* and the *DSM-V* provide a way of classifying all types of mental disorders, most agencies have a copy of the *DSM–IV–TR* and will have copies of the *DSM-V*. Training related to the new *DSM-V* will help case managers and other professionals use the new manual.

The diagnosis section of the report may be followed by a prognosis section—a statement about future behavior. Recommendations conclude the report and suggest some possible courses of action that would be beneficial in the psychologist's opinion, based on the psychological evaluation. For an example of a psychological report, see **Figure 8.3**.

Psychological evaluations differ according to the client's needs. The client profiled in Figure 8.3 was referred for assessment of his reading problems and to determine his eligibility for special services. The tests administered and the final report would have been different if the client had been referred for other reasons (e.g., behavioral problems).

 # Social History

For a complete case file, the client's past history and present situation must be investigated. The person's past adjustment can give indications of how he or she will adjust in the future. A **social history** also provides information about the way an individual experiences problems, past problem-solving behaviors, developmental stages, and interpersonal relationships. Some of the information in a social history may duplicate what has been gathered during the intake interview. In the social history, however, the client can relate the story in his or her own words, with guidance from the helper.

A social history has a number of advantages. Often the informal history contains gaps, but a carefully taken social history completes the picture. The case manager can then plan the appropriate integration of services and provide better information for future referrals. The social history often includes a better assessment of the client's need for services; this is especially helpful for clients who have multiple problems. A social history can also fulfill legal requirements. Finally, the process of taking a social history can help build the relationship between the case manager and the client.

The social history also has limitations. History taking is a preliminary activity in case management, but the client may perceive it as a phase in which solutions are put in place. Unfortunately, categorizations and judgments made at this stage may be premature. The process of taking the history can also give an inaccurate view of what will happen between the client and the case manager. Excessive questioning by the case manager may lead to a dependent role for the client, and culture-bound questions can create barriers to the development of the helping relationship. In addition, an exhaustive history is not absolutely necessary to develop a plan of services; it may be helpful, but the information gathered may not be relevant to service delivery. Spending too much time on history taking can also be harmful. The client may use the process to resist significant facts. Other clients may construe it as therapy, but it is not intended as such and may not even be therapeutically valuable. Despite these limitations, the social history still has the important function of completing the case file. Moreover, the case manager can use certain strategies to mitigate the limitations.

CONFIDENTIAL PSYCHOLOGICAL REPORT

NAME: Scott Garrett
AGE: 7 years, 6 months
DATE OF BIRTH: 3/2/XX
GRADE: 2
SCHOOL: Pineview Elementary
PARENTS: Mr. and Mrs. Scott N. Garrett
EXAMINER: Claudia Zimmerman
DATES OF ASSESSMENT: 9/30/XX

Reason for referral and background information
Scott Garrett was referred by his mother, Mrs. Sue Garrett, who notes that Scott doesn't enjoy reading and isn't good at it. She notes that he doesn't appear to invest in classwork, particularly seatwork. His teacher, Ms. Cole, is also concerned about Scott's progress. She notes that he is behind in reading but is on grade level in math. According to the developmental history, Scott accomplished developmental milestones in a typical fashion, with the exception of speech. His first words were at about 18 months. There was no history of physiological problems except for an eye muscle imbalance problem; he is wearing corrective lens for that problem. His hobbies are typical for his age. Birth history is unremarkable. Labor and delivery were normal.

Assessment procedures
Wechsler Intelligence Scale for Children III (WISC-III)
Woodcock–Johnson, Revised, Tests of Achievement
Brigance Comprehensive Inventory of Basic Skills

Test behavior
Scott was evaluated during two sessions. In both sessions, he was willing to work but would often ask, "Are we finished yet?" His problem-solving attack skills seemed typical for a child his age, and he displayed no signs of hyperactive or excessive distractibility in the one-to-one sessions. When confronted with tasks requiring reading, he would sometimes exclaim, "Oh, no!" In appearance, Scott is approximately average in height for his chronological age, but somewhat overweight. He is an engaging youngster, with close-cropped blond hair and round facial features; he could be described as "cute."

Figure 8.3   Psychological report

Test results

On the WISC-III, Scott obtained a Full Scale IQ of 106 + or –3, a Verbal Scale IQ of 100, and a Performance IQ of 112. His Full Scale score of 106 is at the 66th percentile rank and is slightly above average for his chronological age. Chances are 2 out of 3 that the range of scores from 103 to 109 contain his true score. (A true score is the hypothetical average score a child would obtain on repeated testing with the same instrument, minus the effects of practice, fatigue, etc.). Scott's individual subtest scores are as follows. (Scores can be compared to a population mean of 10, a standard deviation of 3.)

VERBAL-SCALED	SCORES	PERFORMANCE-SCALED	SCORES
Information	11	Picture completion	11
Similarities	8	Coding	15
Arithmetic	12	Picture assessment	10
Vocabulary	10	Block design	11
Comprehension	9	Object assembly	12
Digit span	13	Symbol search	14
Mazes	12		

Scott's profile can be presented using a number of "factor scores." These factor scores have the same psychometric properties as a Full Scale score. That is, the population mean is set to 100 and standard deviation to 15 for the factor scores. The factor scores include Verbal Comprehension, 98; Perceptual Organization, 107; Freedom from Distractibility, 115; Perceptual Speed, 124. These scores reflect average to above-average performance in general, but there is some variability. For example, the Verbal Comprehension scores are considerably lower than the Perceptual Speed score and the Freedom from Distractibility score. In general, his Verbal Fluency, fund of general vocabulary words, and Verbal Reasoning and Judgment appear to be about average compared to chronological age-mates. He is able to focus attention when directed and maintain that attention and concentration to a degree significantly better than chronological age-mates. His nonverbal reasoning and synthesis/analysis scores appear to be average to slightly above average compared to age-mates. This intellectual profile is typically predictive of average-to-better classroom performance.

(continues)

Figure 8.3   *(Continued)*

On the Woodcock–Johnson Revised Tests of Academic Achievement, the following scores were obtained:

	GRADE EQUIVALENT	SCALE SCORE	PERCENTILE RANK
Letter and word identification	1.6	94	35
Passage comprehension	1.7	97	43
Word attack	1.6	96	39
Reading vocabulary	1.7	95	37
Math calculation	2.8	97	43
Applied problems	1.8	97	43
Science	3.1	111	76
Basic reading skills	—	97	42
Reading comprehension	—	98	44

In general, Scott's scores on the Woodcock–Johnson ranged from slightly below average to slightly above average. His math- and science-related scores were slightly better than the reading scores. His language arts/reading abilities are approximately one grade level below his current grade level, which is somewhat consistent with teacher observations. His math scores were approximately grade appropriate, although applied problem-solving skills are depressed relative to straightforward calculation. Applied problems are compounded by a language component, which is not his forte. His best performance is in science, which he acknowledges as his "best subject." This suggests that Scott's relatively poor performance in language arts and reading may be motivational. Because his birthday comes late (September), he started the first grade young compared to other first-graders. Consequently, early language arts acquisition may have been particularly difficult for Scott, and there may be negative affect associated with reading skills currently. It should be mentioned that Scott's scores on the Woodcock–Johnson were compared to his chronological age-mates, not grade-mates. His scores would have been reduced by approximately 4 to 7 standard score points had age norms been applied. Results from the Brigance reveal some specifics associated with Scott's language arts problems. That is, he missed 3 words out of 10 on the primer level, 2 out of 10 on the grade 1 level, and 5 out of 10 on the grade 2 level. Scott seemed particularly adept at calling beginning consonants but often added a sound to the consonants. For example, rather than "mmm" for M, he responded with "mu." The same was the case for the letters B, H, J, G, R, S, D, M, and F. His greatest difficulty occurred with ending sounds. For example, RIX was pronounced as RIC and LIN as LINE. Other examples of ending problems included the following: SAT for SIB, TIDE for TID, PEN for PIN, TOX for TAX, and OX for OC. Obviously, Scott needs considerable work to master basic word lists.

Summary and recommendations
Although Scott is in the second grade, his language arts and reading skills are not consistent with that grade placement. Performance on the WISC-III

Figure 8.3   *(Continued)*

indicates average to above-average intellectual ability. Results from the Woodcock–Johnson are consistent with classroom performance, which reflects relatively poor language arts and reading skills, but relatively stronger math and science skills. His medical history is normal, and there are no obvious physiological impairments. (Glasses apparently are correcting the eye muscle imbalance, and exercise as prescribed by the physician should be continued.) Results from various tests suggest good reasoning and judgment, ability to concentrate and sustain attention, and average nonverbal reasoning and judgment. One purpose of the evaluation was to rule out the presence of a developmental reading disorder, which would be a Diagnostic and Statistical Manual IV, Axis I 315.00 diagnosis of Reading Disorder. There is not sufficient evidence to warrant this diagnosis. However, the evidence does seem to be compatible with a general developmental delay. Scott started the first grade at a disadvantage relative to other first-graders. That is, his birthday comes in September and, in general, boys develop at a slightly reduced rate relative to girls. Consequently, it is likely that his reading difficulties began because of developmental immaturity relative to age-mates, which is possibly confounded by early muscle imbalance problems of the eyes. It is possible to rule out the presence of specific learning disabilities/dyslexia, poor educational environment, low intelligence, and visual-auditory processing problems. The following recommendations are offered.

1. Scott's to-be-learned material should be individualized as much as possible to produce maximum gain and maximum motivation. Later, low-vocabulary/high-interest reading material can be assigned and a contract system developed to maximize reading and development automatized reading skills.

2. Because Scott's word-calling skills are poor, he should practice basic sight words, those most common in the reading content for his grade placement.

3. Scott would profit from instruction from a tutor. To-be-learned material should be coordinated with his classroom teacher.

4. Retention is not an appropriate option for Scott. There are considerable negative implications, primarily social, interpersonal, and self-esteem related. In addition, the literature is controversial regarding long-term academic gains associated with retention. Tutoring is a more appropriate solution. Tutoring will likely be phased out during the fourth and fifth grades.

5. Scott would profit from having stories read to him. Any activity that would increase his desire to read independently is appropriate.

6. Scott is overweight, and a low-calorie diet would be beneficial. A children's weight-loss program is available at University Hospital.

Claudia Zimmerman, Ph.D.
Licensed School Psychologist

Figure 8.3 *(Continued)*

WANT TO KNOW MORE?
SOCIAL HISTORIES

Different organizations or agencies use a variety of formats to collect information for social histories. Use the Internet to search for various formats. Note the strengths of each one. The following example, used by the state of North Carolina, structures an intake interview to determine eligibility for services and to begin service planning. Compare other forms you find with this one. What are the strengths and limitations of each of the forms you find?

STATE OF NORTH CAROLINA DEPARTMENT OF HEALTH AND HUMAN SERVICES SOCIAL HISTORY SUMMARY FOR THE DISABLED

_____ County Department of Social Services   Date _____

Claimant _____ SSN _____

County Case # _____ District # _____

Telephone # or a number you can be reached _____

Person Providing Information and Telephone # (if different from claimant)

_____

Nature of Disability (based on claimant's description or statement)

_____

_____

I. Onset of Impairment

A. Date illness or injury began

_____

B. Date claimant stopped work

_____

C. Date the illness or injury became disabling

_____

D. If still working:

   Name of Employer

_____

   Supervisor's name and telephone number

_____

Hours worked

_____

Gross earnings _____ weekly _____ monthly _____

II. Claimant's Description of Impairment

A. Indicate how the claimant describes the symptoms of the disability and how they affect his ability to work.

_____

_____

_____

_____

_____

B. Describe claimant's daily activities and explain how the impairments affect him such as seeing, hearing, speaking, reading, walking, writing, standing, breathing, sitting, using hands, arms, and other joints. Describe how his impairments limit what he can do.

_____

_____

_____

_____

_____

C. Worker's Observation of Difficulties

_____

_____

_____

_____

_____

(continues)

(continued)

III. Vocational Information (include self-employment)

A. Principal Job (job done the longest in 15 years prior to onset)

1. Job Title _____   4. Hrs./day _____

2. Industry _____   5. Days/week _____

3. Beginning date _____   6. Rate of pay/average earnings

Ending date _____ $_____ per _____

Other Jobs – List of jobs done in last 15 years prior to alleged onset date. Give approximate dates of employment (use additional sheet if necessary).

_____

_____

_____

_____

_____

_____

B. Education/Highest Grade Completed

_____

High School Graduate? _____

Name and address of school if known _____

Additional education _____ Type _____ Is claimant currently at-

tending school? _____

Name of school and address if known _____

Can claimant read and write? _____

IV. List all medical sources (physicians, hospitals, emergency facilities, health departments, therapists, nursing homes, clinics, mental health centers), including names and dates seen in the twelve months prior to and including application month, plus any future medical appointments. Give hospital or clinic number, which is on hospital or clinic card or hospital bills.

Medical Source Condition Treated Dates Seen at Dr.'s office, Name, Address, Ph. # EKG, X-rays clinic, hospital

_____

_____

_____
_____
_____
_____
_____
_____
_____
_____
_____
_____
_____

Is claimant still being treated? Yes _____ No _____

V. **VR Referral** _____ **Yes** _____ **No**

    Date last seen _____ **VR Office** _____

    Counselor's Name _____ **Phone #**_____

VI. **If a mental impairment is alleged, if there is evidence of drug or alcohol abuse, or if the person is homeless, in a shelter or in a halfway house, please give name, address, and phone number of a third-party contact.**

_____
_____
_____

_____ Signature _____

Title _____ Telephone # _____

Form retrieved from the Department of Human Services, North Carolina, http://info.dhhs
.state.nc.us/olm/forms/dma/dma-5009.pdf

There are several strategies that can make history taking, social or otherwise, a positive experience for both the client and the case manager. First, remember that the main concern is the client, not the completion of a form or a survey. So, it is important to make sure that the client understands the reasons and benefits of the data gathering. Second, use this time to continue to build the relationship with the client. Being sensitive to the client's wishes for privacy or need to discuss some aspect of his or her history will move the relationship forward. Last, remember that it is important for the case manager to guide the interview, so maintaining a balance between relationship building and completing the interview is critical.

Using these strategies, the case manager gathers pertinent information about what appears to be the client's problem. The primary source of information is the client, who is encouraged to tell the story in his or her own way. The helper listens carefully to what is said, how it is said, and what is not said. The sequence of events, reactions, feelings, and thoughts are all taken into consideration as the client relates the history. Note taking should be kept to a minimum so that important nonverbal information is not missed.

Social history is taken within the context of the culture of the client. For example, interviews with individuals who belong to a collectivist culture must be treated with cultural sensitivity. In a collectivist culture, the focus is on the importance of the group rather than the individual. In a collectivist context, individuals must fit into the group: There is a focus on group values, beliefs, and needs, and the group influences individual behavior.

Because of group influences, social history may hold very different meaning to an individual from a collectivist culture than it would to a person in the American mainstream. As the client responds to questions and tells his or her story, there may be much more emphasis on the family and the community. The client may not be able to clearly define personal characteristics or personal problems, but instead will describe them in terms of the group or family. It may appear that the client is avoiding answering the questions or not taking responsibility for his or her own behavior, but the client's experience of history may be that of the group or the family. It is also possible that the client may not wish to share his or her story. In many collectivist cultures, this information stays in the family or in the group.

There is no set form or procedure for taking a social history. Some agencies use forms to guide information gathering, such as the social data report shown in **Figure 8.4**. Others just provide guidelines for their case managers; as a result, the length and detail of social histories may vary. In all cases, the social history is prepared when a comprehensive picture of a client's situation is desired. The outline used for writing it depends on what the agency wishes to emphasize, but certain topics are almost always included: identifying data, family relationships, and economic situation. Which other areas are emphasized depends on the focus of the agency and the presenting problem. For example, a social history of a couple involved in marital counseling might target such areas as family relationships and psychosocial development. For someone seeking economic assistance, important areas might be financial status, income, expenses, and work history. In general, the following information may appear in a social history.

> *Identifying information:* Name, address, date and place of birth, Social Security number, military service, parents' name and address, children's names and ages.
>
> *Presenting problem:* Brief description of the problem.

SOCIAL DATA REPORT

Student's full name: _____ SS#: _____

Address: _____ Phone: _____

Age: _____ D.O.B.: _____ Race: _____ Sex: _____ Hair: _____

Height: _____ Weight: _____ Eyes: _____

Distinguishing marks: _____

Offense: _____

Date of offense: _____ Prior court record: _____ Yes _____ No _____
            (If Yes, show details under Additional Information.)

School attending: _____ Grade: _____

Have you been suspended from school, given detention, in-school suspension, or had truancy problems? Yes _____ No _____ (If Yes, show details under Additional Information.)

Health: _____ Are you on any prescription medications? Yes _____ No _____ (If Yes, give type, amount, reason under Additional Information.)

Are you seeing or have you seen a mental health counselor? Yes _____ No _____

If Yes, where and when? _____

Have you ever used drugs or alcohol? Yes _____ No _____

Type, amount: _____

Are you employed? Yes _____ No _____ Where & hours: _____

Use of prescription medication: Yes _____ No _____

If Yes, type and amount: _____

Parents' marital status: Living together, divorced, separated, widowed, non-spousal with child in common

(continues)

Figure 8.4   Social data report

If separated or divorced: When? _____

Discipline in home: Type:_____

Does it work? _____

Who provides discipline: Mother, father, both, neither?

Children in family:

Name   Age   Lives at Home? (Yes or No)   Court/Arrest Record (Yes or No)

_____

_____

_____

_____

Are you receiving any financial assistance from anyone?

Type: Food stamps   Amount:

         Social Security   Amount:

         Child support   Amount:

         Other   Amount:

Do you have medical insurance? (Name, policy #, address, and phone)

ADDITIONAL INFORMATION: _____

_____

_____

_____

_____

Figure 8.4 *(Continued)*

PARENT INFORMATION

Father's name:_____ Age: _____

Address: _____ Phone: _____

Employment: _____ Phone: _____

Educational level: 1 2 3 4 5 6 7 8 9 10 11 12 GED, college attendance, professional training

Court record: Yes _____ No _____ If Yes, for what and when? _____

Use of drugs or alcohol/Type of Use: excessive, moderate, little, none

Have you had any type of counseling? Yes _____ No _____

If Yes, when and where? _____

Use of prescription medication: Yes _____ No _____

If Yes, type and amount? _____

If deceased, give date and cause: _____

Mother's name: _____ Age: _____

Address: _____ Phone: _____

Employment: _____ Phone: _____

Educational level: 1 2 3 4 5 6 7 8 9 10 11 12 13 14 15 16 GED

Court record: Yes _____ No _____ If Yes, for what and when? _____

Use of drugs or alcohol/Type of Use: excessive, moderate, little, none

Have you had any type of counseling? Yes _____ No _____

If Yes, when and where? _____

If deceased, give date and cause: _____

Figure 8.4 *(Continued)*

Referral: Source and reason.

Medical history: Relevant hospitalizations, illnesses, treatment, and effects. Written permission is needed to obtain copies of medical records, if necessary.

Personal/family history: Family life, discipline, parenting, and personal development.

Education: Highest grade completed, progress, records.

Work history: Training, type and length of employment, ambitions.

Present family relationships and economic situation: Family members, ages, relationships, lifestyle, and income.

Personality and habits: Interests, disposition, social activities, personal appearance.

The client provides most of the information for a social history, but other sources may also contribute. When the case manager has gathered material from sources other than the client, he or she should insert it under the appropriate headings, with the source identified. Direct knowledge is the main source, as in the following examples.

- She did not come for her first appointment.
- The client drummed his fingers on the table throughout the interview.
- He states that his goal is to receive a high school diploma and get a job.
- The client stated that during the past week she and her husband had three fights.

The next examples are statements of information from other sources.

- Educational records indicate that the client completed the sixth grade.
- Her parents report that the client lived with them until her marriage two years ago.
- He was fired from his job for absenteeism.
- A psychological evaluation indicates a mildly retarded thirteen-year-old with possible hearing loss.

The social history shown in **Figure 8.5** combines two approaches. The Identifying Information section is a form that the case manager completes. The remaining sections are a narrative based on information compiled from several sources (listed at the end of the report). At this agency, a social history may be compiled by more than one case manager, and all who are involved in the writing of the social history sign the written report.

Another way social information appears in a case file is illustrated by the court report shown in **Figure 8.6**. It was prepared for juvenile court, based on social information gathered by a caseworker at the state department of human services (DHS). DHS caseworkers frequently prepare court reports, for example, if parental rights are being terminated or if the court asks DHS to investigate a petition for custody. All juvenile court reports have certain things in common, such as the reason for the referral to the department and the circumstances of the

SOCIAL HISTORY

I. IDENTIFYING INFORMATION

Date: 0/0/XX   Name: Joe Billy Smith   Date of birth: 0/0/XX

SS#: 000-00-0000   TennCare/Insurance provider: Blue Cross/Blue Shield

Policy number: 000000000   Marital status: Never married

Number of children: None reported

Address: 1111 Dogwood Drive

Atlanta, GA

Telephone number: (123) 777-7777

Prevention resource: Community Health Center

Resource contact: Jim Therapist

County court: _____

Custodial dept.: Department of Youth Development

Custodial dept. contact: Jerry Officer

Date of state custody: 000000   County court: Cobb

Age: 16 years   Race: Caucasian   Height: 5' 2"

Weight: 120 pounds   Eye color: Blue   Hair color: Blond

Unusual markings: Tattoo on left bicep

Allergies: Penicillin   Current medications: None reported

Current medical problems: Ingrown toenail

(*continues*)

Figure 8.5   Social history

Special circumstances: Ms. Smith (mother) reported that Joe was abducted from school at gunpoint and was kidnapped for five days. During the time Joe was gone, he was allegedly emotionally and physically harassed and assaulted. This took place the first two weeks in March.

II. *PRESENT PROBLEMS* (Current charges with dates and circumstances)

Joe appeared in Juvenile Court the following February on the charges of violation of a valid court order and disobeying his probation by using cocaine and alcohol. Joe pled true to these charges and was placed in the custody of the Department of Youth Development. On this same date, the court ordered an assessment of Joe and his situation.

III. *PREVIOUS PROBLEMS* (Past charges, adjudications, and placements)

According to Joe's court file, Joe was petitioned to court in October for the charge of running away from home. Joe pled true to the charge and was placed on probation with the County Probation Service. Ms. Smith (mother) also reported that when Joe disappeared in April, she went to the Police Department to file a missing persons report. Ms. Smith stated that she was informed that a missing persons report could not be filed, but that she could sign a paper declaring Joe a runaway juvenile. This action would allow the police to search for Joe. Ms. Smith reported that she signed the form only out of concern for Joe, and that this is now in Joe's court record. No charges were ever brought against the 20-year-old male who allegedly abducted and abused Joe.

IV. *FAMILY HISTORY* (Name, social security number, current address, phone, date of birth, marital status, employment, educational level, court record, alcohol and drug problems, mental and physical health problems, possible placement resource)

A. Father: Tom Smith is Joe's biological father. His date of birth is 0/0/XX, and his social security number is 000-00-0000. Ms. Smith reported that Mr. Smith lives in Dalton, Georgia, but she does not know his address. Ms. Smith stated that Mr. Smith's phone number is (123) 000-0000. Ms. Smith reported that Mr. Smith is currently remarried and is employed delivering bottled water. Ms. Smith does not know Mr. Smith's delivery route.

Figure 8.5   (*Continued*)

Ms. Smith reported that Mr. Smith obtained his GED and has no prior court history, to her knowledge. Ms. Smith stated that Mr. Smith used to drink alcohol frequently—five times per week—but he did not use any drugs. Ms. Smith is unaware of any physical or mental health problems that Mr. Smith may have. Mr. Smith would not be an appropriate placement for Joe because of his sporadic interest in Joe's life. Ms. Smith stated that Mr. Smith has let Joe down many times in the past.

B. Mother: Betty Smith is Joe's biological mother. Her date of birth is 0/0/XX, and her Social Security number is 000-00-0000. Ms. Smith's current address is 1111 Dogwood Drive, Atlanta, Georgia, and her phone number is (123) 777-7777. Ms. Smith is divorced and works at Kroger's in South Atlanta. Ms. Smith reported that she completed the 10th grade and then earned her GED. Per Ms. Smith, she has no alcohol or drug problems or court history. Ms. Smith stated that she deals with mild depression and seeks professional mental health services when depression sets in. At this time, Ms. Smith reports that she is taking no prescription psychotropic medication for her depression.

C. Stepparents: Not applicable.

D. Siblings:

SIBLING NAME	GENDER	DOB	FULL/HALF SIBLING
Jeff Smith	Male	0/0/XX	Unknown
Ty Smith	Male	0/0/XX	Full

Sibling interaction:

Ms. Smith stated that Joe gets along well with his older siblings, but he is closest to Ty. Ty's wife has a baby girl, and Joe likes to help take care of her. Joe's siblings are older than he is, and Ms. Smith reports that Joe does not get to see them as much as she would like.

E. Other: (Grandparents, relatives, boyfriends, girlfriends, etc.)

Joe's last remaining grandparent died prior to this first introduction to the court. Ms. Smith reported that Joe was very close to this grandparent. Currently, Joe reports no girlfriend involvement.

(continues)

Figure 8.5 *(Continued)*

V. *FAMILY INTERACTION* (Family dynamics/relationships, current issues, financial resources, needs, risks, etc.)

Ms. Smith reported that she and Joe are very close. Ms. Smith appeared to be very protective of Joe and defended him and his actions. Ms. Smith stated that the only current issue in the home is Joe's opposition to house rules. Ms. Smith is angry about Joe's placement in the Department of Youth Development, and she is threatening to sue the state. Ms. Smith reported that she earns minimum wage and usually works at Kroger's 30–40 hours a week. Ms. Smith stated that she pays rent and utilities. Ms. Smith reported that she receives food stamps, but she is not sure how much she will be getting now that Joe has gone into custody. Ms. Smith did not feel that all of her needs were currently being met.

VI. *HOME AND NEIGHBORHOOD* (Date of home visit, type of home, adequacy of space, housekeeping standards, hazardous conditions, neighborhood description)

A home visit was made by the case manager on 0/0/XX. Ms. Smith said they have been renting their two-bedroom apartment since July, prior to Joe's trouble, and that both she and Joe have plenty of room. This case manager observed that housekeeping standards were very good. No major inappropriate housing conditions were noticed. Their apartment is in a low-level crime area of Atlanta.

VII. *CHILD*

A. Early development history (Problems with pregnancy or delivery, planned/unplanned pregnancy, parental A&D use during pregnancy, developmental milestones, serious illnesses or accidents, diagnosis of hyperactivity, etc.)

Ms. Smith reported that she had no problems with the pregnancy or delivery of Joe; she delivered Joe herself, with the help of a midwife. Ms. Smith reported that she did not use alcohol or drugs while she was pregnant, and that Joe reached all of his appropriate developmental milestones earlier than most children. Ms. Smith reported that the only serious illness Joe had as a child was pneumonia.

Figure 8.5 (*Continued*)

B. Peer interaction (Relationships with peers, age of friends, activities of friends, does or does not have friends)

Ms. Smith reported that Joe does not have many age-appropriate peers. Joe's friends are usually older. Per Ms. Smith, Joe does not do much in his free time except sleep. Ms. Smith was not certain what Joe does when he goes out with his friends, but she stated that he does not go out often.

C. Education (Last school attended, grade level, major school problems; accelerated, remedial, or special ed classes, truancy history)

Joe was last enrolled at Greenbriar High School in the ninth grade. Ms. Smith stated that Joe attends regular classes. Joe has had some truancy and tardiness problems. Since the abduction from school in April, Joe has become increasingly paranoid about attending school. It takes full cooperation with the school system to make him attend.

D. Psychological (Current and prior psychologicals, stating examiner's name, location of testing, and test dates)

Joe had a psychological evaluation, conducted at Lakeside Mental Health Institute in April. The examiner was Dr. John Doe.

(continues)

Figure 8.5 *(Continued)*

VIII. *AGENCY CONTACTS AND SOURCES OF INFORMATION*

Name: Tom Casemanager   Relationship: Case Manager

Address: 201 Center Park Drive, 1100

        Atlanta, GA 12345

Phone: (123) 000-0000

Agency name: Lakeside Mental Health Institute

Agency contact: Dr. John Doe

Address: 5900 Lakeside Drive

        Atlanta, GA 12345

Phone: (123) 000-0000

Agency name: Department of Youth Development

Agency contact: Jerry Officer

Address: 222 Jail House Drive

        Atlanta, Georgia 12345

Phone: (123) 000-0000

Prepared by: _____

        (Name)   (Title)   (Date)

        (Name)   (Title)   (Date)

        (Name)   (Title)   (Date)

        (Name)   (Title)   (Date)

Figure 8.5 *(Continued)*

REPORT FOR JUVENILE COURT

Child: Lydia Maza, date of birth 0/0/XX   Petitioner: Jorja Mitten

Address: 100 Washington Pike,   Address: 100 Washington Pike,
         Chicago, IL   Chicago, IL

Mother: Leyla Mitten L   Father: Lloyd Maza

Address: Unknown   Address: P.O. Box 18
                                    Hot Springs, AR

REFERRAL

The petitioner has been the primary caretaker of the child. The legal custodian is incarcerated in Shelby County Jail at this time. Petitioner asks that legal care and custody be given to her and her husband.

CIRCUMSTANCES OF THE CHILD

Lydia came to Chicago when she was 10 years old, to visit her maternal grandparents. During this time Leyla Mitten was arrested for drug trafficking, possession, and dealing and sent to Shelby County Jail, where she will be eligible for parole in two years. Jorja Mitten received a letter from her daughter asking her to care for Lydia until she is able to do so.

When asked if she would like to go back to Arkansas, Lydia stated that she would rather stay in Chicago. Lydia is a shy girl who states that she has more friends in Hot Springs, but has friends in Chicago too and enjoys playing with them. Lydia stated that she likes living with her grandparents, and she seems happy there.

Lydia's teacher at Ritta School says she is doing very well in class but seems emotionally fragile. The school records show that Lydia has missed four (4) days of school this year, all of which have been excused.

(continues)

Figure 8.6   Court report

CIRCUMSTANCES OF PARENTS

Mother—Leyla, date of birth 0/0/XX, was arrested in Chicago on several drug charges. She will be eligible for parole in two years. Reportedly, Leyla expects Lydia to be returned to her then. Leyla has alcohol and drug issues.

Father—Lloyd Maza is a spa owner in Hot Springs. He has had no contact with Lydia in recent years.

CIRCUMSTANCES OF THE PETITIONER

Jorja Mitten, maternal grandmother, age 58, is the petitioner in this matter. She has been married to Gus Mitten, age 59, for 30 years. The Mittens live in a three-bedroom, one-bath home, which they own and have resided in for 18 years. The Mittens have two children, Leyla and her twin brother Boyd, who lives in Red Springs.

Ms. Mitten is currently employed by a local utility. She stated that she has worked there for 19 years. She reportedly earns $3,500 per month.

Ms. Mitten denies any alcohol or drug abuse problems. She is a smoker, as is Mr. Mitten, and they use air filters in their home. Ms. Mitten states that she is in good health and takes Oruvail and Adalet daily under doctor's orders.

Mr. Mitten is currently employed. He stated that he has worked at Wind Industries for 22 years. He reportedly earns $4,000 per month.

Mr. Mitten denies any alcohol or drug abuse problems. He reports that he is in good health. Mr. Mitten takes Oruvail daily.

Mr. and Ms. Mitten are members at the YMCA, where they exercise weekly.

Ms. Mitten stated that she is concerned that Leyla is not emotionally or financially stable and is therefore in no position to care for Lydia. She stated that she is afraid that Leyla, after being released from jail, will "drag Lydia down" with her.

All references speak very highly of the Mittens and hold them in the highest regard. All stated that they have no concerns about the Mittens caring for Lydia.

A check with local law enforcement agencies revealed no prior record in Cook County on Jorja or Gus Mitten.

Figure 8.6 *(Continued)*

child, of both parents, and of the petitioner. Also included is the recommendation of the department, which the court may or may not follow. Although the format of this report is determined by the court, you will see content similarities to the social history in Figure 8.5. In this court report, a grandmother is asking for full

RECOMMENDATION

At this time, the department would recommend that custody be granted to Jorja Mitten. The legal custodian is currently unable to provide for the child, and the child states her reluctance to leave her grandparents' home. This worker knows of no reason why Lydia should not remain with Ms. Jorja Mitten.

As always, we will respect the court's wishes in this matter. We hope this information will be of assistance to the court.

Submitted by:

Illinois Department of Human Services

Tina Rachael
Caseworker 1

Figure 8.6   (*Continued*)

custody of her granddaughter. A caseworker has been out to the home, completed a social history of the family, and obtained a signed release of information from the petitioner. The caseworker has also consulted with the law enforcement agencies, checked references, and obtained as much information as possible from other sources. The caseworker then writes a report, informing the court as succinctly as possible of all the relevant information gathered.

 # Other Types of Information

Other types of information may be relevant to the case file, depending on the agency's mission and services as well as the client's problem. Educational and vocational information, the most commonly needed, is discussed here.

Educational information can have many parts: test scores, classroom behavior, relations with peers and authority figures, grades, suspensions, attendance records, and indications of academic progress such as repeated grades or advanced work. The sources of educational information are just as varied: school records, teachers, guidance counselors, and principals. Often, the particular information that the case manager obtains depends on which source is contacted. Rarely is it gathered in a single report, as medical information might be. In many cases, the case manager decides what information is needed and contacts the source or sources most likely to have that information. For example, a teacher is probably the best source of information about classroom behavior, whereas school records provide test scores and indications of past academic performance. The contact may occur formally (in writing) or orally (by telephone or personal interview).

Vocational information can be important for several reasons. People seem to be happiest when their activities are satisfying and fulfill their needs. There is also the need to earn a living, and self-support often engenders self-respect. Ways of gathering vocational information range from asking the client about his or her work history to arranging for a formal vocational evaluation. The types of information gathered include jobs previously held, the ability to get along with coworkers, work habits (e.g., punctuality and reliability), and reasons for frequent changes in employment. How much more information is needed depends on the client's problem and the agency's mission. For example, if the client has no work experience, an exploration of vocational interests and aptitudes may be in order. For the client who has had varied employment, the focus may shift to attitudes toward work and the skills developed. The client who has an extensive work history may need help in reviewing his or her experience and skills to establish a vocational objective.

 # Deepening Your Knowledge: Case Study

Let's return to Roy Johnson's case, discussed in Chapter 1. Roy's counselor requested a period of vocational evaluation at a regional center that assesses an individual's vocational capabilities, interests, and aptitudes. Roy and the counselor, Tom Chapman, attended a staffing to hear the vocational evaluation report. Mr. Chapman later received a written report (**Figure 8.7**). The report illustrates

VOCATIONAL EVALUATION REPORT

ADMISSION DATA

To: Tom Chapman

From: Dan Howard

Re: Roy R. Johnson

D.O.B.: 0/0/XX

S.S.#: 000-11-2222

Sex: Male

Marital status: Single

No. of dependents: None

Date: 0/0/XX

Date reported: 0/0/XX

Dates attended : 0/0/XX

Scheduled hours: 8:00 A.M.–3:00 P.M.
                      Monday–Friday

Work history: Bartender, plumber's helper

Medication: None

Education: Two quarters SSCC

Transportation: Own vehicle

Program manager: Jo Singletary

TESTS ADMINISTERED

Purdue Pegboard

Valpar Component Work Samples

    Size discrimination

    Upper extremity range of motion

    Simulated assembly

    Eye–hand–foot coordination

    Bennett hand tool

Reason for referral: Mr. Johnson was referred to the Vocational Training Center for vocational evaluation to assess his manual and finger dexterity. This evaluation was to consist of paper/pencil testing and situational assessment. The client completed his paper/pencil testing, but did not complete his situational assessment. This report is based on the results of the tests he took and the limited time he spent in situational assessment.

BACKGROUND INFORMATION

The information in this section was based on statements made by the client on his evaluation interview.

    Social/educational: Mr. Johnson was a 30-year-old white single male who lived with his mother and sister. The client disclosed during his

(continues)

Figure 8.7   Vocational evaluation report

evaluation process that his mother has been severely hearing impaired since birth. Because of this, he learned sign language before he could talk. He stated that he maintained this skill.

He had recently received a settlement of $40,000 due to his injury on the job and was building a home with these funds. He did not have any other source of income. The client completed two quarters at Silver State Community College. He quit when his grandmother became ill. He also received plumber's training for 6 months at T.A.T. and attended the Georgia School of Bartending for 5 weeks. The client has held 3 jobs in the past 3 years and has not worked since April. The client's stated vocational interest was in working as a mechanical engineer. He said he would choose this because it was within his physical capabilities, and that he had a background in mechanics.

WORK HISTORY

The client's work history was obtained from the client during his evaluation interview.

The client worked as a bartender at the Ramada Riverview for 10 months. He left to go to work at Joe B's Restaurant.

The client worked at Joe B's as a bartender for 2 months. He left this position because of a personality conflict.

The client last worked at Rock City Mechanical as a plumber's helper. He left following his injury.

MEDICAL INFORMATION

General: The client stated that he was in general good health. The client's general medical examination revealed that his only problems were related to his back injury.

Handicapping condition and functional limitations: The client was handicapped by a back injury while working on a plumbing job. It is stated in a report from Dr. Alderman that the client also had a history of previous lumbar disc disease. It is also stated that the client's recovery will be slow and that he will have a permanent impairment of 10% to the body as a whole.
The client's limitations were taken from his general medical examination, done by Doctor Jones. These limitations include walking, standing, stooping, kneeling, lifting, reaching, pushing, and pulling.

Figure 8.7 *(Continued)*

Vocational implications: The client would work best in a sedentary job, one where he can get up and move around when needed to relieve his back pain. The client was still recovering, and these limitations may change with time.

BEHAVIORAL OBSERVATIONS

General: The client was cooperative during testing and attempted all tasks asked of him. In situational assessment, the client could not work on all the contracts that we had. He wanted to wait until we had something he was able to do. When we did not get work in that was easier than what we had, he quit coming in spite of efforts to make the work easier for him. The reason stated by this client for not attempting this work was that he feared injuring his back again.

Vocational: The client did not complete the situational assessment, and the observations made were not complete.

OBSERVATIONS OF WORK BEHAVIORS

The client was observed in 13 different areas while in situational assessment. Ratings were as follows:

Rating scale for work behavior		
	NA	Not acceptable
	1	Not acceptable; needs improvement
	2	Satisfactory; meets criteria
	3	Excellent; employment strength

Attendance	1
Punctuality	2
Co-worker relations	3
Supervisor relations	2
Work quantity	
Contract	3/1
Cleaning	NA

(continues)

Figure 8.7 *(Continued)*

Work quality	3
Work tolerance	1
Job flexibility	1
Follows work instructions	3
Use of work time	2
Works without close supervision	3
Observes all safety procedures	3
Care of tools/equipment/materials	3

Comments

Attendance: The client attended the center 43% of the time. He did not come in because he could not do the work we had (sanding and painting dumpsters).

Work quantity/contract: The client's production on the nuts and bolts contract for Smithfield Industries was 95% of industrial norms. His production on the labeling contract for A.P.S.U. was 25% of industrial norms. It is noted that this contract required 100% quality.

Work quantity/cleaning: The client could not do this.

Work tolerance: The client could work a full day but complained of pain in his feet. He said that this was because of the concrete floor.

Job flexibility: The client could not do the work on the dumpsters or the cleaning duties. He worked on the A.P.S.U. labels, the tape-recycling contract for Triad Corporation of Tennessee, and the nuts and bolts contract for Smithfield Industries. These were all sedentary tasks.

Figure 8.7   *(Continued)*

TEST INTERPRETATION

Purdue Pegboard
    Right hand—58%ile—industrial applicants
    Left hand—56%ile—industrial applicants
    Both hands—72%ile—industrial applicants
    Right+left+both—59%ile—industrial applicants
    Assembly—93%ile—industrial applicants

Valpar Component Work Sample #2—Size discrimination

	TIME	ERRORS
Assembly	115%	<100%
Disassembly	80%	<100%

Performance—64%
Norm group: Seminole Community College Vocation Assessment Center

Valpar Component Work Sample #4—Upper extremity range of motion

	%ile	MTM%
Assembly: Dominant hand	85	70
Assembly: Other hand	80	75
Disassembly	70	90

Performance: 87%ile
Norm group: Seminole Community College Vocational Assessment Center

Valpar Component Work Sample #8—Simulate assembly
    Performance—95%ile   MTM—75%
    Norm group: Seminole Community College Vocational Assessment Center

Valpar Component Work Sample #11—Eye–hand–foot coordination
    Points Performance—25%ile   MTM—50%
    Time Performance—85%ile   MTM—140%

Bennett Hand-Tool Dexterity Test
The client's performance on this instrument was 75%ile when compared to boys at a vocational high school.

(continues)

Figure 8.7   (Continued)

SUMMARY/RECOMMENDATIONS

Summary: Mr. Johnson was a 30-year-old man who was handicapped by a back injury, with limitations in walking, standing, stooping, kneeling, lifting, reaching, pushing, and pulling. The counselor is referred to the medical information section of this report for details.

The client's performance in testing could be considered strong, with the exception of his accuracy in performing the eye–hand–foot coordination work sample. The client's scores (except eye–hand–foot, as mentioned) were above average when compared with the norm group.

The client's strengths in situation assessment were in co-worker relations, work quality—nuts and bolts contract, working without close supervision, observing all safety procedures, and care of tools/equipment/materials.

The client's weaknesses in situational assessment were attendance, work quantity—labeling contract, work tolerance, job flexibility.

The client had acceptable performances in punctuality, supervisor relations, and use of work time.

Recommendations: The results of the testing done here at the center to assess the client's dexterity show that he could perform a sedentary, small, hand-assembly job. He expressed no interest in pursuing this line of employment. As noted in the Medical Information section, he would need to be in a job that would allow him the freedom to get up and move around when he needed to relieve his back pain. During situational assessment, he was observed to have the need to get up and down as he worked.

As stated in the Social/educational section of this report, the client is proficient in sign language for the hearing impaired. The client could be employed as an interpreter for the hearing impaired if he wanted to pursue certification. The client said that he had never considered this, but was open to the idea. It is recommended that the counselor investigate the possibility of having the client obtain his certification.

Figure 8.7   (*Continued*)

This client expressed an interest in attending the state university to pursue a degree in mechanical engineering. This evaluator cannot make a recommendation regarding this pursuit, since no academic testing is performed at this facility.

_____

Dan Howard, Evaluator

_____

Jo Singletary, Center Manager

Figure 8.7 *(Continued)*

two important points. First, information about a client is integrated with other new information to complete the picture. In this report, you will read about work history, medical information, and test scores, as well as the results of the vocational evaluation. Second, this report is a vocational evaluation report. Vocational evaluation is a process of gathering, interpreting, analyzing, and synthesizing all data about a client that has vocational significance and relating it to occupational requirements and opportunities.

Vocational and educational information add other dimensions to the client record, making the case file more complete. This information rounds out the case manager's understanding of who the client is—strengths, weaknesses, abilities, and aptitudes.

Discussion Questions

1. In addition to the reports mentioned in this case, what other reports might Tom require to understand Roy's circumstances and needs?
2. What steps should Tom take to help Roy prepare for and understand these assessments?

Chapter Summary

The information about the client that is gathered from other professionals helps the case manager see a more complete picture of the client. This information includes medical reports, psychological evaluations, social histories, and educational and vocational information. When the case manager requests the information from other professionals, the goals must be clear, and it is helpful if the client's problems are identified. The case manager asks culturally sensitive professionals to conduct the assessments. Once the information is received, the case manager reviews it and integrates the results with the information previously gathered.

Medical information is critical, especially when a client has disabilities or mental illness. It is important for the case manager to understand medical terminology and be familiar with medications. A psychological evaluation is also an important part of a client file because it contributes to the understanding of the client as an individual. Often the case manager needs a psychological evaluation to establish eligibility for services, to justify a service, or to screen for criteria to determine the need for services.

Social histories provide information about the way an individual experiences problems, past problem-solving behaviors, developmental stages, and interpersonal relationships. The social history can help to complete the picture about the client and assists the building of the relationship between the case manager and the client. The file may include other information, such as educational and vocational information. Relevance is determined by the agency's mission and services, as well as the client's problem.

Chapter Review

Key Terms ....

Medical diagnosis
Medical consultation
Physical examination
Diagnosis
Medical terminology

Word root
Psychological evaluation
Psychological report
DSM–IV–TR
Social history

Reviewing the Chapter ....

1. Identify the resources that will help you understand medical reports.
2. How does medical information contribute to a case file?
3. In what situations would a medical consultation help a case manager?
4. Describe a general medical examination.
5. Describe the guidelines under which a culturally sensitive physician would conduct a medical exam.
6. Define each of the elements that form medical words.
7. What are the three basic steps in working out the meaning of a medical term?
8. Why is keeping current with medical terms a challenge for case managers?
9. List reasons to refer a client for a psychological evaluation.
10. How does a case manager make a good psychology referral?
11. What types of information does the *DSM-IV* or *DSM-V* diagnosis provide?
12. Describe a psychological report.
13. What is a social history?
14. Describe the advantages and limitations of a social history.
15. How will the guidelines for history taking help you do a social history?
16. Complete a social data report (Figure 8.4) on yourself.
17. Write a social history on yourself, using the nine content areas of a social history.
18. Describe the three ways in which a social history may appear in a case file.
19. What do educational and vocational information add to a case file?

Questions for Discussion ....

1. Why do you think it is important to have medical information in a case file?
2. What difficulties do you expect to have in understanding a psychological report?
3. Develop a plan to gather information for a social history of a client who is in prison for armed robbery.
4. Do you believe that you can have too much information about a client? Why or why not?

● *References* ....

American Psychiatric Association. (2000). *Diagnostic and statistical manual of mental disorders: Text revision* (4th ed.). Washington, DC: Author.

American Psychiatric Association (2012). *APA DSM-5 Development.* Retrieved from http://www.dsm5.org

Dabatos, G., Rondinelli, R. D., & Cook, M. (2000). Functional capacity for impairment rating and disability evaluation. In R. D. Rondinelli & R. T. Katz (Eds.), *Impairment rating and disability evaluated* (pp. 73–94). Philadelphia: Saunders.

Felton, J. S. (1992). Medical terminology. In M. G. Brodwin, F. Tellez, & S. K. Brodwin (Eds.), *Medical, psychosocial, and vocational aspects of disability* (pp. 21–33). Athens, GA: Elliott & Fitzpatrick.

Tervalon, M. & Murray-Garcia, J. (1998). Cultural humility versus cultural competence: A critical distinction in defining physician training outcomes in multicultural education. *Journal of Health Care for the Poor and Underserved* 9(2):117–124.

University of Washington. (2009). *Medical interviewing: Communication skills benchmark.* Retrieved from http://courses.washington.edu/icmweb/icm2/sites/default/files/communications_toolbox.pdf

CHAPTER NINE

Service Coordination

Thriving and Surviving as a Case Manager

Jessica

My company does something a little bit different than case management. It's case management and then it's more. We are actually called resource coordinators. My agency is the premier provider for therapeutic foster care in the region. What that means is if a child goes into state's custody the state does a CANS— a Child and Adolescent Needs and Strengths assessment—and if they determine that the child has therapeutic needs or is medically fragile due to birth defects or abuse, they refer the child to us. They call us first and if we have room for the child, we place him or her. The reason our agency was started was that a man who was working for human services figured out that a lot of these kids who have pretty serious needs aren't getting their needs met by the department and needed specialized, intensive treatment and support.

So what do I do? I don't have any medically fragile children right now, but if I did have a medically fragile child, the coordination of services would include help from early intervention specialists who work with in the home with parents, foster parents, or guardians. They provide occupational and speech therapy, and any other kind of therapies children need to be as successful as they can be. We like a lot of those services to be provided in the home. This is what we do for our medically fragile children. And if we are referred a child who is in the hospital, we pick a foster parent whom we know is very nurturing and we send them to the hospital while the child is in ICU to start building a bond, because reactive attachment disorder is prevalent and scary and we want to prevent that if we can.

—Permission granted from Jessica Brothers-Brock, 2012, text from unpublished interview.

In my job you have to know about the resources in your community. You can really help your clients get to the right place.

> —Director and case manager, intensive case management services,
> Los Angeles, California

When I worked as a case manager in the downtown hospital, I needed networks in the hospital, in the medical community, and in the human service community. In fact, since I was doing discharge planning, this knowledge was a critical part of my job. I could not meet all of my clients' needs. And the range of needs was so great, from detox services, to Social Security and government services, to food stamps and housing. I also had to help my families deal with the bureaucracy.

> —Case manager, urban hospital, Atlanta, Georgia

It is incredible how important community is with other service providers. In my job I am a broker and I have to be able to meet people, establish relationships with them, and work well with them.

> —Care coordinator, health system, Pima County, Arizona

One of the most important roles in case management is service coordination. Rarely can a human service agency or a single professional provide all the services a client needs. Because in-house services are limited by the agency's mission, resources, and eligibility criteria—as well as by its employees' roles, functions, and expertise—arrangements must be made to match client needs with outside resources. Case managers must know which community resources are available and how to access them.

The preceding quotations reflect the knowledge and skills that a case manager uses to meet client needs. According to the case manager at the intensive case management service provider, knowledge of resources is important, as is knowing how to negotiate the service delivery system to gain access to those resources for the client. A key to using resources, according to a hospital case manager, is to have networks in place so that the case manager knows both the resource agency and the name of a contact. Perhaps the one indispensable skill in using resources is communication; the care coordinator of a health systems organization states that working well with other service providers depends on the case manager's communication skills.

Today's service delivery environment imposes new roles and responsibilities on the case manager. In the past, many services were provided directly by the case manager, but service delivery has become more specialized. Professionals must be careful not to provide direct services in areas in which they are not trained or lack the necessary resources. Case management has thus come to mean providing selected services, coordinating the delivery of other services, and monitoring the delivery of all services. This shift in job definition calls for skills in advocacy, collaboration, and teamwork.

This chapter explores service coordination as a critical component of modern case management. We examine the coordination and monitoring of services as well as the skills that will help you perform these roles. After reading the chapter, you should be able to accomplish the following objectives.

COORDINATING SERVICES
- Describe a systematic selection process for resources.
- Make an appropriate referral.
- Identify the activities involved in monitoring.
- List ways to achieve more effective communication with other professionals.

ADVOCACY
- Explain why advocacy is important.
- Name several client problems that are appropriate for advocacy.
- Identify ways to be a good advocate.

TEAMWORK
- Describe the purpose of a treatment team.
- Define departmental teams, interdisciplinary teams, and teams with family and friends.
- List the benefits of working in and with teams.

 ## Coordinating Services

If a client needs services that an agency does not provide, it is the case manager's responsibility to locate such resources in the community, arrange for the client to make use of them, and support the client in using them. These are the three basic activities in coordinating human service delivery. In coordinating services, the case manager engages in linking, monitoring, and advocating, while building on the assessment and planning that have taken place in earlier phases of case management. The case manager continues to build on client strengths or emphasize client empowerment within the context of the client's cultural background and basic values.

Coordinating the services of multiple professionals has several advantages for both the case manager and the client. First, the client gets access to an array of services; no single agency can meet all the needs of all clients. The case manager can concentrate on providing only those services for which he or she is trained, while linking the client to the services of other professionals who have different areas of expertise and have the necessary resources. Second, the case manager's knowledge and skills help the client gain access to needed services. Often, services are available in the community, but clients are unlikely to know what they are or how to get them. The success of service delivery may depend on advocacy by the case manager. Also, service coordination promotes effective and efficient service delivery. In times of shrinking resources, demands for cutbacks in social services, and stringent accountability, service provision must be cost effective and time limited. In addition, customer satisfaction is important. Clients have a right to receive the services they need without getting the runaround or encountering frustrating confusion among providers.

Service coordination becomes key once the client and the case manager have agreed on a plan of services and determined what services will be provided by someone other than the case manager. For services that will be provided by others, a beginning step is to review previous contacts with service providers. What services do they provide? Is this client eligible for those services? Can the services be provided in-house? What about the individual's own resources and those of the family? Family support may be critical for the success of the plan, and the client's own problem-solving skills and strengths may be helpful. This mean a thorough case manager does not ignore the resources of the client, the family, or significant others.

The next step is referral—the connection of a client with a service provider. The final step is monitoring service delivery over time and following up to make sure the service has been delivered appropriately. These steps may vary somewhat, depending on whether the services are delivered in house or by an outside agency, but the flow of the process is likely to be the same. Before examining these steps in detail, let's review the documentation and client participation aspects of service coordination.

Documentation is critical in this part of case management. Staff notes must accurately record meetings, services, contacts, barriers, and other important information. During this phase, reports from other professionals are added to the case file. Any progress that occurs in the arrangement of services must be recorded by the case manager.

Client participation is important throughout the service coordination process. This entails more than just keeping the client informed; his or her involvement should be active and ongoing. First of all, the client participates in determining the problem that calls for assistance. Second, the values, preferences, strengths, cultural perspectives, and interests of the client play a key role in selecting community resources, and of course client participation is critical in following up on a referral. Clients also have the right to privacy and confidentiality. Without the client's written consent, the case manager must not involve others in the case or give an outsider any information about it.

Sometimes service coordination does not go well. Problems in the coordination process include, but are not limited to, the following:

- Professionals do not agree about the nature of client problems.
- Professionals would rather work alone.
- Professionals break policy rules or governmental regulations.
- Professionals do not deliver the services they promise.
- Professionals provide less-than-effective services.
- Professionals act unethically.

Regardless of the difficulties case managers encounter during service coordination, remembering that meeting client needs is of primary importance is an excellent guideline. Of course, maintaining good professional relationships is also important. Case managers often represent their agencies or organizations during the service coordination effort. Relationships among agencies sometimes hinge on the working relationships between individual direct service providers.

Resource Selection

Once client needs and corresponding services have been identified, the client and case manager turn their attention to **resource selection**—selecting individuals, programs, or agencies that can meet those needs. Paramount in this decision are the client's values and preferences. The information and referral system the case manager has developed (see Chapter 7) is useful in this regard.

Rube Manning is a fifty-three-year-old white male who is on parole for aggravated rape. He had sexual relations with his twelve-year-old niece; she later gave birth to his son. Both parties claim that the intercourse was consensual; the severity of the charge and conviction were due to the girl's age. The girl and the family seem to harbor no animosity toward Rube, and even went so far as to write a letter on his behalf to the department of corrections. Rube was sentenced to three years in prison and is now eligible for parole. Angela Clemmons is the parole officer assigned to this case. She and Rube must develop a plan of services for him to pursue once he is released. Among the conditions of Rube's parole are completing a mandatory sex offender program, supporting his son, and finding employment.

There are no options for the mandatory sex offender program; only one program is available in this community. Angela senses that Rube is motivated to do everything in his power to comply with the conditions of parole. Although he does not talk much about his prison experience, he does say that he didn't like it. Angela suspects that he was abused by other inmates. Sex offenders are usually on the lower rungs of the prisoner hierarchy unless they are very strong or charismatic; Rube is neither.

Finding employment and supporting the child are tied together. Checking her information and referral computer file, Angela advises Rube that there are three short-term training programs that can provide him with job skills. The first two are at the vocational school and would give him a certificate in either horti-culture or industrial maintenance. The third one is on-the-job training in food services, with a modest salary until training is finished. Rube's preference is hor-ticulture, because he grew up on a farm and thinks he would feel more comfort-able outdoors. He knows that industrial maintenance is a fancy term for janitorial work, and he's not interested. The location of the food services training is not on the bus line, and Rube has no transportation of his own, but this option offers a salary immediately. Angela notices that Rube sounds interested—even a little excited—about horticulture, so she checks her addresses and e-mail file for the phone number of her contact. (**See Figure 9.1.**)

In this case, resource selection is systematic, which has advantages for both the client and the case manager. The client and the case manager proceed objec-tively and deliberately, taking into account Rube's values, beliefs, and desires. The rationale for the choice is articulated, and it reinforces his motivation to fol-low through with the referral. Rube Manning and his parole officer have chosen the horticulture program: It is on the bus line, it builds on Rube's previous farm-ing experience, and it is something he wants to pursue.

The selection process can also accommodate many alternatives and can tai-lor services to the client's unique circumstances. The conditions of Rube's parole include work, and he does want the independence, salary, and respect that come with employment. However, he is not willing to do just anything. Being a janitor doesn't appeal to him, and he does not want to work indoors. Had the parole officer ignored his feelings at this point and decided to steer him toward janito-rial work, Rube would probably not be motivated to do well. At the very worst, he would do nothing, and his parole would be revoked. In addition, the relation-ship between Angela Clemmons and Rube Manning would not develop as a partnership.

Service Coordination Agency: Lincoln Vocational Technical School

Address: 30512 Townview Parkway

Contact Persons: Lynda Johnston, Admissions
                 Robert Griffin, Student Services

Phone: 555-1516

Services: Short-term training programs in auto mechanics, cosmetology, horticulture, industrial maintenance, printing, and secretarial services.

Comments: Good student services and advising: Janet Evans 7/1/XX

Figure 9.1   Entry in information and referral file

Being aware of the client's preferences, strengths, and values is critical to the success of the selection step in service coordination. There must be a strong partnership between the participants. In our example, the decision to try the horticultural program takes into account Rube's wishes, along with his need for training and employment.

Making the Referral

As mentioned earlier, no helper can provide all conceivable services. Therefore, arrangements must often be made to match client needs with resources. This is done by referring the client to another helping professional or agency to obtain the needed services. Referral is the process that puts the client in touch with needed resources. According to a case coordinator, working with parents of students at an urban high school, "We know that many of our families and students do not have money to provide the basics of food, housing, clothing, and medical services. We work hard here to help families and students find reliable sources within the community to provide help in these areas .... [I]n this city there are emergency services and long-term services ... often we refer to both of these types of services. Churches and local and state government services provide the most support for families. Sometimes we also need shelters—these are available in the community. During the economic downturn the needs have more than tripled."

A **referral** connects the client with a resource within the agency structure or at another agency. In no way does referral imply failure on the case manager's part. Limitations on the services a case manager can personally provide are imposed by policy, rules, regulations, and structure, and reflect his or her own expertise and personal values.

The case manager assumes the role of broker at this point in service coordination. The broker knows both the resources available in the community and the policies and procedures of agencies. He or she acts as a go-between for those who seek services and those who provide them. Consider the following case with regard to the referral process and the broker role.

Bethany's first client on Tuesday is Anna, a young woman who has just discovered that she is pregnant. This pregnancy has caused a crisis in Anna's family. Her parents are first-generation immigrants from El Salvador, they are Catholic, and they are very opposed to both the pregnancy and abortion. Although the agency that employs Bethany specializes in career development services, Anna feels comfortable with Bethany and wishes to discuss her options for the pregnancy with her. On the other hand, this is a difficult subject for Bethany, because her sister had an abortion three years ago and still feels guilty and upset about her decision. In fact, the whole family is still having difficulty with it, since the sister is living at home. Bethany also knows that her training is in career development, and she has never worked with anyone dealing with an unwanted pregnancy.

The encounter illustrates a situation that is appropriate for a referral. Bethany has some personal feelings that may impair her objectivity; she recognizes that she has no professional experience with this problem; and her agency's purpose is career development. For these reasons, she decides it is best to make a referral to someone who can help Anna explore options related to the pregnancy. Bethany will continue to support Anna's career development efforts. In the referral process, Bethany's role is that of a broker.

Making a referral may seem like a fairly uncomplicated process, but it often results in failure. If a case manager believes that all that is necessary is being aware of client needs and making a phone call, the referral is likely to be unsuccessful. In fact, it is common for clients referred to other community resources to resist making the initial contact. Clients may also fail to follow through after the first interview and drop out before service provision is complete.

A referral can fail for three reasons. The first is insensitivity to client needs on the part of the case manager. Identifying the problem but failing to grasp the client's feelings about it contributes to an unsuccessful referral. The client may not be ready for referral at this point, feeling only that he or she is being shuffled among workers or agencies. Second, if the case manager lacks knowledge about resources, the client may be referred to the wrong resource. This makes him or her feel lost in the system, think that it is all a waste of time, and believe (sometimes correctly) that the case manager is incompetent. A third reason for failure is misjudging the client's capability to follow through with the referral. Suggesting to an involuntary client that she call to make an appointment for a physical examination may not work, perhaps because she is new to town, is unsure of who to call, doesn't have a phone, or does not actually want the exam.

How can the case manager make the referral process a successful one? Two areas of focus are assessing client capabilities and understanding the case manager's role in making the referral. Assessing clients' capabilities means finding out how much they can do on their own. It is good to encourage independence and self-sufficiency in clients, but some of them will prove unable to identify what they need and take the steps to obtain it. The nature of the problem, the feelings the client has about it, and the energy required for action may all contribute to feelings of being alone, an inability to act, and a lack of motivation to follow through.

In addition to assessing the client's capabilities, the referring case manager must form a clear idea of what role he or she will play in the referral process. In this, the case manager should be guided by what the client needs and what relationship the case manager has with the other professional or agency. The case manager's degree of involvement in the referral can fall anywhere on a continuum, from discussing several resources with the client, who then takes responsibility for selecting a resource and following through, to giving concrete assistance with details, such as making the appointment on the client's behalf and having an agency volunteer accompany him or her to the appointment.

Bethany approached the referral process in the following way. She acknowl-edged Anna's concern about her situation and recognized her desire for some help. She also shared with Anna her reservations about being able to assist her, explaining that her training was in career development and she had limited knowledge about options for an unmarried pregnant woman. How-ever, she did know of two agencies that offered just the services Anna was seeking. Anna wanted to know about these, so they discussed the services they provide and their geographic locations. Anna was concerned about the cost of services, and Bethany was unsure about the agencies' charges. She checked her computer file and found that both agencies charged fees on a sliding scale. Anna didn't know what that meant, so Bethany explained that such a scale determined the fee in accordance with the individual's income. Anna was unsure how to get an appointment—who to call, how to explain the problem, and so forth. She also wondered whether she would be able to continue working with Bethany on career development. Bethany discussed all these concerns with Anna. Together, they decided on one of the agen-cies, and Bethany agreed to make the initial contact. Her previous work with Anna led her to believe that once the initial anxiety of making contact was over, Anna was capable of showing up for the appointment and getting the services she needed.

BETHANY:	Hello. This is Bethany Douglas at Career Development. I am working with a client who needs help identifying her options for dealing with an unplanned pregnancy. Will someone at your agency see her?
RECEPTIONIST:	We do provide counseling. Let me connect you with one of our counselors.
COUNSELOR:	Hello, this is Carol Fong. May I help you?
BETHANY:	Yes, Bethany Douglas here. I am a career counselor at Career Development. My client has just found out she is pregnant and would like to talk with someone about her options. She is nineteen and single. Could we set up an appointment for her to come see you?
COUNSELOR:	Yes, I would be glad to see her. Would Monday at 11:00 o'clock be okay?
BETHANY:	(*Checks with Anna, who nods*) Yes, that would be fine. Her name is Anna Rodriguez. She will see you at 11:00 o'clock Monday. Thank you.

Bethany used several strategies to ensure that Anna's referral was a successful one.

1. *Discuss with the client the services that are provided by the resource.* The discus-sion should include why the referral is needed, how it will be helpful, how

the client feels about it, and what information should be provided. If client information will be shared, then the client or guardian signs a release form at this time.

2. *Make the referral.* Making a referral may entail just providing the client with a telephone number and an address; helping him or her with the initial contact, as Bethany did; or taking the initiative to contact the resource. The interaction may involve scheduling an appointment, telling what the client knows about the resource, and finding out what information the resource needs. Of course, before any information is released, the client's permission must be obtained. Finally, if there are cultural dimensions to be considered, the case manager should communicate this to the referral source. Any language or cultural barriers should be addressed before the client arrives at the point of referral.

3. *The referral does not go as planned.* When Bethany made the call, Carol Fong might have responded differently—perhaps she couldn't see Anna until next month, or her agency didn't do that kind of counseling anymore. Bethany would have two options. She could return to her file to locate another agency that provides the services Anna needs. However, suppose that Anna lives in a small town or a rural area where there aren't any other agencies to call. Bethany's second option would be to become a **mobilizer**—one who works with other community members to get new resources for clients and communities. Bethany could try to mobilize Carol Fong and other professionals so that Anna could gain access to needed services.

4. *Share the referral information with the client.* He or she needs to know the appointment time, the location, and the name of the person to see on arrival. It is also appropriate to find out what support the client might need to follow through with the appointment.

5. *Follow up on the referral.* The service coordinator can do this by talking with the client and the helper who received the referral. Did the client show up? What happened? Was the client satisfied with the services? With the worker? Service coordinators with thorough information and referral systems make a habit of noting such information in their files. Information from the worker who saw the client may be conveyed in a phone call, a written report, or not at all.

Bethany followed up on Anna's referral by talking with her about it the next time they met. She discovered that Anna had had no trouble finding the agency, liked the worker immediately, and felt positive about exploring her options with her. Bethany received no official report from the other agency and did not request one.

The referral process is a flexible one that can be adapted for use with any client, but client participation is vital to good service coordination. Clients participate in the decision to refer and the choice of where to refer. Their capabilities determine the extent of their involvement in the steps of the referral process, including making an appointment, getting to the agency, and so forth.

The case manager's role in the referral process varies from little involvement to integral involvement, depending on the client's capabilities. The case manager's responsibilities include knowing what resources are available for the client, how to make a referral, and how to assess the client's capabilities accurately. His or her involvement does not end after the referral; the next step, monitoring services, is also the case manager's responsibility.

Monitoring Services

Once the referral is made, monitoring service delivery becomes the focus of case management. Monitoring services is more than following up on the contact; it may mean offering information, intervening in a crisis, or making another referral. The case manager continues to work in the roles of broker and mobilizer throughout this phase of service coordination. In **monitoring services**, the case manager reviews the services received by the client, any conditions that may have changed since the planning phase, and the extent of progress toward the goals and objectives stated in the plan. This review can occur as often as once a day or three times a week or as little as once a month or once a year, depending on the goals of the program, caseload of the helper, and resources available.

REVIEW OF SERVICES

Once a case manager has made a referral, delivering the needed service becomes the responsibility of the resource—the agency or professional that has accepted the referral. However, the case manager does not relinquish the case completely. He or she remains in contact with the client to ensure that the services are being delivered, that the client is satisfied with them, and that the agreed-on time frame is maintained. As you remember, all these are specified in the plan of services.

When checking with Rachel Vasquez after her visit to the health clinic, the case manager heard about the generous time a volunteer had spent with Rachel in making out a balanced nutrition plan for her diabetic son. Rachel was excited about knowing what to buy, how to prepare it, and why it made for a good meal. Most of all, she was impressed by how much time the volunteer spent with her.

If there are problems with service delivery, the case manager has ultimate responsibility to intervene. Problems may be caused by the agency, the client, or both. For example, the agency may be unable to see the client for several weeks, or may neglect to do what the client has been promised. The client, on the other hand, may fail to show up for appointments or refuse to cooperate (e.g., be reluctant to give needed information). The case manager must be aware of the situation if he or she is to know that intervention is required. The intervention in such a case involves identifying exactly what the problem is and working with the client and the resource to resolve it.

Sam Miller received a call from the VA hospital where twenty-two-year-old Raymond Fields (who is developmentally disabled) had been placed as an orderly just two weeks before. Both the supervisor and Raymond had been pleased with the match. This morning, the supervisor reported that twice in the past three days, Raymond had been seen unzipping his pants and playing with his penis in the hallways. Sam hastened over to the hospital to talk with Raymond about the behavior. He told Raymond to keep his pants zipped. There was no more trouble afterward.

CHANGING CONDITIONS

Often there is a time lag between plan development and the provision of services. During this period, the case manager seeks agency approval, if necessary, and arranges for services either within the agency or at another one. It is also likely that there will be changes in the client's situation during this time. Living arrangements, relationships, income, and emotions are some of the factors that may change. The presenting problem may show some alteration, or additional problems may surface. Any such changes may necessitate review and revision of the plan.

Alma Justus is raising two granddaughters and one grandson with the help of her son Zack. The mother of the children, Alma's daughter, lives in another state with her boyfriend and his two children. Alma and the children are receiving assistance from a case manager at the local office on aging. Last week Alma was placed in the hospital, and after extensive testing, it was confirmed that she had had a series of slight strokes. The case manager will work with the family to determine the changing need for services.

The client's circumstances may also change during service delivery. Part of service monitoring is staying informed of changes that occur in the client's life. Some changes may occur as a result of service delivery; for example, a client might learn more appropriate ways to express anger than hitting his spouse. Other changes may have nothing to do with service delivery, yet they influence it. For example, a client might decide to marry while halfway through service delivery—an action that could well affect her economic eligibility for services. Again, monitoring of services helps the case manager stay abreast and be ready to intervene if necessary.

EVALUATING PROGRESS

Monitoring services also entails continually checking progress toward the goals and objectives set forth in the plan of services. Continual evaluation may lead the case manager to modify the plan to improve its effectiveness or deal with new developments. In monitoring services, the case manager repeatedly asks the following questions:

- Has the identified problem changed?
- Was the referral made correctly?
- Were the desired outcomes achieved?
- Were cultural differences acknowledged and attended to?
- Should the plan be altered?
- Should the case be closed?

Monitoring of services goes most smoothly if close contact is maintained with the client. Outcome measures focus on the client, so he or she is a key source of information about service delivery. Did the client use the resource? Was the goal of the referral attained? The case manager's responsibility continues until the client's problem is resolved. Follow-up and monitoring are performed to make sure that referrals result in the desired outcomes. Client satisfaction is important in a successful referral.

The following case illustrates how a case manager monitors services by reviewing the services received, considering any changes in conditions, and evaluating progress toward goals and objectives.

It was chilly on February 17, but the Naylors were happy. It was Presidents' Day weekend, and they were going to have three days off. Everyone gathered in the den in front of the fireplace. Jennifer, the younger daughter, was wearing a tank top and shorts to be comfortable, since she had just come down with chickenpox. Johanna, the older daughter, went to look for the kitten her grandparents had given her for Christmas. It appeared to be just another quiet evening.

Johanna came in the back door about 6:30 p.m. and said, "Mom, there's a fire in the garage!" Mrs. Naylor looked out the door that led to the garage and saw flames at least ten feet high. Calmly she said, "Everybody out," and headed for the front door. Mrs. Naylor and both daughters made it out safely. As the Naylors stood watching the fire consume their home, they wondered what they were going to do and where they were going to go. Would they be able to salvage anything at all?

The Naylor family lost everything. The burn shelter in their community immediately stepped in to provide the many services that fire victims need. Bettyjean Fleming, a case manager at the burn shelter, was assigned to work with the Naylor family. Once notified of the fire, Ms. Fleming went to the site of the fire to help the family members with their immediate needs. Comfort, clothing, a meal, transportation to the hospital, and temporary lodging are among the services the shelter provides. Case managers work with the victims to cope with any losses they may have suffered (including family members, pets, and possessions). Once the immediate needs are met, the counselor and the victims develop a plan to meet long-term needs, which will be met by other agencies.

The Naylors had a number of needs: transportation, housing, clothing, household furnishings, and counseling. One of the most serious needs was counseling for the two girls. Not only had they lost everything they owned, but they also lost every picture and memento of their father, who had died a year before. They desperately needed help in coping with the loss of their father, as well as loss of their home, their possessions, and their pet. Ms. Fleming learned of these circumstances from Mrs. Naylor, and they developed a plan for long-term services. She was supportive of counseling for the girls; together, they discussed several alternatives, using the file of community services at the burn shelter. They decided on counseling services at the local mental health center, which is known particularly for its children's services. Mrs. Naylor agreed to call for an appointment. Unfortunately, she was told that the earliest appointment was next month; she didn't see any alternative, so she took that appointment.

The following week, Ms. Fleming called to see whether Mrs. Naylor had received her insurance check. In the course of the conversation, she learned about the delay in getting counseling. She alerted Mrs. Naylor to the emergency services that were available and volunteered to call and talk with a counselor. Ms. Fleming often worked with staff at the mental health center, so she called one of her contacts, Frances Lane. Ms. Lane agreed to see the girls that week and requested some background information, which was provided. Mrs. Naylor was grateful for the intervention, since there had been no improvement in the girls' mental state.

At the next meeting with Mrs. Naylor, Ms. Fleming asked about the visit to the mental health center. Were they seen on time? How did the girls feel about Ms. Lane? Were they feeling better? What were the next steps? At the same meeting, Ms. Fleming asked again if the insurance check had arrived, how the apartment was working out, and whether Mrs. Naylor or the girls had other needs.

In this case, the case manager's responsibility did not end after the referral. Because she monitored service delivery, Ms. Fleming became aware that there were problems. She was able to intervene, using her contacts to expedite service delivery. She also monitored the delivery of the other services, verifying that the identified services were delivered within a reasonable time and by the professionals designated to deliver them. Not keeping abreast of developments in a case can result in delays in needed services (or even nonperformance). Then clients become frustrated and dissatisfied with both the resource and the case manager.

Working with Other Professionals

Clearly, effective service coordination depends to a certain extent on the case manager's relationship with other professionals. Both referrals and service monitoring are more easily achieved when the case manager has a relationship with the resource. The professionals on whom a case manager relies have a wide variety of cultural backgrounds, academic achievements, and job descriptions. Often barriers to service coordination appear; they may be rooted in turf issues, competition for clients, or concern about confidentiality. Communication, sometimes a challenge among those with different perspectives, is one way of addressing these barriers.

Good communication skills are critical when working with personnel from other agencies. These skills can be the deciding factor in making effective use of resources on the client's behalf. Here we present certain suggestions for enhancing communication with other helpers. First, avoid stereotyping other professionals. You may have encountered one nurse who was rude, but it is unreasonable to think that all nurses are that way. Second, don't hesitate to ask for clarification or a definition of terminology that you don't understand. It is better to ask than to pretend you know. Third, you can help others learn your own terminology by using it and explaining its meaning. Finally, be aware that other professionals may well have different styles of communication. For example, a clinical style may be more comfortable for psychiatrists. Other styles that have been identified are legal (equal adversaries), political (unequal adversaries), and pedagogical (teacher–student).

PRE-MEETING HOMEWORK

It is important to prepare for any encounters with other professionals, either within your own organization or outside your organization. Here are steps to help you prepare.

> **Gather any background information** you need and research any issues you believe might arise.

> **Learn about the other professionals,** their values, their roles, and what their positions might be. You will want to listen and to demonstrate respect.

> **Understand your own role and the context** of the specific situation you are in. Think about how you will communicate with others in a way that fits into their organizational culture.

> **Be aware of your strengths and limitations.** Don't overcommit without the ability to follow through. Consult often.

MEETING WITH OTHERS

> **Identify commonalities.** Use this time to identify mutual concerns and common goals of all parties. Build bridges with others who may support your position, understand your concerns, or be willing to compromise.

Define the issue. Keep focused on the challenges, concerns, and stated agenda for the work. Don't get sidetracked.

Approach each conversation or meeting with a listening ear. This is the time to hear the other, make sure the other knows he or she was heard, and state your position in a fair minded way.

Be concrete. Don't use jargon. Provide numerous examples.

Be sure to say, "I don't know" and "I'll find out" when that is the case.

Use your humor to deflect anger. Direct the humor to yourself, not others.

Expect to compromise. Expect to revisit your position as you learn more about the issue from others.

POST-MEETING HOMEWORK

Review the past meeting. Outline what happened, what decisions were made, and what responsibilities you have, and think about how you could have improved the communication and outcomes.

Follow up with others if need be.

Make informal contacts and have conversations or exchange e-mails with those involved in decision making. Keep the channels of communication open.

Using these checklists can facilitate your work with other helping professionals. Being a skillful communicator and having good working relationships with other helping professionals enhances the case manager's role as a client advocate.

 # Advocacy

Advocacy is speaking on behalf of others, pleading their case, or standing up for their rights. When case managers act as advocates for their clients, they are supporting, defending, or fighting for another person or group. Advocacy is also related to client empowerment and participation, since case managers support clients' involvement in decisions about their own treatment and welfare. A caseworker at a family services center in the Bronx describes a situation faced by AIDS patients that illustrates the need for advocacy.

> We provide temporary housing for clients in need and the family members that live with them. The family members are in jeopardy sometimes. When the client dies then the family has to move out. Usually these families have few resources and we have to scramble to try to help them, even though officially they are no longer our clients. This is a systems issue, but there are no solutions as yet.

Advocacy is very important for case managers. Often, clients are unable to articulate what they need, and do not understand what choices are available to them. They may not have the necessary information or the skills needed to present their positions. In some situations, what clients want or need is in direct conflict with what is being provided. Clients may then be too intimidated to speak for themselves, or the people in power may refuse to consider client wishes. Terry is an example of such a client.

Terry is a fourteen-year-old who has just been diagnosed as HIV positive. She is terrified of her medical condition and will not talk to anyone about it. She just sits and stares when the caseworker, her mother, or any other member of her family tries to address the subject. Her mother wants to send her away to a residential school, but Terry will not speak of this or any related issue.

In some agencies, the people in charge may not wish to hear client complaints or grievances. There may be no way for clients to appeal decisions or discuss methods of treatment they do not support. Clients, by definition, are usually not in a position to act as advocates for themselves. Advocacy requires confidence, a feeling of control, and an understanding of the system. Most clients are in the human service delivery system because they do not have these characteristics. One caseworker at a school for the deaf tells how she remained an advocate even after her official responsibilities were completed.

There are times when we simply advocate for our clients. We encounter situations where the children we serve return home to live with their parents, the Department of Human Services closes the case, and we have to ask them to reopen it. We can see from our interaction with these children and from our home visits that the children are in danger. This case is unique because I maintained the connection with the foster family after the children were removed from the home. I knew that I would be participating in future legal affairs relating to these children.

The following are some common client problems that case managers may consider appropriate for advocacy.

- Client has been denied services, or services have been limited.
- Client has expressed interest in one method of treatment but has been given another.
- Client's family has made decisions for him or her.
- Client has little information about the assessments gathered or the decisions made.
- Client has been denied services based on factors such as race, gender, or religion.
- Client has been given treatment contrary to his or her cultural norms.
- Client has been treated with disrespect by human service professionals.

- Client is caught in the middle of a conflict between two agencies or professionals.
- Client is being given unsafe or indifferent care.
- Client does not know what his or her rights are.
- Rules and regulations do not serve the client's needs.
- There are no available services to meet the client's needs

Case managers are well positioned to provide advocacy for clients they serve. In the first place, case managers know their clients. They are most familiar with their skills, values, wishes, and the treatment they have received. A measure of trust has built up between the client and the case manager, so the case manager is likely to hear from the client if there has been unfair treatment. Case managers are also in contact with family members, friends of the client, and other professionals with whom the client is involved. The case manager therefore has immediate access to anyone who may be involved in any unfair practice. Finally, in the process of monitoring service delivery, the case manager may discover situations that warrant advocacy.

Advocacy is not an easy role to perform. Before an advocacy effort is begun, several tasks must be completed. First, the case manager gathers the facts of the situation and determines whether there is a legitimate grievance—whether making requests on behalf of the client would be justified. The case manager then decides whether advocacy is more appropriate than other methods, such as problem resolution or conflict resolution. Third, the case manager discusses the need for advocacy with the client and shows willingness to speak for the client. He or she needs the client's approval before any advocacy takes place.

We note that new avenues for advocacy arise as the social, economic, and political landscape shifts. Clients and their needs are dynamic. When thinking about the effects of the long-term economic downturn in the United States, the demand increased for services to meet basic needs of adults and children such as housing, food, and vocational assistance. Addressing issues related to foreclosures and job loss required advocacy on the part of most human service agencies. The following three case managers talk about the influence of the economic downturn on the clients they serve.

> We are seeing so many more families in the homeless shelter. In the past two years, our numbers doubled. We are just now seeing families in lower numbers. These were not what we would typically say fit the profile of regular clients. In these families, the adults were educated and the children did not have any outstanding problems by mental health status exam standards. But the families were in crisis with nowhere to go.
>
> Case coordinator, homeless shelter, Houston, Texas

> I coordinate care of veterans. Life is very difficult for them when they return to the United States. Many have PTSD or traumatic brain injury and some suffer from anxiety or depression. All of the issues arising from their military experiences are exacerbated by the economic pressures the country faces. For example, even if they can work, they have trouble

Want To Know More: Advocacy

Find out the range of social advocacy occurring internationally, nationally, by state, and locally using the Internet. First search for "advocacy." A variety of sources provide examples and definitions of advocacy. Second, search for "social advocacy." This term provides you with an incredible range of services available and work underway to improve the lives of others.

getting work. This means they cannot support their families, they lose a way to find meaning, and generally feel unappreciated and rejected.

> Case manager and counselor, veterans' services, Columbia, South Carolina

Immigrants we see are affected by the current economic conditions. What do I mean by this? Citizens are so protective of what jobs exist that there is a prejudice against giving work to a non-U.S. citizen, even if that individual has a green card.

> Case manager, settlement services, San Antonio, Texas

These three quotes illustrate the impact of the context in which services are delivered and the advocacy needed. The focus of advocacy continues to shift. As a case manager, look to define trends that occur globally, nationally, by state, and locally. This will help you know the direction your work should take.

How To Be a Good Advocate

Many case managers do not like conflict, and they fear that, as advocates, they will not be successful in meeting the needs of their clients. Advocacy is complicated because case managers have loyalties divided among the agency, the supervisor, and the client. The case manager's personal values and beliefs may complicate the advocacy process. The following guidelines support the work of effective advocacy. These guidelines build on the knowledge, skills, and values presented in our previous discussion of the assessment, planning, and implementation phases of the case management process.

Know the environment in which the conflict takes place.   This is important when planning how to act as an advocate for the client. It is helpful to know the appropriate person to whom to make the appeal. Having a good referral network and a good relationship with many agencies is useful in understanding the environment.

Understand the needs of the client and of the other parties involved.   It is helpful if the advocate can understand why the other parties are in conflict with the client or seem not to respect the client's wishes. If the advocate has this information before the appeal, strategies can be developed to meet some of the needs of all involved parties or, at the very least, to articulate the common ground.

Develop a clear plan for the client.   The advocate must be clear about what the client needs and what the other parties must give or give up to meet those needs.

Use techniques of persuasion when appropriate.   Many situations are suited to persuasion. Persuasion is used to support client needs while respecting the rights of other parties. It includes stating the problem clearly, presenting critical background information and facts, explaining why the situation needs to be changed, and detailing an acceptable solution.

Once there is agreement, it needs to be stated or written.   Remember that all parties need to agree.

Use more adversarial techniques when persuasion proves ineffective.   When using techniques that challenge the system directly, the case manager makes a formal appeal or takes a case through legal channels in an effort to promote change on behalf of the client. In a grievance procedure or a legal challenge, the client has an opportunity to be heard, either verbally or in writing. Usually an impartial person or group hears the client's case, and a ruling is made. Another adversarial technique is to use the media to take the client's case to the public.

Terry's mother wants a solution to the problem right now. She is determined to send Terry away and has already made contact with three residential programs. She plans to move Terry somewhere as soon as possible. In reality, Terry's mother is frightened by what lies ahead for Terry and for the family. The only way she knows how to cope is to distance herself from the problem. The caseworker feels strongly that Terry should have a voice in the decision; she begins her advocacy work by trying to talk with Terry about what she wants.

It is not necessary for special circumstances to be present for case managers to assume the advocacy role. In the normal course of the job, a case manager encounters many opportunities to act as an advocate for the client. The case manager then integrates advocacy into other case management responsibilities. Each of the actions listed below represents the basic goals and values of case management discussed in Chapter 1. Not only do they characterize effective advocacy, but they are also good standards of practice.

- Provide quality services by involving a team of professionals and the client (or, when appropriate, a member of his or her family).

- Interact with the client to plan treatment that is congruent with his or her values, strengths, and cultural orientation.
- Monitor the case and set goals and outcomes based on quality standards of professional care.
- Continually communicate with other professionals about issues relating to client rights.
- Plan treatment that takes into consideration the client's preferences, strengths, and limitations, and provide additional support if you anticipate that he or she will have difficulty.
- Speak for the client only when he or she gives permission.
- Educate the client about the agency's policies and procedures.
- Create an environment that facilitates decision making by the client.
- Educate the client about options in treatment and about the process of developing the treatment plan, and discuss the barriers that may be encountered during the implementation phase.
- Work within the system to support, modify, and create policies. Know which professionals are involved in this work, and discuss your opinions with them. Volunteer for committees that consider policy issues.
- Become involved in the political process. Contact public-sector policymakers.

Advocacy work is difficult, and the case manager must exercise good judgment in choosing when to attack barriers to the client's cause. At times, the need for advocacy may not be clear-cut.

Martha Severn has just been asked by one of her clients not to report the $10 a week that she makes taking in laundry. Reporting this income would cause the client to lose some of her scholarship aid for school. Instead, the client wants Martha to try to change the eligibility rule. Although changing the rule might seem to be a special favor for this particular client, Martha happens to believe that changing this policy is a good idea.

In other instances the case manager may have to deal with competing interests—those of the agency and fellow staff members, as well as the client's.

James Dowling is a student intern working in a transitional housing unit for clients receiving community-based mental health services. He believes that one of his clients is being abused by a member of the night staff. His client tells him of beatings during the night shift and asks him for help.

Ms. Wise is an elderly client who receives attendant care at home. The agency that coordinates Ms. Wise's care has a policy of keeping clients on home services for as long as possible. Cheryl Santana, a case manager at the agency, believes that clients remain much too long in home care. Cheryl makes recommendations for residential care, and the agency routinely rejects those recommendations.

Voices from the Field: Processes of Advocacy

Ritu Sharma developed an advocacy training guide for those working with the U.S. Agency of International Development, African Bureau, Office of Sustainable Development. In this document Sharma describes the following process for those who wish to be good advocates.

Selecting an Advocacy Objective

Problems can be extremely complex. In order for an advocacy effort to succeed, the goal must be narrowed down to an advocacy objective based on answers to questions such as: Can the issue bring diverse groups together into a powerful coalition? Is the objective achievable? Will the objective really address the problem?

Using Data and Research for Advocacy

Data and research are essential for making informed decisions when choosing a problem to work on, identifying solutions to the problem, and setting realistic goals. What data can be used to best support your arguments?

Identifying Advocacy Audiences

Once the issue and goals are selected, advocacy efforts must be directed to the people with decision-making power and, ideally, to the people who influence the decision makers. . . . What are the names of the decision makers who can make your goal a reality? Who and what influences these decision makers?

Developing and Delivering Advocacy Messages

Different audiences respond to different messages. . . . What message will get the selected audience to act on your behalf?

Building Coalitions

Often, the power of advocacy is found in the numbers of people who support your goal. Especially where democracy and advocacy are new phenomena, involving large numbers of people representing diverse interests can provide safety for advocacy as well as build political support. . . . Who else can you invite to join your cause? Who else could be an ally?

(continues)

(continued)

Making Persuasive Presentations

Opportunities to influence key audiences are often limited. A politician may grant you one meeting to discuss your issue, or a minister may have only five minutes at a conference to speak with you. Careful and thorough preparation of convincing arguments and presentation style can turn these brief opportunities into successful advocacy.

Fund Raising for Advocacy

Most activities, including advocacy, require resources. Sustaining an effective advocacy effort over the long term means investing time and energy in raising funds or other resources to support your work. How can you gather the needed resources to carry out your advocacy efforts?

Evaluation Advocacy Efforts

How do you know if you have succeeded in reaching your advocacy objective? How can your advocacy strategies be improved? Being an effective advocate requires continuous feedback and evaluations of your efforts. (pp. 5– 6).

Sharma, R. R. (n.d.). An introduction to advocacy: A training guide. Retrieved from http://dat.acfid.asn.au/documents/an_introduction_to_advocacy_-_training_guide_(full_document).pdf

Case managers must be aware of how their advocacy efforts are perceived by others. Many case managers believe that their first loyalty is to the institution or agency; others feel that they need to support the efforts of the team. Vigorous advocacy, at the expense of team camaraderie, may jeopardize the client's trust in the team. Another factor is the difficulty in acting as an advocate for certain clients: Some people *are* dishonest, greedy, or troublemakers. With such clients, the case manager must think clearly about the legitimacy of any demands the clients make and approach the issues with fairness in mind.

 ## Teamwork

As stated earlier, working with others—**teamwork**—is a key component of the case management process. Professionals find themselves forming relationships with clients, families and friends of clients, coworkers, other professionals, and other agencies. These relationships include working on teams, counseling families, and forming partnerships with other agencies, businesses, or governmental units and departments.

Treatment Teams

In meeting the needs of clients who have multiple problems—children, those with developmental disabilities, the elderly, and many other client populations—a coordinated team approach is necessary because several professionals are involved. Sometimes referred to as the **treatment team**, this group of professionals meets to review client problems, evaluate information, and make recommendations about priorities, goals, and expected outcomes. The director and care coordinator, who works at a shelter for runaway girls, describes the shelter's staffing procedures.

Every Tuesday the entire staff meets, and we discuss every client. We begin with the newest girls who have entered the residence. There we talk about the initial intake and assessment. For the others we talk about any new information, how the girls are progressing with their goals, what other services we might provide, how might we enhance their success; then we think about changing goals, transition, and perhaps termination.

Using a team to make decisions has numerous advantages. Working as a team, professionals can share responsibility for clients as well as the emotional burdens of working with clients who have difficult problems. When a group of professionals is involved, all the dimensions of the client's situation are more likely to be considered, and team members can get one another's viewpoints about the advantages and disadvantages of each decision made. Furthermore, in the team setting, helpers can share their expertise and knowledge as they focus on each client's unique set of needs and circumstances.

Types of Teams

One type of team used in case management is the **departmental team**, made up of a small number of professionals who have similar job responsibilities and support each other's work. Colleagues bring their most challenging cases to the team, and their coworkers help identify client problems and generate alternative approaches to treatment. One case manager, who works with adults with disabilities, talks about her work in the departmental staff meetings. "When we began our services, we were so excited to move clients from the institution to group living and then to independent living. . . . Honestly, we didn't think through the process very well and we were optimistic about how well clients would do. We now have a team in charge of the transition. We support our clients better and they are more stable."

Departmental teams are particularly useful when decision making is difficult or client problems create a stressful situation for case managers. The departmental team shares information, offers opinions, and often makes group decisions about how to work with clients. A case manager, who works with children, describes the importance of teams: "My team saved my life; it is so important to have relationships and support with other members of your team."

Within the departmental team, the case manager can fill either of two roles: leader or participant. When assuming the leadership role, the case manager presents a case for review. He or she describes the client and the current status and summarizes conclusions and decisions made to date. The material is presented in such a way as to encourage feedback and dialogue with other colleagues. When the case manager is in the participant role, another colleague presents a case, and the case manager must listen, study the situation, and provide advice and counsel.

An **interdisciplinary team** is a different way to work with other professionals to provide services to the client. As the name suggests, this team includes professionals from various disciplines, each representing a service the client might receive. Often it includes the client or a member of his or her family. Early in the treatment of a client, the interdisciplinary team gathers and shares data, establishes goals and priorities, and develops a plan. In the later stages of treatment, the interdisciplinary team monitors the progress of the client, revises the plan, and often makes decisions about aftercare. The case manager is often the team leader, giving other team members a holistic view of the client, ensuring that the client or a member of the family is heard or representing that individual's viewpoints, and conducting an assessment of client problems. In addition, the case manager is expected to monitor client progress between meetings, set the agenda for the meeting, give a summary of client progress, discuss next steps and any dilemmas, and help reconcile differences of opinion. At times, the team leader is placed in a difficult situation, since each helper at the table has credentials in a specialized area of professional expertise. Many of them are accustomed to managing cases, being leaders, and making assessment and treatment decisions on their own.

One outcome of initial interdisciplinary team meetings is an organized, well-integrated plan designed to meet the goals that have been established to meet client needs. A good example is the Family Service Plan (FSP) for Carl, Martha, and Redana Young presented in **Table 9.1**. This plan supports Redana, who is developmentally delayed, and her parents. They are involved in the Hope Project, supported by a coalition of providers including a local hospital, an early childhood support program, and the mental health advocacy center. This program involves families as care managers for infants and toddlers with chronic illnesses and developmental disabilities. In this project, infants and toddlers are identified and receive comprehensive services to meet their own and the families' needs. These include the provision of services during hospitalization, assistance in transition from hospital to home, and follow-up when children return home.

The service coordination functions include coordinating assessments and coordinating the FSP Plan. One component of the plan was the designation of responsibility to the professionals providing each of the services listed. Besides the parents, there is a team of four professionals involved in developing the comprehensive plan. Tasks are clearly stated, and an individual is assigned to oversee

TABLE 9.1   FAMILY SERVICE PLAN FOR REDANA YOUNG

Carl and Martha Young	Parents
Sofia Roberts	Nurse Specialist
Christina Cho	Child Life
Eleanora Howard	Human Services
Suzi Toutzel	Education

Identified	*Plan*	*Person Responsible for Positive Outcomes*
1. The Youngs will receive developmental assistance before Redana leaves the hospital.	a. Referral to Jones School. b. Give Youngs ideas about how to work with Redana. c. Youngs will decide when they need help from team.	a. S. Toutzel b. S. Toutzel c. C. Cho
2. The Youngs will be given information about resources to help family.	a. Give the Youngs information about resources. b. The principal at Jones Elementary will be contacted by the Youngs.	a. S. Roberts b. The Youngs
3. The Youngs will look at financial needs and support.	Arrange for consultation with Eleanora Howard.	C. Cho E. Howard
4. The Youngs will receive information about Redana's condition.	a. Educator will discuss interventions and visitation. b. The Youngs will check out material about how Redana's condition affects the family.	a. C. Cho b. S. Roberts
5. Help Redana communicate with others.	Set up ways to play games for young children. See if Redana responds.	The Youngs Pediatric nursing staff
6. Help Redana by sestablishing routines and games.		S. Toutzel E. Howard The Youngs
7. Reinforce Redana's development using material provided.	Use interventions provided by staff.	The Youngs

or perform each task. Once the treatment plan begins, interdisciplinary team meetings are scheduled to help monitor the client's progress. Each of the professionals involved presents a progress report, and together they make a decision about how to proceed. The case manager often meets with each of these professionals one-on-one before the meeting. In crisis situations, he or she may have to revise the plan without team approval. Interaction with other professionals can sometimes be difficult if they have not provided the services for which they are responsible or if there is some question about the quality of service delivered. The case manager is most often not their supervisor, so when such issues arise, he or she must manage by persuasion and collaboration.

Two other benefits of the team are support, described earlier, and challenge. Other team members bring multiple perspectives to each case management situation. These members may support a case manager's work, but also may provide critical feedback that allows case managers to see situations in different ways. For example, others may see different goals as a priority, may suggest alternative approaches, and may challenge your assumptions about the clients or families. They may ask for evidence for conclusions, helping you separate facts from intuition or identify biases. The challenge and critical feedback that occurs in a supportive and trusting context can enhance case manager skills and improve work with clients.

Teams with Families and Friends

Because case management is a viable model for serving clients with long-term, complex needs, families and friends are often an important part of the team. When working with clients such as the elderly, people with limited mental capacities, children and youth, the mentally ill, and other populations who depend on family and friends to help make decisions, provide care, or both, the case manager must recognize that the caregivers expect to be involved in the planning for services. There is also a trend to include families in treatment of pregnant teens, single mothers, parolees, immigrants, and others who were traditionally given individual treatment. Input and participation by families and friends is viewed as important at each phase of the case management process.

Working with families and friends can be rewarding and challenging. The benefits include expanding the network of support and care for the client, adding another perspective on the environment and needs of the client, and engaging those in the immediate environment as part of the solution. Not all family and friends facilitate the teamwork process: Family members may not agree with each other, may not support the client's commitment to change, or may try to sabotage the case management process. Most case managers would still rather have family members on the team even if they are not totally supportive.

One example of broad inclusion of family is the Relationships Australia Family Integration project, which provides support to individuals who are members of stepfamilies or are considering joining a stepfamily. Members of families in this situation have multiple needs and, at times, few resources or support. The family is at the center of the service delivery, which includes counseling, group activities, child care services, and other services as needed. Service delivery is based upon identified needs of stepfamilies, including adjusting to the breakup of previous families or the death of a family member, supporting members of families as they explore new roles and relationships, helping children in vulnerable situations, and expanding the ways that a new couple can relate to each other (Relationships Australia, 2012).

Benefits of Teams

Working on a team is an exciting experience for most case managers. They welcome the creative thinking and the support that comes from a collaborative effort. However, building an effective team requires the efforts of all the members, especially the case manager. In most interdisciplinary teams, the case manager has the responsibility for leading the team and developing an atmosphere conducive to good **collaboration**. A positive atmosphere in which a team functions well occurs when there is a common goal. In the case of teams involved with case management, the goal is the successful development and implementation of a plan that meets client needs. Team members must have respect and trust for the others. Respect is important because teams share in decisions that can radically alter clients' lives. Mutual respect is especially important in interdisciplinary teams because each member is relied on to bring knowledge and skills in a particular area of expertise. In the case of departmental meetings, the participating colleagues continue to work side by side, so it is important to maintain respect for each other.

For the case manager who is involved with teams, several aspects of teamwork can directly improve the services provided and the work environment of the professionals involved. First, the clients receive services from several professionals working together. The greater the sum of expertise, creativity, and problem-solving skill the team applies, the more effective the planning and delivery of services will be. Each professional can perform his or her responsibilities better because of having participated in the process of setting goals and priorities as well as planning. Because of the team, the professionals also have a better sense of the client as a whole person and a clearer conception of how their own expertise and treatment fits into the larger plan.

Teamwork also enhances members' skills in making decisions and solving problems creatively. Not only do they have opportunities to practice those skills, but they can also learn from the other professionals. The environment fostered by teamwork is valuable to any helping professional, but especially to a case manager who is coordinating services. Good communication skills are developed in a team atmosphere; members learn to listen well and to speak to the group when appropriate. The opportunity to share responsibility—to ask for assistance, to volunteer or give it, and to receive it—helps all team members by increasing their sense of community and reducing isolation.

At a home for the elderly, one case manager used team meetings to discuss difficulties, and it brought positive results.

> In our team meetings we talked about the difficult situations we were facing. It is really difficult to work with some of our clients. Sometimes they are not nice, they scream, yell, complain, and are violent. And often

they demand much of the staff. We give each other room to say, "I've had it!"

Teamwork is central to performing the case management role of service co-ordination. Linking clients to services, monitoring client progress, and communicating with other professionals are important components of effective service delivery. And team members can provide support and critical feedback to one another. The team is—or can be—a safe place.

 # Deepening Your Knowledge: Case Study

Karen is a case manager with a juvenile justice program in Phoenix, Arizona. She works with a wide range of adolescents who have become involved in the juvenile justice system and require additional treatment and service options after they spend time in the mandatory municipal program for first-time offenders. She recently took on case management services for Ashley, who at fifteen was placed in the municipal program for grand theft auto, driving under the influence, and drug possession after stealing her neighbor's car to run away from home with her older boyfriend. Ashley is ten days from completing the program and has been recommended for a probationary period that includes additional services after her discharge.

In meeting with Ashley, Karen learns that she has experienced repeated physical and sexual abuse from her stepfather. She has a strained relationship with her mother whom, she says, "doesn't care about me. She always puts him (stepfather) first and if I have to live in that house then I'll keep running away until I make it or get killed." When asked about her boyfriend, Ashley states that "he's a lowlife; I'm only with him because he knows how to get pot." Karen asks Ashley if she has any other options for places to live and informs her of referral options with local teen abuse centers. Ashley reveals that she has wanted to move in with her aunt, but hasn't seen her in a few months and isn't sure that it would be a possibility. Karen spends the rest of the interview assessing Ashley's strengths and weaknesses, documenting her substance abuse history and listening to Ashley's thoughts and concerns.

When the interview concludes, Karen informs Ashley that she will have to report the physical and sexual abuse and lets her know that they will meet two more times before her discharge. During those meetings, Karen agrees to update Ashley on her post-discharge expectations and possible follow-up treatment and living options for Ashley after the program. Ashley seems skeptical about Karen's commitment, though she states that the abuse should be reported. Karen reminds Ashley that she will remain the case manager in charge of monitoring her experiences in future inpatient or outpatient treatment and that she knows of many caring facilities that can assist Ashley in working through these issues.

Morgan, C. (2012). Unpublished manuscript, Knoxville, TN. Used with permission.

Discussion Questions

1. Even though Karen works with a court-mandated program, she attempts to include Ashley in the decision-making process. What are some of the ways in which she does this? Why is this important?
2. What steps might Karen take to advocate for this client?
3. How might a treatment team approach affect the course of Ashley's treatment?
4. How might Karen's ability to communicate with and make referrals to external resources affect the course of Ashley's treatment?

 # Chapter Summary

Service coordination is a key to the case management process. Because it is the case manager's responsibility to locate resources, make arrangements for clients to use them, and monitor client use and progress, coordinating services is a critical phase of case management. First the helper and the client agree on a plan of service; then they determine what services can be provided by the case manager and what services need to be referred. It is important for the client to participate during resource selection so that his or her values and preferences and unique circumstances are considered. Because the case manager often does not provide all the services needed, referral is an important component of service coordination. Effective referrals match available services with client needs. Readiness of the client to be referred, appropriateness of the referral, and readiness of the new agency to receive the client are all important factors in making a successful referral.

Once a referral is made, it is the case manager's responsibility to monitor the services provided and the client's progress. This includes periodically reviewing the services, noting changing conditions either on the part of the client or in the services being provided, and evaluating client progress. Evaluation of client progress is an ongoing responsibility and includes investigating the status of problems, verifying client satisfaction with the referring agency and staff, monitoring the status of outcomes, and determining whether to continue or close the case. There are guidelines that can facilitate working with others on this assessment. Case managers are encouraged to use good communication skills, to know their own limits and the limits of other helpers and agencies, to listen well, and to encourage dialogue about issues of disagreement. Referral and monitoring among helping professionals, at its best, occurs in an atmosphere of understanding and mutual trust.

Advocacy, a key element in service coordination, focuses on providing the best services available to the client. Advocacy often requires speaking and acting on behalf of clients when they are not able to speak for themselves. Sometimes organizations and individuals must respond more fairly to the clients they serve; clients may be denied services; services may be limited; or the client may be caught in a conflict between two professionals or two agencies. Whatever the situation, part of service coordination is helping clients negotiate the system, working to ensure that they receive the most effective services available.

Meeting client needs often requires working in teams. Teamwork can take place within departments, across departments and agencies, and with clients and their families. The benefits of using teams include being able to serve clients with multiple needs and increasing the number of resources available.

Chapter Review

Key Terms

Resource selection   Teamwork
Referral   Treatment team
Mobilizer   Departmental team
Monitoring services   Interdisciplinary team
Advocacy   Collaboration

Reviewing the Chapter

1. Name the three activities of service coordination.
2. What are the benefits of service coordination?
3. How would you use a systematic resource selection process when making a referral?
4. Under what circumstances does a case manager refer a client?
5. Discuss three reasons why referrals fail.
6. What are the steps in a successful referral?
7. Describe how the roles of broker and mobilizer apply to monitoring services.
8. What are the three components of monitoring services?
9. State some guidelines for working with other professionals.
10. Define advocacy.
11. Name some common client problems that reflect the need for advocacy.
12. What are some guidelines for good advocacy?
13. How does advocacy relate to the goals and values of case management?
14. What different types of teams are encountered in case management?
15. What are the benefits of teams?

● *Questions for Discussion* . . . .

1. After reading this chapter, what evidence can you give that coordinating services is a critical component of the case management process?
2. Why do you think advocacy is an important responsibility for a case manager?
3. Suppose you were a leader of a case management team. How would you conduct your first meeting?

● *References* . . . .

Relationships Australia. (2012). *Stepfamilies are different.* Retrieved from http://www .relationships.org.au/relationship-advice/relationship-advice-topics/second -chances-remarriage-and-repartnering/stepfamilies-are-different

. . . .

CHAPTER TEN

Working within the Organizational Context

Thriving and Surviving as a Case Manager

Jessica

I used to work for an agency where my values weren't the same as my employer's. It was all about the dollar as the bottom line. I wanted to work for my current agency for a long time—I knew its reputation and I supported its mission and goals and objectives. I had worked with its resource coordinators and had always been impressed. So when I got the job there I was ecstatic. And still, two years later, I'm ecstatic to have the job. The values that the company operates on reflect, "Do what you say and say what you do." What this means is if we say that we are going to provide this, if we say that we are going to do this, then we are going to do it. Another value is child advocacy. I think the most important thing about case management is that we work for people who don't have a voice and if you are a case manager or a resource coordinator or whatever label you want to put on it, if you chose to be an advocate for somebody without a voice, then you need to be passionate about it. This agency really is passionate about that and so am I.

There is a case that has been very trying. It involves four children who were sexually abused by many people, one of whom was their father, and for some reason he's always just gotten away with it. We had to send the children back to their parents. So I've been trying to figure out ways to keep these kids safe and then human services dropped the ball on something and told me something that they shouldn't have and then said, "We lost the file. This is why they are going home. You didn't hear that from me."

I went straight to my supervisor and I said, "This is what this person just told me. I can't keep that to myself." And my boss said, "You know what? You've got to keep on fighting. As long as you are advocating for the safety of the child you are always going to have a job, so just do what you need to do."

> —Permission granted from Jessica Brothers Brock, 2012, text from unpublished interview.

*O*utreach is important to us. In our work we plan for toward more than one outcome and we think of outreach in terms of levels of success. There are some individuals we work with who, when we move them from homelessness to a shelter, we consider that outreach a success, even though it is a small step. Also, some people are homeless but they don't want to live in a shelter or a home. They are not interested in housing but want very temporary shelter and food for the day. We consider it a success if these individuals want our help at all. We try to expand their vision of services available and another way of living when we interact with them. We try to meet the client's needs as identified by them.

　　—Outreach worker, youth services, Missouri

*I*n our school program we are able to meet the needs of our high school students and their families because we established partnerships and teams with other agencies. We write grants together; we receive state funding together. We evaluate our outreach programs together.

　　—Case manager, Full Service Schools, Riverside, California

*Y*ears ago we just provided services. When our clients left we never knew what happened to them. We are trying to change that and to be more accountable; we follow our clients six months out and a year out, just to track whether they maintain the status they leave with. We want to be accountable for our outreach efforts. This helps us change our programs and fine-tune them. We also use this information to determine if we need to offer new types of services.

　　—Director, supervisor, and case coordinator, community services center, Missouri

The case management process takes place in the context of an agency, and client services often involve more than one organization. Effective case managers understand the organizations to which they belong. This requires mastery of three key concepts: organizational structure, agency resources, and improving services. The preceding quotes illustrate the importance of these three concepts in case management.

The organizational structure of an agency reflects the agency's mission, goals, and policies. Knowledge of these aspects allows the case manager to understand his or her actual job responsibilities and working environment. According to the outreach worker at community services, the goal for providing outreach to the homeless is to respond to client needs as expressed by the client, not just to find the client housing. The wide range of services given the clients, including meals, clothing, bathing facilities, and recreational opportunities, reflects this mission of the agency. All these services reflect a strong commitment to letting clients determine the services they want.

An agency's budget constraints are also important. Even though the Healthy Start budget is limited, the Full Service School program offers a wide variety of services to students and their families because they work with other agencies in California. With these partnerships, the Full Service School Program is able to provide parenting classes, childcare, welfare-to-work family case management, mental health care, and many other services. Without these partnerships, the services would be limited to basic health care.

Case management work also involves trying to improve services, with an emphasis on meeting client needs. This often means evaluating outcomes or periodically reviewing services and cost effectiveness. The outreach program in Missouri changes its service delivery after it collects data about the success clients have six months and one year out. Establishing measurable outcomes helps increase the effectiveness of service delivery.

 # Deepening Your Knowledge: Case Study

The Deepening Your Knowledge: Case Study section of this chapter is presented as an ongoing story. During our discussion of organizational factors, you will meet Carlotta Sanchez, who works for the Sexual Assault Crisis Center, an agency located in a city of 400,000. She has just begun her responsibilities as a case manager for the agency. Along with Carlotta, you will learn about the organizational structure of the Sexual Assault Crisis Center, the budgeting process and her role in it, the agency's informal structure, and the organizational climate in which the center's work is performed. You will also read about her participation in agency programs designed to improve the quality of services.

For each section of the chapter, you should be able to accomplish the following objectives.

UNDERSTANDING THE ORGANIZATIONAL STRUCTURE
- Name three ways that knowledge of organizational structures benefits a case manager.
- List some sources of information about an organization.
- Differentiate between formal and informal structure.

MANAGING RESOURCES
- Trace the planning and budgeting process of an organization.
- Identify other ways in which the case manager is concerned with resource allocation.
- Name the components of a budget.
- List sources of revenue.

IMPROVING SERVICES
- Identify four processes for improving the quality of services.

 # Understanding the Organizational Structure

To be an effective case manager, one must understand the organization and its structure. This knowledge is helpful in three ways. First, it gives the case manager a better understanding of his or her job responsibilities and how they relate to the goals of the agency and the specific objectives of the unit. Second, case managers can help meet client goals more easily if they use agency procedures correctly. Finally, once case managers understand the organizational climate, their work can fit appropriately into the context of the agency. They know what is expected of them, how much autonomy they have, and who can help with the difficult situations they face. Case managers encounter obstacles in any setting; knowledge of the particular working environment helps to identify barriers and develop strategies to cope with them.

The Organization's Plan

Several documents may shed light on the way an agency is structured. A certificate of incorporation defines the legal parameters of an agency's responsibilities and actions. In addition, agency bylaws provide process and procedures for agency action (National Council of Nonprofits, 2012). In other words they operationalize agency practice, including the daily activities of the case manager. Another document is the agency's mission statement. Other important statements are its goals, objectives, and policies and procedures. As you examine these documents it is helpful to look for the following information:

Does the organization have a mission statement? What does it say? What does it mean?

Can you identify the values embedded in the mission statement? Are they explicit or implicit?

Do the values in the mission statement influence how case management is delivered to clients?

Does the agency have publicly stated goals and objectives? What are they?

How do these relate to the work of the case manager?

Where can you find the policies and procedures of the agency? What do they say? Do you understand them?

What impact do these policies and procedures have on the case manager?

Let's look at one of the documents described to see how the document influences case management responsibilities and actions. Emphasizing values, a **mission statement** provides a summary of the guiding principles of the agency. It usually states the broad goals of the organization very concisely and describes the populations intended to benefit from the work of the organization. It may also specify the values that guide all decisions, the agency structure, sources of funding, agency priorities, and the work of the staff (Covey, 2012). The mission statement for the National Institute of Mental Health, a federal agency responsible for research in mental health and mental illness, follows. Notice that it is succinct, has a clear focus on research, and targets the public as the recipient of services.

> The mission of NIMH is to transform the understanding and treatment of mental illnesses through basic and clinical research, paving the way for prevention, recovery, and cure. For the Institute to continue fulfilling this vital public health mission, it must foster innovative thinking and ensure that a full array of novel scientific perspectives are used to further discovery in the evolving science of brain, behavior, and experience. In this way, breakthroughs in science can become breakthroughs for all people with mental illnesses. (http://www.nimh.nih.gov/about/index.shtml)

Each organization has a set of policies and procedures, some of which pertain to the behavior of all employees. Such documents are often very specific, describing procedures in great detail. Many reflect a standard of practice that is determined by the agency, as well as legal requirements established by the federal, state, and local governments. Also included are standards of practice established by a professional code of ethics or by a professional accrediting body. For example, an adjustment training center in South Dakota has a policy on the composition of the review board that approves applicants for services: It must include the executive director, a service coordinator, a nurse, and one representative from each of the other three units of service. At many agencies, the speed with which clients move through the system is regulated by stated policies and procedures. An Arizona health system has such a timeline. "Once we determine the eligibility of our clients, we need to see them before a 48-hour period of time elapses. We want them to become clients, and schedule a meeting with a case manager by the end of five working days. This meeting is face to face and used to develop a treatment plan of care."

Once the case manager has a clear picture of the agency's direction, guidelines, and rules, it is important to know the job description, the job's relationship to the work of the department or unit, and the agency's expectations. Case managers should ask the following questions of their supervisors or the people who are responsible for hiring new employees.

Is there a job description of a case manager? Where is it? Will you review it with me?

Do we have a list of goals for our department? How does these relate to the
goals of the agency?

How does case management fit into those goals and objectives?

Does our working unit have policies and procedures that are unique within
the agency context? Which are the most difficult policies and procedures
to follow?

Do these policies dictate how a case manager functions?

Rarely does the description of a position accurately reflect the actual work,
but it does serve as a guideline. The **job description** does define the work for
which the case manager is held accountable, but it often changes with
reorganization, economic pressures, and the changing needs of the client
population. One case manager describes how her job as case manager has evolved.
"Our agency was small when I began my case management work. We had a director who also was a case manager and then I was the full-time case manager . . .
since the setting was a hospital I want working with every unit in the hospital . . .
and clients were sent to us by the doctors . . . since that time, because of government funding for health care, our office has expanded to twenty individuals and
all patients need to have a plan when they leave the hospital. . . . Right now my
job has completely changed even though my job title is still the same."

Structure of the Organization

Another way to understand the structure of the organization or agency is to discover
the relationships among people and departments. The following questions are helpful.

Who is in charge of the agency?

Who makes the policies and procedures?

Is there a reporting structure? What is it?

Is this the evaluation structure? If not, what is the evaluation structure?

Who supervises whom? What authority does a supervisor have? To whom
does the supervisor report?

How does information get communicated in the agency?

In applying this information to your work as case manager, consider how the
relationships define authority, accountability, and communication. The following
questions will be helpful.

Who is my supervisor?

Does my supervisor evaluate me?

Am I supervising anyone?

Do I evaluate those I supervise?

Are there any policies and procedures related to evaluation? What are they?

What is the best way for me to gain information about agency work?

If I have information how to I convey this information to others? Who should
receive information for me?

Two terms important in understanding the structure of the organization are *authority* and *accountability.* **Authority** refers to assuming responsibility for resources and action. The lines of authority in the organization become clearer when departments are examined. Responsibility to others in the organization for what you do and how you use resources is **accountability**. Simply stated, a major component of authority is holding individuals accountable for the jobs they perform. Each person in authority also has to be accountable for that which he or she controls. Accountability also includes responsible fiscal management, external reviews, budgets, services provided, and outcomes for clients.

The **chain of command** of an agency stretches from the position with the most authority to the one with the least. Policy information is often passed along the chain of command, from the top down.

At a mental health center in Tennessee, two case managers explain how the chain of command is used to change an implementation plan for a client.

> If a case manager is not in agreement with a treatment plan and wants to try to have a team reconsider the plan, the first step is to contact the treatment team leader. This person is also the primary clinician. Since we work with clients who have mental health issues, the clinician understands the diagnosis and client needs. If the clinician approves the request for services for the client, then the information is shared with the treatment team, that is, the day treatment team. The coordinator of day treatment would also conduct an assessment and then implement the new intervention or services.

Many layers of professionals may be involved in making this decision, each person having authority for part of the decision and accountable for that part.

A document that is helpful when determining the chain of command is the **organizational chart**. (**See Figure 10.1.**) This chart is a symbolic representation of the lines of authority and accountability, as well as the information flow within an organization. An organizational chart usually consists of an arrangement of boxes that represent offices, departments, and perhaps individuals.

The boxes at the top of the chart represent the positions with the most authority, supervisory experience, and control of resources. Boxes that are connected with solid lines represent departments, offices, or individuals for which there is a formal communication pattern. For example, any communication from the executive director would go to the four directors of the agency.

Boxes that are connected with vertical solid lines represent lines of authority. The executive director is responsible for the entire organization and is accountable to the board of directors. The director of programs and services supervises the associate director of counseling and the associate director of day programs. These two associate directors in turn supervise the counselors and the activities staff. Boxes connected by dotted lines represent departments, offices, or individuals who communicate with one another but do not have supervisory authority or control of resources. For example, all the directors communicate with one another as they plan and implement the work of their departments.

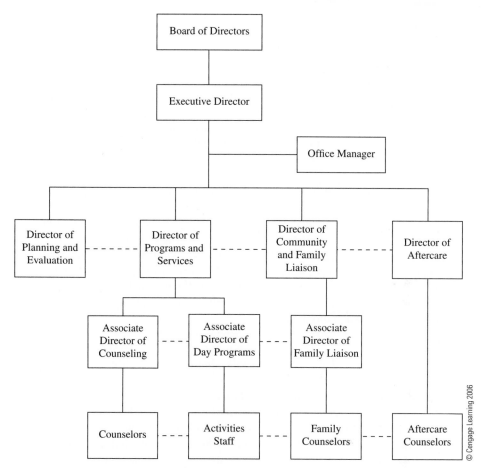

Figure 10.1   Organizational chart

Final authority in many human service agencies and organizations rests with the board of directors. The primary responsibility of the board is the financial health of the organization. Under normal conditions, board members do not work directly with supervisors, staff, or clients, but they do interact with the highest-ranking individual in the organization. The board focuses much of its attention on the budget (National Center for Nonprofit Boards, 2012). Members of the board analyze statistics such as the characteristics of clients served, the staff–client ratio, the sources of funding, and the relationship between expenditures and outcomes. The board is also involved with public relations, helping the agency establish its mission and goals, and working for a positive public image.

Let's see how Carlotta Sanchez is adjusting to the organizational structure of her agency.

Carlotta has been in her job as a case manager for six months now. It has been quite a learning experience for her. She believes that she was hired so quickly in part because she had worked as an intern with a women's shelter during her senior year. Carlotta is not really sure how many women she has been able to help in the past six months, but she knows that she has been learning a lot. She is finally able to work with clients and families without constantly asking for help from her friend Sally, who is also a case manager. Sally has been in her job for three years and was a volunteer before that, so she understands the work of the agency well.

During the first month, Carlotta had difficulty in understanding how to prioritize her work. She was coming in an hour early and staying two hours later than the regular staff just to complete the work she thought she should be doing. During those first months, everyone seemed to think they knew what she should be doing, but Carlotta herself was confused about her role. Her supervisor, Ms. Ludens, who oversees the work of the five case managers, is rarely available to talk with Carlotta about the job, although she is pleasant and supportive. At the end of that first month, Sally pulled Carlotta aside and asked if she needed any help.

Carlotta asked Sally if she could help her plan her work schedule. Carlotta also called her college advisor to talk about the difficulties she was experiencing. Both Sally and her teacher asked Carlotta several questions that she could not answer, ones that focused on the agency's policies and procedures. Carlotta barely remembered her orientation, which had consisted of two days of lectures on agency policy and rudimentary training in working with the target clients (primarily women who had been victims of sexual assault). During that orientation, she also received information about the legal system. At that time, everything was so new that she didn't know what questions to ask.

Carlotta asked whether she could attend orientation again the next time a new case manager was hired. She reread the mission statement and examined the goals and objectives. Some of them looked really familiar, but there were others that she did not remember, and she was not sure the agency ever addressed them in its programming. According to the mission statement and the goals and objectives, the Sexual Assault Crisis Center is dedicated to prevention and education. Carlotta had never contributed to those goals, and she did not see such activities going on in any other unit. Everyone seemed to be too busy working with clients who had already been assaulted or who were in danger.

After reading the policies and procedures, she was confused by discrepancies between what they said and what her colleagues did. She noted one particular serious violation of policy. All case managers are to follow a timeline after receiving a referral. Intake is to occur within forty-eight hours and is followed by a staffing, but Carlotta has never been able to meet that deadline. She talked to Sally about her questions and concerns. Sally just looked at her and told her to ask the supervisor; Sally would not say another word. Carlotta thought that was a strange reaction, so she decided to ask Manuel, another case manager; he had never heard of the policy that she was questioning. She decided that Ms. Ludens must be the person to turn to.

Carlotta scheduled a meeting with Ms. Ludens. She came up with a list of questions that concerned her, including her job in relation to the mission, the timeline that she had not been able to follow, and her need for more supervision. In preparing for the meeting, she found an organizational chart that showed a Director of Case Management, reporting to the Executive Director.

Carlotta felt that the meeting with Ms. Ludens went well, although she still had unanswered questions afterward. Ms. Ludens very carefully reviewed the job description for the case manager and prioritized the responsibilities of case managers. Carlotta discovered that Ms. Ludens expected her to conduct intake interviews, coordinate a team that included all the professionals working with a client, work with the family, and provide long-term yearly follow-up with clients. She did not expect her to participate in educational or preventive activities. At times it seemed that Ms. Ludens did not understand the job of the case manager. Ms. Ludens provided a curt answer to the question about the mysterious Director of Case Management: "We no longer have that position." She gave Carlotta an equally short response when asked about the timing of intake and staffing: "That is not my policy."

Although she was still unclear about the conflict between policy and practice, Carlotta left the meeting knowing a bit more about the organization.

An organization's planning documents, the written job description for a case manager, and the chain of command (as illustrated by an organizational chart) give a picture of the formal structure. However, as in Carlotta's experience, the formal organization does not give any case manager all the necessary information about the agency. Each agency also has an informal structure that influences the work that is done and how it is accomplished.

The Informal Structure

The **informal structure** of an agency supports many of its functions. The agency's informal structure also helps meet the needs of supervisors, staff, and case managers. The case manager must distinguish between the formal structure and the informal structure. The major differences between the two are as follows (Tsai, 2002).

FORMAL	INFORMAL
Who is the documented supervisor?	Who provides informal advice about the work?
What responsibilities are listed in the job description?	What has the job evolved to be?
What are the communication lines of the agency?	How does communication actually occur?
What is stated policy?	How is policy followed?

An informal structure develops within every organization as a way of meeting the needs that are unmet by the formal structure. It enables workers to use relationships and alternative ways of getting the job done. This structure facilitates the work of the case manager and helps meet social and professional needs; coworkers celebrate successes and support each other in times of trouble. They form a network of support and cooperation based on mutual respect and long-term relationships. As discussed earlier, such a network helps the case manager locate services and monitor the progress of clients.

The informal structure of an agency typically evolves long before the case manager assumes responsibilities; he or she then becomes a part of that structure. For example, through informal channels, coworkers give the new case manager advice on defining client problems, addressing issues of client motivation, finding services, and using agency policy to help meet client needs. Often, coworkers provide support to each other when one of them encounters difficult clients or stressful situations, and they rejoice together when their work goes particularly well. An interesting change in how agencies communicate reflects the growth of electronic communication and social media. Now members of an organizational community view each other both formally and informally within this new context. Jue, Marr, and Kassokatis (2009) suggest that tools such as wikis and blogs, and social media such as Facebook and Twitter, provide opportunities that are "rapidly changing how we build professional relationships and work collaboratively." Many organizational specialists see how technology can improve an organization by facilitating communication, involving more individuals in decision making, and making the workplace more personal (Jue, Marr, & Kassokatis, 2009; Sjoberg, 2010). In Chapter 4 we discussed issues related to technology and guidelines for operating in that environment. Although those guidelines are relevant to this discussion, we emphasize careful attention to organizational and personal communication practice.

At one outreach program, staff members always go out in pairs on bikes or in the van. "It helps for us to know our clients if we meet them on the streets. That way we can tell them about the program." This is a particularly helpful strategy when working with clients outside the office.

The informal structure also involves a different pattern of communication flow. More individuals in the organization thus receive useful information outside of channels. Having access to information supports the decision-making process. For example, case managers may hear about changes in policy or procedure or budget cutbacks before they happen. Advance notice gives them time to adjust

Want to Know More? Organizational Social Media Guidelines

As you know, social media and other Web-based communication are important ways in which we communicate with others. The strengths of communicating using technology within an organizational context include ease of access, quick response time, informal sharing of information, and creative problem solving. However, using the technological tools requires thoughtful practice. Use the Internet to conduct a search related to organizational social media policies. Find at least five policies and compare and contrast them. Which guidelines do you believe are the most helpful?

their work for the coming changes or seek to modify them, serving as advocates for clients' interests.

Although the informal structure contributes to the strength and the success of an organization, it can also contribute to difficulties. For example, the informal structure tends to represent tradition, preservation of the status quo, and resistance to change. This hinders case managers from advocating change to help meet the needs of their clients. In addition, communication within the informal structure may be unreliable or even hurtful. Such informal information must be considered carefully and not always regarded as truth. Several roles, such as advocate, broker, and planner, depend on the case manager's status as a representative of the agency. Passing on unreliable or inaccurate information or making decisions based on such information may damage his or her professional credibility. This type of communication is made more difficult within the context of social media. There is little opportunity for discussion and often is complicates rather than clarifies understandings.

Since there is no official map or diagram of this informal structure, case managers must construct their own. Sometimes case managers make diagrams of the interaction patterns that they observe. The following questions about one's own case management activities provide insight into one's interaction with the informal structure.

Which colleagues provide me with information?
Which colleagues do I help?
Who is my supervisor?
Whom do I supervise?

There can be many combinations of answers to these questions. Drawing a chart based on the answers and then comparing it with the organizational chart can help illuminate the differences.

One case manager, who works in a day care setting for the elderly, describes an informal way of sharing information about clients with other staff that begins the moment the clients arrive at the agency. "The individuals who transport the clients to and from the day care provide us valuable information about our clients, based upon their morning or evening observations. Sometimes the clients are not feeling well. At other times they talk about the difficulties of the evening before. As the case manager, I make sure I follow-up with these informal reports." This every-day exchange of information is very different from the formal weekly staff meetings, care plan meetings, or the charting that is completed at the end of each day.

Another helpful way of looking at the organization is to analyze the organizational climate.

The Organizational Climate

The conditions of the work environment that affect how people experience their work comprise the **organizational climate**. Such a climate is difficult to describe, for it is based on the values, attitudes, and feelings of people in the work setting. Case managers need to understand the climate so they can clarify their work expectations and their degree of autonomy. Ultimately, the organizational climate influences how they perform their jobs and how they relate to their clients. Several factors, such as policies and procedures and supervision, influence the organizational climate.

As discussed earlier, each organization has a set of policies and procedures that guides the behavior of employees. These can provide different messages to employees, indicating one or more of the following (Robbins, 2012):

1. We trust you will do a good job if we supervise you closely.
2. We trust you to do a good job because you want to do a good job.
3. We trust you to do a good job because you are loyal to the agency, which is like a family.

Case managers within the first, highly supervised environment have job responsibilities that are very clear. The policies and procedures are narrowly defined and well articulated. In the second, less structured environment, policies and procedures allow more autonomous decision making, and case managers have less rigid job descriptions and responsibilities. In the third, "family" framework, an agency invests much time and energy in its case managers and expects loyalty from them in the form of long-term commitment and a focus on excellence.

How a case manager is supervised also contributes to the organizational climate, and the supervision is not always consistent with the agency's stated policies and procedures. The supervisor can establish a tone that encourages independent thinking and activity, or he or she can narrow the range of responsibility and decision making. Even within an organization that has strict policies, the supervisor can foster a less rigid climate. For example, if the supervisor encourages autonomy, case managers may have the freedom to determine their own schedules, make final recommendations for client treatment, and establish agreements with other organizations, all without prior approval. Two case managers who work for the same agency described how the supervision structure of their agency affects their work: "We share an office and we divide our responsibilities in terms of who is the coordinator, who is the supervisor, and who are the case managers . . . although these roles seem clear and well-named, it is a very loosely governed office. . . . The agency gives to us lots of flexibility to use our own style in working with clients and families. Although it is a new program, throughout the state there are marked individual differences between each of the five state programs."

Let's go back to the organization in which Carlotta works.

After Carlotta met with Ms. Ludens, she decided that she really needed to work more closely with her colleagues if she was going to improve her performance and help her clients more effectively. Sally had been very helpful, and Carlotta decided that she would also work more closely with Charlotte Jones, the legal liaison for the center. Ms. Jones has been with the agency for ten years and has a wealth of knowledge about the law, relations with the police and the community, and the welfare of clients. In staffing, everyone listens when she makes recommendations. When she seeks Ms. Jones's advice, Carlotta noticed that many of her colleagues were doing the same. It also seemed that Ms. Jones was willing to answer any questions that Carlotta asked.

Carlotta broached the subject of the forty-eight-hour turnaround from intake to staffing. She honestly felt that she was not doing her job when she had to schedule staffing after the forty-eight-hour period. Ms. Jones smiled when she heard Carlotta's question, and she answered it straightaway: "So you have noticed. Let me tell you why we have that policy. Our board of directors feels that forty-eight hours is a reasonable time to begin to work on a case. They also believe that we have not really begun our work until the staffing takes place. We have explained many times why intake is often a longer process than forty-eight hours, but they refuse to change the policy. Our last director lost her job over that battle. Ms. Ludens just refuses to acknowledge the policy. We have all decided to do the best we can. We know the policy; we try to follow it. Sometimes it is just not possible."

Carlotta also asked Ms. Jones about the Director of Case Management position that she had seen in the organization chart. Ms. Jones shook her head before she spoke. "That was another battle we lost with the board. Our budget was cut two years ago, and the board recommended that we reduce our indirect personnel costs. They wanted us to put most of our resources in direct services to clients. With the difficulty of this work, the turnover for case managers is high, and training our new case managers is critical. Your questions are evidence of that need for supervision."

Several months later, Carlotta was meeting Sally for dinner after work. Carlotta had actually left work at the conclusion of office hours that day. She told Sally that she was beginning to understand her work better, and she said, "I am even feeling more at home in the agency."

Providing case management within the context of an agency is an exciting and challenging activity for Carlotta and other case managers. Another challenge is managing resources. The next section discusses how resource allocation influences the work of the case manager.

 # Managing Resources

Understanding the basic concepts of budgeting is important for a case manager. Much of the planning within the agency relates directly to the allocation of resources. Resource allocation determines the number of staff, the degree of operating support, the number of clients served, and the programs and services available for clients. Allocation of resources even covers the amount of time case managers spend on specific tasks.

Because costs are rising, and the number of clients who have complex needs is simultaneously increasing, there is an imperative to spend every dollar well. Allocating resources wisely requires an understanding of budgets and their applications to case management work.

What Exactly Is a Budget?

A **budget** describes agency income and expenditures. It is also a plan of action, representing anticipated expenditures during the next months or years. Budgeting has two purposes: planning and controlling. Sometimes the planning and controlling functions of a budget have opposing goals; planning represents the creative and expansive approach to human service delivery, whereas control is restrictive and measured. However, both are important facets of the budgeting process and have great relevance to the case management process.

Experts agree that planning and budgeting should be linked and that planning should guide the budgeting process. A human service agency must be clear about goals, objectives, and priorities before it begins budget preparations. A two-step process links planning and budgeting. The first step is to set up three categories: goals and objectives, planned activities, and the costs of those activities (The World Bank, 2009). For example, an initial part of the budgeting process for a training adjustment center was to submit a request to the state for "ten new case managers to coordinate the care of veterans applying for services." Planners at the center believe that without ten new case managers, the agency will not be able to meet the standard of care outlined by the federal government."

The second step in the budget process considers the goals and objectives and the resources available to provide the services. Revision of the budget then occurs within the parameters of the available resources. The result of the second step is a budget with costs revised to link the goals, objectives, programs, and services to the funds actually available. This process allows planning to determine the budget. Once the training adjustment center receives its allocation from the state, it will complete the second step of the planning and budgeting process—revising the goals for clients on the basis of the state's response to the request for ten new case managers.

The case manager's work in the planning and budgeting process is determined by the agency's policies and procedures. There is now a management trend toward participatory management and budget responsibility at the unit level, so case managers may be asked to provide input into the process (Mullahy, 2010). Because case managers have knowledge of client needs, departmental goals and objectives, and new professional trends in research and service delivery, they can offer helpful recommendations for priorities and resource allocation.

Many case managers are involved directly in budgeting for their individual clients. Some have responsibility for managing an account of services for clients; they generate and monitor a budget for each one of them. This includes defining needs, setting priorities, procuring services, and paying for services rendered. Usually this begins with a fixed allocation of funds and a list of services that may be purchased for clients who have a specified set of needs. Case managers help set priorities and determine how best to use the funds. A case manager at a local behavioral health system in Tucson has these responsibilities: "Each client has a limit on the amount to be spent. We study each client to make sure we provide cost-effective services. One of our guidelines is that the cost of home and/or community-based services can not exceed 75% of inpatient care costs."

Resource allocation extends to how case managers use their time. Time is a valuable resource of the case manager, who allocates it among various activities such as intake, planning, coordinating resources, monitoring client progress, advocacy, providing counseling, educating, interdisciplinary teamwork, building networks, and completing paperwork. Time management of each day is a resource decision, and the use of time should be linked to outcomes.

Another purpose of a budget is to establish control of expenditures. The budget provides a baseline from which to judge and project obligations. If an organization adheres to the budget during the year, the budget controls spending. It is important to maintain a balanced budget (one in which expenses do not exceed expenditures); this is one standard of good fiscal management.

The control function of the budget has a relationship to case management work in terms of how programs are implemented and how needs are met. Planning takes into account the resources available and any restrictions on their use. There may be policies governing what services clients are eligible to receive. For example, a counseling center in Texas has multiple funding sources: Social Services block grants, Title III, the Older Americans Act, and county and city funds. The agency's policy is to ask clients to pay for some of the services if they can. Limited resources may also restrict the number of clients that an agency can serve. Sometimes, planners must develop eligibility criteria that consider the prevailing limitations on resources.

Features of a Budget

A budget has several components: projected expenditures, actual expenditures, and balance. (**See Table 10.1.**) *Projected expenditures* are cost estimates made at the beginning of the budget cycle; they designate the money that has been set aside for expenses in a given category. *Actual expenditures* represent the money that has been spent in a given category to date. At any time in the budget year, it is clear what has been spent in each category. The *balance* is the amount allocated to a given category minus the amount already spent.

In Table 10.1, expenditure categories are listed in the left column.

Salaries are money paid to employees, either for providing direct services to clients or for providing indirect services such as administration and supervision.

Supplies are necessities for maintaining the activities of the agency. Usually, such items have a short life or are used within a year. Examples are paper, pens, gasoline, and essentials for repair and cleaning.

Equipment includes machines that are bought for a specific purpose, such as computers, a copy machine, and recreation equipment. These are usually one-time expenditures.

Travel represents the cost of transporting personnel and clients. Some agencies cover clients' travel expenses to and from sites where services are delivered. Case managers may make home visits or transport clients to other agencies. The travel budget is also used for purposes of professional development.

Training costs often include payment to consultants who conduct staff training for professional development, as well as the cost of certification classes, books, and professional journals.

Communications includes expenditures for using telephones, sending faxes, mailing, and any other expenditures incurred in transmitting information.

Total designates the whole amount that has been committed or spent.

TABLE 10.1   BUDGET FORMAT FOR A HUMAN SERVICE AGENCY

Category of expenditures	Projected expenditures	Actual expenditures	Balance
Salaries			
Supplies			
Equipment			
Travel			
Training			
Communications			
TOTAL			

Often the budget is simplified by using just three categories: salaries, operating, and capital. The *salaries* category is the same as that for the budget presented in Table 10.1. *Operating* includes supplies, most equipment, travel, training, and communications. The *capital* budget includes major equipment and building projects.

Sources of Revenue

Funding for a human service agency or organization determines if an agency is public or not-for-profit (voluntary). **Public** or governmental **agencies** exist by public mandate and receive funding from one or more of the following sources: the federal, state, regional, county, and municipal government. State departments of human services or children's services are examples. Individual contributions, fundraisers, foundation grants, state and municipal debt securities, and corporate donations fund **not-for-profit agencies** or organizations. An elected board of directors also governs them. Today the distinction between public and not-for-profit agencies is not always clear, because not-for-profit agencies are increasingly providing services for public agencies on a contractual basis.

Recently, a third category—**for-profit agencies**—has proliferated. Reasons for their increase are the limited resources for voluntary agencies, reduced governmental funding, and changing political and economic times. Two functions guide service delivery by a for-profit organization: (1) providing a service and (2) making money. Managed care organizations, health maintenance organizations, and private corporations are examples.

Whether public, not for profit, or for profit, agencies and organizations often receive funding from multiple sources. Four main revenue streams fund human service programs: (1) federal, state, and local government-sponsored funding, (2) grants and contracts, (3) fees, and (4) private giving.

Federal, state, and local government funds   Government funds are available for programs and services. As in the case of Medicare and Medicaid, they may be direct reimbursements for client care. In the past, this was considered stable long-term funding. In today's climate of rising costs, balanced government budgets, and the diminishing role of government in human services, such funding is less secure. There is a movement to allocate financial support to local levels of government in order to increase accountability and to develop programs closer to where the problems occur.

Grants and contracts   Many agencies write proposals for grants and try to develop **contracts** to secure funding from both government agencies and private corporations. Staff members are sometimes assigned to write proposals to acquire grants and contracts. Sponsoring agencies often establish priorities and set restrictions on resource allocation. If human service agencies and their target populations are to benefit from such external funding, the priorities of the granting or contracting body must be similar to those of the agency.

Fees   Revenue from **fees** is generated in two ways. First, many clients have insurance that covers the costs of the services they receive. The insurance is managed through a third party, and the payment is made according to the policies of the insurance agreement. In the managed care environment, there may be restrictions on the range or quantity of services available. Fees for services are also collected from individuals who choose not to use their insurance, who have exceeded the amount the insurance will pay, or who are ineligible for government or insurance support. In such cases, fees are often assessed on a sliding scale: The higher the person's income, the higher the fees; the lower the income, the lower the fees.

Private giving   Private giving is becoming an important source of revenue. Human service agencies solicit individuals, businesses, and corporations for donations of money, equipment, and professional expertise. The current trend is to establish relationships with such donors in the hope that funding can be stabilized by a long-term partnership.

How an agency is funded influences the case management process. Government funding and insurance reimbursements have direct effects on the services provided. Any change in funding has the potential to expand, restrict, or alter services. Many programs have changed over time as funding patterns have shifted. For example, a parole officer who manages a caseload of adult parolees describes changes in an adjustment counseling program for parolees. "When I first began working, we had a therapist who focused on adjustment counseling for the paroles. We lost our funding. Now we are expected to support the adjustment process."

Case managers may be involved in grant writing or may serve as consultants to those who are making applications; working on grant-writing projects gives them an opportunity to speak for the needs of clients. In addition, case managers may be involved in fundraising. For many years, private fundraising was the sole responsibility of the board of directors and the executive director. As private giving becomes a more important part of agency revenue, case managers can expect to increase their participation in fundraising activities. The case managers at a family services agency in the Bronx solicit private donations to enhance the services they provide. For one particular program, clients receive an allowance to buy a bed and a kitchen set for a new apartment. With private donations, caseworkers are able to get their clients additional furniture. Many clients are grateful for this extra support. According to a case manager, one client expressed her appreciation. "She was so happy to be able to have her apartment completely furnished. She could just move it and begin living there immediately. It was the first time in her life she had been able to do this."

Carlotta Sanchez, the new case manager at the Sexual Assault Crisis Center, has little experience with budgets. Let's see how she begins to learn about the resource allocation process at the agency.

To Carlotta, it seemed that the more she talked with her colleagues, the more she discovered the importance of the board of directors. She knew that Ms. Ludens spent quite a lot of time with board members, and during the week of a board meeting, the whole staff was busy guessing what changes the meeting might bring for the agency's functioning. Ms. Jones had also told Carlotta that funding was very tight for the coming year. Carlotta had discovered that although the budget year for the agency began January 1 and ended December 31, budget preparations began very early in August.

Carlotta still did not have a supervisor, and she had no way of getting key information about the agency unless Ms. Ludens spoke at the large staff meeting or sent an "all-staff" memo. Sometimes, Carlotta heard news from the other case managers or Ms. Jones. By the middle of the summer, several of her colleagues were expressing worry about the budget for the coming year. Several funding sources were cutting back, and the staff had no way of guessing how the agency would be affected. The state government had experienced a shortfall for the second year in a row and was talking about reducing support to social services. United Way had increased the list of agencies it was supporting and therefore might decrease its funding to the center. Donations from the public sector were also down from the previous year.

In September, Ms. Ludens called the entire staff together and asked them to prepare the budget for the coming year. They were to consider three scenarios: a budget increase of 2%, a budget decrease of 2%, and a budget decrease of 4%. She established a budget team and asked them to describe what impact the increase scenario and the two decrease scenarios would have on programming and staffing. She also asked the budget team to consult with everyone in the agency. Carlotta was chosen to represent the case managers on the team.

Carlotta learned a lot during the two months she worked with the budget team. They talked about priorities, the cost of programs, the target population, and how to set and measure expected outcomes. Carlotta was amazed at how her own responsibilities would change in each scenario. It was a stressful period, and involving the entire staff in the budget planning took a lot of time and effort. The group had the three scenarios ready for Ms. Ludens by the deadline, and she prepared to present them to the board at its meeting the following week. The whole staff was holding its breath; the actual budget total would not be available until final word about state funding arrived.

One other effort related to organization influences the roles and responsibilities of case managers in the delivery of human services. The next section discusses how agencies involve their case managers in the process of improving services.

 ## Improving Services

In recent years, there has been an increasing emphasis on providing quality services. This has resulted in the use of four processes: linking outcomes to cost, conducting a utilization review, planning quality assurance programs, and promoting continuous improvement. Each process addresses the issue of quality in a different way.

What Is Quality?

Quality is a term that is very difficult to define. In *Zen and the Art of Motorcycle Maintenance,* Pirsig (1974) eloquently expressed the dilemma this way:

> Quality . . . you know what it is, yet you don't know what it is. But that's self-contradictory. But some things are better than others, that is, they have more quality. But when you try to say what quality is, apart from the things that have it, it all goes poof! There's nothing to talk about. But if you can't say what quality is, how do you know what it is, or how do you know it even exists? (p. 163)

In the fields of health and mental health care, *quality* is understood to mean that the delivery system meets its objectives with a particular emphasis on client needs (Department of Health and Human Services, 2011). Defining and measuring quality is only one component in providing quality services. There must then be an identification of any problems and a corresponding change in service delivery. This section presents two approaches to defining and measuring quality: linking objectives to outcomes and conducting utilization reviews. We also discuss two approaches to improving quality: conducting quality assurance programs and implementing continuous improvement programs.

One way to identify the quality of services provided is to evaluate the outcomes of programs in relation to the resources spent. To do this, programs and the necessary resources for implementation must be weighed against expenditures. Outcomes such as the number of clients served, the programs delivered, and client progress are matched to actual expenditures. Experts in evaluations have suggested areas that need to be explored (Department of Health and Human Services, 2011; World Bank, 2009):

Look at the objectives of the organization.
Outline what programs are linked to each objective.
Describe the cost of each of the programs. Include both human and financial expenditures.

Voices from the Field: Department of Health and Human Services Commitment to Quality

News Release

FOR IMMEDIATE RELEASE 21, 2011   Contact: HHS Press Office
(202) 690-6343

National Quality Strategy will promote better health, quality care for Americans

Created under the Affordable Care Act, first-ever strategy will guide local, state and efforts to improve quality of care

The U.S. Department of Health and Human Services (HHS) today released the National Strategy for Quality Improvement in Health Care (National Quality Strategy). The strategy was called for under the Affordable Care Act and is the first effort to create national aims and priorities to guide local, state, and national efforts to improve the quality of health care in the United States.

"The Affordable Care Act sets America on a path toward a higher quality health care system so we stop doing things that don't work for patients and start doing more of the things that do work," said HHS Secretary Kathleen Sebelius. "American hospitals, doctors, nurses and other health care providers are among the best in the world. With this ground-breaking strategy, we are working with local communities and health care providers to help patients and improve the health of all Americans."

The National Quality Strategy will promote quality health care that is focused on the needs of patients, families, and communities. At the same time, the strategy is designed to move the system to work better for doctors and other health care providers—reducing their administrative burdens and helping them collaborate to improve care. The strategy presents three aims for the health care system:

- **Better Care:** Improve the overall quality, by making health care more patient-centered, reliable, accessible, and safe.
- **Healthy People and Communities:** Improve the health of the U.S. population by supporting proven interventions to address behavioral, social, and environmental determinants of health in addition to delivering higher-quality care.
- **Affordable Care:** Reduce the cost of quality health care for individuals, families, employers, and government.
- To help achieve these aims, the strategy also establishes six priorities, to help focus efforts by public and private partners. Those priorities are:
- Making care safer by reducing harm caused in the delivery of care.
- Ensuring that each person and family are engaged as partners in their care.
- Promoting effective communication and coordination of care.
- Promoting the most effective prevention and treatment practices for the leading causes of mortality, starting with cardiovascular disease.
- Working with communities to promote wide use of best practices to enable healthy living.
- Making quality care more affordable for individuals, families, employers, and governments by developing and spreading new health care delivery models.

(continues)

(continued)

The strategy was developed both through evidence-based results of the latest research and a collaborative transparent process that included input from a wide range of stakeholders across the health care system, including federal and state agencies, local communities, provider organizations, clinicians, patients, businesses, employers, and payers. This process of engagement will continue in 2011 and beyond.

The National Quality Strategy is designed to be an evolving guide for the nation as we continue to move forward with efforts to measure and improve health and health care quality. HHS will continue to work with stakeholders to create specific quantitative goals and measures for each of these priorities. In addition, as different communities have different needs and assets, the strategy and HHS will empower them to take different paths to achieving these goals.

The National Quality Strategy is just one piece of a broader effort by the Obama Administration to improve the quality of health care, and will serve as a tool to better coordinate quality initiatives between public and private partners. For example, the Affordable Care Act established a new Center for Medicare and Medicaid Innovation that will test innovative care and service delivery models. These new models are being tested to determine if they will improve the quality of care and reduce program expenditures for Medicare, Medicaid, and the Children's Health Insurance Program (CHIP).

SOURCE: Department of Health and Human Services. (2011). National quality strategy will promote better health quality care for Americans. Retrieved from http://www.hhs.gov/news/press /2011pres/03/20110321a.html.

The results of this analysis are helpful to a case manager, for one of the most frustrating elements of case management is the difficulty of determining one's own effectiveness and efficiency (Lewis, Lewis, Packard, & Souflee, 2012). If case managers can begin to link their activities to the objectives of the organization (progress made by the client) and to the cost of the service delivered (the costs of direct service to the client plus the cost of any support services), they begin to see the relationship between their work and the benefits to their clients.

One type of quality analysis consistently used in the managed care environment is utilization review, which has been developed to oversee service provision and monitor costs. It is a response to accusations that managed care substitutes cost effectiveness for quality care. In the area of mental health, managed care has also been criticized for underfunding necessary treatments. Let's examine utilization review and its impact on the case management process.

Conducting a Utilization Review

Peers within the human service delivery system conduct a utilization review. Two approaches are used during utilization review: pretreatment review and second-opinion mandates (Mullahy, 2010). In **pretreatment review**, a managed care professional must review and approve a treatment plan before service delivery. **Second-opinion mandates** apply to certain diagnoses; consultation with a second professional must take place before a treatment plan can be approved. Utilization review is designed to determine the proper treatment for each of the client's problems or conditions, the appropriate course of the treatment process as a whole, and how the submitted plan must be revised.

Utilization review is often applied because the treatment is provided on a prepayment basis. In other words, since limited funds are allotted for each client, it is critical to assess client needs and the cost of relevant services before providing such services. The responsibility of people who perform utilization reviews is to provide for the treatment measures that represent the best use of resources in light of the anticipated outcomes.

In many managed care PPOs, health care and mental health care are provided to employees of a corporation, or to members of a designated group at discounted rates, in return for a high volume of clients. For cost-effective management, the diagnosis and treatment regimen must be shortened. The ultimate goal is to spend the least amount of time and provide the lowest-cost treatment that restores the client to good health or mental health (Mullahy, 2010). Another goal is to keep professionals out of legal difficulties. To meet these goals, another process has developed, one that emphasizes quality care. The process of quality assurance documents that the care provided has met quality standards (Department of Health and Human Services, 2011).

Planning Quality Assurance Programs

The focus of **quality assurance** programs in the human services is on developing standards of client care. Relevant to the work of the case manager are two quality-assurance measures: the standard treatment plan and the client satisfaction survey.

An agency develops a standard treatment plan to ensure common practices in intake, assessment, goal setting, planning, implementation, and evaluation of outcomes. Agency practice must establish standards of care for common assessments within the scope of the agency's activities. The work of the case manager and others must then be compared with the standards established. Thus, each case manager provides the same set of services for each client. When services do differ, reasons must be provided. For the quality assurance process to be of real value, the case manager must be able to link assessment with both implementation and positive outcomes.

Client satisfaction is another component of quality assurance programs. As a dominant criterion for evaluating their work, professionals view it both positively and negatively. Some would question the client's competence to judge the quality of services. On the other hand, there is evidence that clients who are pleased with their treatment are more likely to follow the plan and to be straightforward in reporting whether their goals have been met. Because case management aims to empower clients and increase the extent of their responsibility for their own lives, client satisfaction can be considered a key element in providing quality service, even though this is only one source of quality assurance data.

Figure 10.2 shows how client satisfaction is measured by one agency, in this case a provider of adolescent substance abuse treatment services, that uses service coordinators to manage cases. Investigating satisfaction is difficult; it requires an atmosphere in which clients believe that someone wants to hear what they have to say. It is also helpful to hear from clients who terminate services prematurely or whose treatment proves unsuccessful. The identification of problems in service delivery is beneficial to agencies that value quality assurance, for this feedback can serve as the basis for needed improvements. Unfortunately, such clients are sometimes difficult to reach.

The final approach to improving the quality of services is **continuous improvement** (also known as total quality management). This movement began in industry but is fast becoming an accepted part of quality assurance in the human service delivery system. Supervisors and case managers each have a role in the continuous improvement process. Supervisors are charged with providing leadership that encourages all staff to participate in improving service delivery; case managers are expected to provide quality work and to collaborate with others in a spirit of cooperation and teamwork.

Early in the quality assurance movement, Edward Deming (1986), a leader in the total quality movement, suggests several guidelines to be observed if work toward quality in an organization is to succeed. First, note that *all* processes and procedures can be improved, so each component of the process is subject to scrutiny and possible change. Another guideline of Deming's is that customer needs (in this case, those of clients) must be considered. A third component of continuous improvement is the requirement of teamwork across units and departments to solve problems and improve outcomes. Clearly, this goal is consistent with the commitment of the case manager to work effectively with colleagues.

If an agency is participating in a total quality program, the case manager may be asked to conduct peer reviews or join a quality circle. In a *peer review* system, professionals establish a process for evaluating one another. Another feature of continuous improvement is the *quality circle*, in which those who serve on case management teams work together to solve problems.

Still another area of focus for the continuous improvement approach is the management of time (mentioned previously in the section on managing resources). Managing time well means knowing how time spent is related to priorities and expected outcomes. Good time-management techniques for the case manager are discussed in Chapter 11.

Let's revisit Carlotta as she becomes involved in her agency's efforts to institute continuous improvement.

1. What is your age?

   15 _____ 16 _____ 17 _____ 18 or older _____

2. What is your gender?

   Male _____ Female _____

3. Which best describes your race?

   American Indian/Alaska Native _____ Asian/Hawaiian/Pacific Islander

   Black/African American _____ White _____

4. What is your education level?

   High school graduate _____ GED _____

   Some college _____ College graduate _____

5. How long have you been in this treatment program?

   Less than 1 month _____ 1–3 months _____

   4–6 months _____ 7–12 months _____

   Longer than 1 year _____

6. Are you being treated for a mental health problem?

   I receive mental health services from this agency. _____

   I receive mental health services from another agency. _____

   I don't currently receive mental health services but I have in the past. _____

   I do not have a mental health problem. _____ I don't know. _____.

7. I value the services I receive here.

   Strongly disagree _____ Disagree _____ Neutral _____

   Agree _____ Strongly agree _____

8. If I had other choices I would still come to this case manager for services.

   Strongly disagree _____ Disagree _____ Neutral _____

   Agree _____ Strongly agree _____

9. I would recommend this case manager and this program to a friend or family member.

   Strongly disagree _____ Disagree _____ Neutral _____

   Agree _____ Strongly agree _____

(continues)

Figure 10.2   Quality assurance questionnaire

10. The services I receive here are helpful.

Strongly disagree _____   Disagree _____   Neutral _____

Agree _____   Strongly agree _____

11. I was able to get services quickly.

Strongly disagree _____   Disagree _____   Neutral _____

Agree _____   Strongly agree _____

12. The case manager responded to me within 24 hours.

Strongly disagree _____   Disagree _____   Neutral _____

Agree _____   Strongly agree _____

13. I was able to get services even though I could not pay

Strongly disagree _____   Disagree _____   Neutral _____

Agree _____   Strongly agree _____

14. Services are available at times that are good for me.

Strongly disagree _____   Disagree _____   Neutral _____

Agree _____   Strongly agree _____

15. The location of services is convenient.

Strongly disagree _____   Disagree _____   Neutral _____

Agree _____   Strongly agree _____

16. I am encouraged to participate in my care.

Strongly disagree _____   Disagree _____   Neutral _____

Agree _____   Strongly agree _____

17. Staff members help me get the information I need so I can take charge of managing my substance abuse problem.

Strongly disagree _____   Disagree _____   Neutral _____

Agree _____   Strongly agree _____

18. I feel comfortable asking questions about my treatment and medications.

Strongly disagree _____   Disagree _____   Neutral _____

Agree _____   Strongly agree _____

Figure 10.2   *(Continued)*

19. Case managers respect me and give me information about my choices.

    Strongly disagree _____   Disagree _____   Neutral _____

    Agree _____   Strongly agree _____

20. Staff members here believe I can grow, change, and recover.

    Strongly disagree _____   Disagree _____   Neutral _____

    Agree _____   Strongly agree _____

    SATISFACTION WITH PRESENT TREATMENT

1. I was given information about my rights as a client.

    Strongly disagree _____   Disagree _____   Neutral _____

    Agree _____   Strongly agree _____

2. Staff members respect those rights.

    Strongly disagree _____   Disagree _____   Neutral _____

    Agree _____   Strongly agree _____

3. Staff members act appropriate and professionally.

    Strongly disagree _____   Disagree _____   Neutral _____

    Agree _____   Strongly agree _____

4. Staff members respect my wishes about who can be given information about my treatment.

    Strongly disagree _____   Disagree _____   Neutral _____

    Agree _____   Strongly agree _____

5. Staff members are sensitive to my needs.

    Strongly disagree _____   Disagree _____   Neutral _____

    Agree _____   Strongly agree _____

6. Staff members are sensitive to my cultural/ethnic background.

    Strongly disagree _____   Disagree _____   Neutral _____

    Agree _____   Strongly agree _____

7. I decide my treatment goals with the help of my case manager.

    Strongly disagree _____   Disagree _____   Neutral _____

    Agree _____   Strongly agree _____

(continues)

Figure 10.2 *(Continued)*

8. The staff members I work with are helpful and knowledgeable.

   Strongly disagree _____   Disagree _____   Neutral _____

   Agree _____   Strongly agree _____

9. My case manager encourages me to talk about and work on many of the problems at the same time.

   Strongly disagree _____   Disagree _____   Neutral _____

   Agree _____   Strongly agree _____

10. I have a good relationship with my case manager.

    Strongly disagree _____   Disagree _____   Neutral _____

    Agree _____   Strongly agree _____

11. Right now I deal better with my daily problems than I did when I began my work with my case manager.

    Strongly disagree _____   Disagree _____   Neutral _____

    Agree _____   Strongly agree _____

12. Right now I feel better about myself than I did when I began my work with my case manager

    Strongly disagree _____   Disagree _____   Neutral _____

    Agree _____   Strongly agree _____

    RESULTS OF CASE MANAGEMENT

1. I am better able to handle the circumstances in my life.

   Strongly disagree _____   Disagree _____   Neutral _____

   Agree _____   Strongly agree _____

2. I am better able to deal with difficult situations.

   Strongly disagree _____   Disagree _____   Neutral _____

   Agree _____   Strongly agree _____

3. My mental health is improving.

   Strongly disagree _____   Disagree _____   Neutral _____

   Agree _____   Strongly agree _____

4. My financial situation is improving.

   Strongly disagree _____   Disagree _____   Neutral _____

   Agree _____   Strongly agree _____

Figure 10.2 *(Continued)*

5. My relationships with family and friends are improving.

   Strongly disagree _____ Disagree _____ Neutral _____

   Agree _____ Strongly agree _____

6. My housing situation has improved.

   Strongly disagree _____ Disagree _____ Neutral _____

   Agree _____ Strongly agree _____

7. I have a better understanding of my problems.

   Strongly disagree _____ Disagree _____ Neutral _____

   Agree _____ Strongly agree _____

8. I am meeting my goals for case management.

   Strongly disagree _____ Disagree _____ Neutral _____

   Agree _____ Strongly agree _____

9. On a scale of 1 to 10, what is your overall rating of the services you received/ are receiving?

   _____

Adapted from Adolescent Client Satisfaction Survey, Office of Substance Abuse, Maine Department of Health and Human Services. (2011). 2011 Client satisfaction survey. Retrieved from http://www.maine.gov/dhhs/osa/pubs/data/2012/CSS-2011-AdolescentStateReport.pdf

Figure 10.2 *(Continued)*

Once she had participated in the budget team, Carlotta's confidence increased, and she felt that she was able to add something of value to the agency. She began to look around for other committees that she might join. She felt that she was too new to the organization to join the Personnel Committee, and she was not very interested in fundraising and development activities (dealing with large crowds and planning events was not much fun for her). However, she was interested in the total quality management (TQM) program that the agency was beginning apply to its volunteer training. The TQM group included two volunteers, the volunteer coordinator, a board member, a client, and a member of the Mental Health Association. The group wanted a member who worked directly with clients, and they invited Carlotta.

In the beginning, Carlotta thought that she had never been to so many meetings. They had training on six consecutive Saturdays just to become familiar with the principles of TQM. Carlotta's father worked for a chemical company

in a neighboring state, and he had told her about his terrible experiences with quality teams. He thought that they were a waste of time and gave workers false hopes that things would get better. Carlotta was more optimistic, and she really enjoyed meeting and working with people she normally had little contact with. In the first sessions, the group learned about TQM and the progress it had fostered in other organizations. By the third Saturday, the group had started focusing on the volunteer program.

First, they had to establish the agency's mission, goals, priorities, stakeholders, and customers. Carlotta thought that the agency only had one customer—the client—but she learned differently. After much discussion, it was decided that the agency had many customers, among them the board, the citizens of the county, the local government, and the police in the city and the county. The list was much longer, but the point for Carlotta was that more people depended on her agency than she had ever realized.

The first six Saturdays were only the beginning of the meetings, and the group spent a lot of time outlining what happens in the agency, exactly how the work flows, and the role of volunteers in that work. They learned about the training and supervision of volunteers, their actual work experiences, how clients responded to working with them, and how the volunteers themselves felt about their work. They sent out a survey and then interviewed clients and volunteers about their experiences to determine what they valued about their experiences with the agency and how satisfied they were. Carlotta was not sure exactly how this work would turn out, but it gave her a different perspective, and she enjoyed working with these new colleagues. She came to be confident that all her work would help provide better conditions for volunteers as well as clients. Carlotta learned so much about her agency when she participated in the quality team effort. She began to have a deeper understanding of the organization, priorities, and client outcomes. The effort required time and effort, but she has a clearer sense of belonging to the agency and better understands her place in it.

 # Chapter Summary

Case managers work within an organizational context and must understand the essentials of agency structure, how to manage resources, and how to improve services. Understanding these fundamentals enables them to use the organization more effectively to meet client needs. They can then see more clearly how their work is integrated into the mission of the agency.

Many case managers are also expected to manage resources. A thorough understanding of the budget process and its effects on the resources available for clients helps case managers serve as effective advocates for their clients' interests. This includes knowing the sources of funding, distinguishing between public (governmental) and not-for-profit (voluntary) agencies, and understanding the impact of for-profit agencies on service delivery today.

Case managers are naturally committed to improving services to meet the changing times, resources, and client needs. Providing quality services occurs in four ways: linking outcomes to cost, conducting utilization reviews, planning quality-assurance programs, and promoting continuous improvement.

Chapter Review

Key Terms

Mission statement	Public agencies
Job description	Not-for-profit agencies
Authority	For-profit agencies
Accountability	Contracts
Chain of command	Fees
Organizational chart	Pretreatment review
Informal structure	Second-opinion mandates
Organizational climate	Quality assurance
Budget	Continuous improvement

Reviewing the Chapter

1. Why is it important for a case manager to understand organizational structure?
2. List sources of information about an organization and describe the type of information each provides.
3. How can a case manager find out about relationships within an organization?
4. Describe the relationship between authority and accountability.
5. What is the purpose of the informal structure? How does it operate, and how is it different from the formal organizational structure?
6. How does the organizational climate affect the work of the case manager?
7. Name the two steps that link planning and budgeting.
8. What role do resources play in preparing a budget?
9. Describe how budgeting for clients takes place.
10. In what other ways does resource allocation affect case management?
11. Name the components of a budget.
12. Design a budget for yourself for one week, using the categories discussed in this chapter.
13. How does an agency get money to support its work?
14. Distinguish between public, not-for-profit, and for-profit agencies.
15. Describe the two ways in which fees are generated.
16. How does an agency's funding affect case management?
17. How do linking objectives to outcomes and conducting utilization reviews relate to defining and measuring quality?
18. How do quality-assurance programs and continuous improvement programs contribute to better services?

19. Describe pretreatment reviews and second-opinion mandates and their roles in utilization review.
20. How does the use of standard treatment plans affect case management?
21. Discuss client satisfaction as an element in quality case management.

Questions for Discussion ·······························

1. Why do you think it is important to understand an organization's informal structure?
2. Can you provide a rationale for the statement, "Case managers should understand the process of managing resources"?
3. What do you think will happen as case managers learn more about how to improve services?
4. Do you think it is a good idea to involve clients in the process of improving services? Why or why not?

References ································

Covey, S. (2012). *The community: Business mission statements.* Retrieved from https://www.stephencovey.com/mission-statements.php

Deming, W. E. (1986). *Out of crisis.* Cambridge, MA: MIT Press.

Department of Health and Human Services. (2011). *National quality strategy will promote better health quality care for Americans.* Retrieved from http://www.hhs.gov/news/press/2011pres/03/20110321a.html

Jue, A. L., Marr, J. A., & Kassokatis, M. E. (2009). *Social media at work: How networking tools propel organizational performance.* San Francisco: Jossey-Bass.

Lewis, J. A., Lewis, M. C., Packard, T. R., & Souflee, F. (2012). *Management of human service programs* (5th ed.). Belmont, CA: Thomson/Brooks/Cole.

Mullahy, C. M. (2010). *The case manager's handbook* (4th ed.). Sudbury, MD: Jones & Bartlett.

National Center for Nonprofit Boards. (2012). *Effective boards.* Retrieved from http://www.nsba.org/sbot/toolkit/EfBoards.html

National Council of Nonprofits. (2012). *What is a membership nonprofit?* Retrieved from http://www.councilofnonprofits.org/resources/resources-topic/membership

Pirsig, R. (1974). *Zen and the art of motorcycle maintenance.* New York: Bantam Books.

Robbins, S. P., & Judge, T. (2012). *Essentials of organizational behavior* (11th ed.). Englewood Cliffs, NJ: Prentice Hall.

Sjoberg, L. (2010). *Social media in organizations: A review.* PsycCRITICS, 55(34). Retrieved from http://psycnet.apa.org/index.cfm?fa=buy.optionToBuy&id=2010-14384-001

Tsai, W. (2002). Social structure of competition within a multiunit organization: Coordination, competition, and intra-agency knowledge sharing. *Organization Science, 13*(2), 179–190.

The World Bank. (2009). *Linking the PRS with national budgets.* Retrieved from http://siteresources.worldbank.org/INTPRS1/Resources/383606-1106667815039/PRS_Budgets_GuidanceNote.pdf

CHAPTER ELEVEN

Thriving and Surviving as a Case Manager

Thriving and Surviving as a Case Manager: Time and Pace

Ellen

A typical day? I would pull into the parking lot and the clients would be waiting to get into the door when the building opened at 8:30. They would be waiting for me. . . . It would start before I got into the building. With our informal family-type environment, the clients knew they were safe there. When people who have different mental illnesses know they are safe, they are free to express things that they don't typically express in other places. So it was always exciting and interesting. We would have, on a typical day, 300 people in and out of the doors at some point during the 8:30 to 5:00 workday. I had a caseload of 100 people. And it was really fast paced.

So the clients are coming in and we have a certain number of doctor hours we have to structure. The entire operation is based on the clients seeing their doctor; that would drive who was coming in and out at any given point, and then we also had a core group of clients that came during the day every day. It was their place and so they'd be there at 8:30 and they'd come in and they would go back to the day treatment room and they would start the coffee and they would get the cereal out and make sure everyone had breakfast. There was a great sense of pride in the community there.

I was pretty much booked solid. I might have four or five scheduled appointments and then have certain periods of time set aside for unscheduled clients that come in. Maybe they've got something in the mail from Social Security and they need help figuring out what to do. There were nine case managers there and we all worked in the same office. It was a great big bullpen and so there were desks all around the perimeter of the office. It was one of the most fabulous working arrangements I have ever been a part of. At times it was frustrating because it got really noisy. Everyone was working, talking on the phone; but it was also an environment of support and care and compassion.

—Permission granted from Ellen Carruth, 2012, text from unpublished interview

In the last four years there have been a lot of cuts. And we need to serve more clients. And those clients have more problems. There is a limit to what we can do and what is manageable.

—Director and case worker, human services center, New York, New York

Culture is an important aspect of all of the work that we do with our clients. For example, we have individuals living here from Iran and Iraq. They don't deal with our culture but truly bring their own culture with them. For example, they have their own doctors and dentists, and they have a different approach to mental illness. And they want to stay within their own niche. They have their own shopkeepers and people who run the small businesses. They even have their own newspapers. They build a country of their own within our country. The Vietnamese also have their own way of doing things. We differ in this culture in so many ways. Teachers think many of the kids are abused or they suspect abuse. We get referrals all of the time from the schools.

—Caseworker, case manager, hospital setting, Houston, Texas

I feel very passionate about my job. Really passionate about making lives better for children. They have so many problems but they can still be so hopeful.

—Caseworker, children's ministries, Nashville, Tennessee

The passage of federal legislation, the development and growth of the managed care industry, and the evolution of the human service professional have contributed to the reemergence of case management in human service delivery. You read in Chapter 1 about the reasons for its reemergence. One reason is a response to the shift in service delivery from large, state-operated institutions to community-based services. Coupled with this shift is the resulting dual nature of service delivery that requires both service provision and service coordination in order to meet client needs. Finally, the managed care industry, in its efforts to control the quality and cost of health care, has had a significant impact on terminology. Its use of the terms *case management* and *case manager* has increased their visibility. In addition, the growth of technology, the rise in varied cultural and ethnic cultures, and the continuing economic downturn increase the need for case management services for clients and families with multiple problems.

You have also read about the evolution of case management as a service delivery strategy that reflects these changes. Some aspects of case management that are different today from thirty years ago are the identity of the case manager, the necessary knowledge and skills of those who deliver services, the types of problems clients experience, and the nature of the bureaucracy. Case management as a service delivery strategy will continue to evolve, making use of technology, recognizing new client problems, and responding to the changing economic and political climate.

The purpose of this chapter is to integrate what you have learned reading and studying this text with the wisdom of case managers across the United States who are engaged in service delivery. They offer real-life lessons, as the director and case worker in a human services center does in her chapter-opening quote, about the reality of case management today. Her words illustrate the struggle with budget cuts, large caseloads, and unresolved problems. Case managers are constantly facing challenges as they work with diverse populations. The case manager in the hospital setting illustrates the difficulties of providing services to diverse cultures. He suggests that services be tailored to fit the unique characteristics of people from another culture. Working for a nonprofit organization that uses volunteers to provide help in the community, the case manager who works with children takes a leadership role as she attempts to persuade the community to address current and future social service issues related to children and safe environments.

We think you will find that the knowledge and experiences of these case managers affirm the case management concepts introduced and discussed in this text, suggest new areas of emphasis, and identify new skills. To help you understand the application of the information shared here, the chapter includes a case study that is a composite of experiences of several case managers. For each section of the chapter, you should be able to accomplish the following objectives.

THEMES IN CASE MANAGEMENT TODAY
- List the themes that case managers identify when they talk about their professional work.
- Define what case managers mean by a Jack or Jill of all trades.
- Illustrate why communication is important in case management.
- Describe the use of decision-making and critical thinking skills in case management.
- List three personal qualities that help case managers do their jobs.

CASE STUDY
- Describe five interactions that take place in the case concerning the guardianship of baby Juan.

SURVIVAL SKILLS
- Describe five characteristics of burnout.
- Describe four characteristics of vicarious trauma.
- List ways that burnout can be identified and prevented.
- Explain why time management is an important skill for the case manager.
- Describe four steps in establishing a time management system.
- Define assertiveness and explain why it is important.
- Illustrate ways in which the case manager can be assertive.
- Describe the importance of seeking and providing supervision in case management.

Themes in Case Management Today

Case managers whose voices you will "hear" as you read this chapter represent social service agencies across the United States. They either identify themselves as having the job title of case manager or they describe their primary job responsibility as case management. They may have different job titles, such as *caseworker, social worker, family advisor, or behavior specialist.* Our analysis of the interviews we conducted with over seventy human service professionals resulted in the articulation of eight common themes. These themes describe what they do and what they need to know how to do to be effective case managers. The themes respond to the complex and sometimes difficult situations case managers encounter as they cope with large caseloads, clients with multiple needs, and scarce resources. In addition, case managers often work with clients who are silent, reluctant, or resistant in a bureaucracy that requires detailed documentation for each interaction. The eight themes are (a) the performance of multiple roles, (b) organizational abilities, (c) communication skills, (d) setting-specific knowledge, (e) ethical decision making, (f) boundaries, (g) critical thinking, and (h) personal qualities.

Performance of Multiple Roles

Every day, as case managers work with their clients, they perform multiple roles, including advocate, broker, coordinator, planner, and problem solver. (You read about these roles in Chapter 3.) It helps to be a **Jack or Jill of all trades**.

Many helpers combine the planner, broker, and coordinator roles into an intensive case management function and assume responsibility for determining the real issues, developing care plans, finding resources, and coordinating care among other professionals. Problem solving is often part of case management, and occurs when everything is going smoothly as well as when there is a crisis. Problem solving requires a plan A, a plan B—and a plan C, if necessary. For many helpers, the final goal of managing cases is self-sufficiency or the resolution of issues for the client.

Advocacy, discussed in Chapter 9, is an example of the complexity of roles in case management. One interviewee suggested that "services begin with advocacy," and another described the advocacy role as a means to "instill in clients what is best for them." Advocacy occurs when case managers are fighting for quality services for their clients, helping families treat their members fairly, working with agencies and other bureaucracies to serve clients better, and supporting clients "when they can no longer even support themselves." One helper's approach to advocacy was "to teach clients how to deal with their problems . . . how to deal with the system." On another level, interviewees defined advocacy for the agency as fighting for resources, attracting clients, and representing the agency. At the beginning of Chapter 10, Jessica talked about being passionate about her work and helping the children with whom she works. She feels incredible support for this advocacy from her supervisor and her agency.

The successful case manager is the professional who has multiple skills and is able to use them as needed, sometimes simultaneously. The complexity of a particular role, coupled with other job demands, often makes this especially challenging.

Organizational Abilities

Several case managers emphasized the importance of **organizational skills**, and they mentioned the disasters that can occur when professionals are not organized.

> The job and the work are incredibly complex: it is like a crystal chandelier, with so many pieces working together to give light—or, in this case, help. Yet it is so fragile. You think that you have stability, but within each case there are ten, twenty, or more pieces—and if you have thirty cases, well you are constantly juggling all of the responsibilities.

They are also aware that if they are not organized, it is their clients who will suffer. For them, being organized means managing time and completing paperwork.

THE CONCEPT OF TIME

The concept of time influences case management in a variety of ways: organizing, budgeting, scheduling, responding, balancing, and slowing down. Case managers recognize the need for time to "let things percolate up," "sit down and remember everything you have done during the week with that client," and "take one step at a time." Even though time management was consistently mentioned as a tool to alleviate stresses and pressures, many admitted that the workload is so "horrendous" that they never gain control over their work situations. They say that they are so busy that they don't have time to seek a resolution to a client's problem. One participant said, "Clients have to come to me." The difficulty with time management arises from several sources, including the unpredictability of the workday, external deadlines, and the ever-changing bureaucracy.

Themes for case managers (Woodside, McClam, Diambra, & Varga, 2012) related to the meaning of time center around pace, change, choices, and service delivery. All four of these themes relate to different facets of the work: the agency, the human service professional, and the client. Case managers talked about how their work on each level is significantly impacted by time as it relates to the "never-ending" pace of demands and pressures in their work; the "now and then" aspects of changes in protocol and policy over time; the "influence of time" on decisions about agency availability, caseloads, and policies; and the "one step at a time" notion of service delivery relating to repeated services and goal setting with clients.

For example, in the life of the case manager there is a "never-ending" tempo, described in the previous quote; the work is constant and the clients and the needs they present are at times overwhelming. Case managers describe their work in terms of how to "fit it all in," having too much to do, and responding to crises. In addition to the workload, there is a constancy of pressure within the job. Again from Woodside et al. (2012), case managers talk about "the work as 'day after day,' 'one right after another,' 'happening over and over again,' and 'never ending.' The scheduling alone is intense. Another case manager decried, "It is horrible. It is awful. It is terrible. It is a slave [driver]. I wish they would have twenty-four hours that I can work. It is not enough."

PAPERWORK AND DOCUMENTATION

Completing paperwork or documentation is another organizational skill that case managers practice daily. They understand the purpose of the paperwork that flows through their offices and its importance in building records, making requests, accounting for expenditures, documenting a "client's whole life on paper," jogging the memory, and providing an audit trail. The documentation they describe includes initial assessments, family histories, psychosocial assessments, contact notes, goals, service plans, and evaluations. Many of these are discussed in this text. The extensiveness of paperwork is illustrated by the following quote.

> We have to keep copies of all of our records and the files become almost unwieldy. Right now by law we keep information on the computer, the server. We also keep hard copies of all official records such as birth certificate, Social Security card, Medicare card, Medicaid card. . . . Last year our server went down and we were without our records. We hired a company to retrieve the data for us and we got back all but one week. It was a crisis and a wake-up call.

These professionals face the dual challenge of knowing how to write reports and how to find time to do their paperwork. In addition, they struggle to "set up a work space where you can find . . . 5,000 pieces of paper needed at any moment."

Organizational skills, then, pervade each day's work providing case management services. Even though many helping professionals choose this field because of the client contact, they are often surprised at the responsibilities of case managers, including the required documentation. Case managers who fail to master organizational skills experience overload, frustration, and eventually burnout, which forces many to leave the profession. Organizing time, paperwork, caseloads, daily schedules, and emergencies is a critical survival skill.

Katie describes how she organizes her work as a case manager.

Katie

I kind of have an organization problem. I feel like I'm borderline obsessive compulsive with stuff like that. So I have my different folders for different things. I have my case management folder, which has a face sheet for all of the kids that I'm working with and some of their basic information in case their school needs it. Things like their medications and contact numbers and stuff like that and also the goals that we are working on. I also have a place to keep things that we are working on together. For example, if we are doing worksheets or whatever the case may be, it's all there in my notebook. I also have separate folders for my paperwork, like my productivity numbers, my mileage. Also I'm very diligent about keeping up with my schedule. If I don't write something down then I will forget to do it and it doesn't happen. I know my schedule a week ahead of time, at least, so I do my schedule for the next week, including when I am going to be at which school, and I already have home visits scheduled. My planner is also color-coded so I have certain colors for school visits and certain colors for home visits and certain colors for office work I have to do. It gets kind of ridiculous but that's what I do.

—Permission granted from Katie Ferrell, 2012, text from unpublished interview.

Communication Skills

The case managers we interviewed claimed that "communication is more important than any other skill" and is tied directly to establishing a helping relationship, assessing needs and situations, and selling and persuading clients. Central to these results are the **communication skills** of listening, questioning, and persuading.

These case managers talked about "really listening" to what is being said by the client in order to establish a helping relationship, identify problems, and move through the case management process. For them, bonding is an important part of the helping process because the case manager must get clients' "trust . . . so they will tell you what the real problems are." Without this relationship, clients are less likely to accept services or to continue to reach out over a long period. Case managers form strong attachments to their clients; several agencies support clients through death, and case managers can be involved in arranging and attending funerals and "sharing grief with the family."

Good listening helps these case managers understand and evaluate what the client is saying (e.g., "Is the client telling you about the real problem?"). Listening also makes a therapeutic contribution to the client's progress: "Often times a client will start talking to you and . . . will ask . . . questions and answer the questions in one breath." Listening helps some clients let off steam. In other instances, clients have multiple problems that are jumbled together. "You can listen so you can see questions that will help them be more clear about what they mean."

Questioning is another communication skill that facilitates the assessment process. It involves "gathering information from lots of people," "assessing people's needs," and "figuring out what is important and what is not." Interviewees discussed the art of questioning, emphasizing asking "the right questions" to gather information for social and client histories, needs assessments, and intakes. Questions enable case managers to "look in every corner," assess the situation as well as client needs, and make appropriate decisions regarding case management. (Questioning was discussed in Chapter 6.)

To persuade clients to become involved in receiving services and to be active in their own self-care is an important communication tool. Sometimes case managers have to "sell the mother on letting her daughter be independent" and "sell parents whose children are involved in the criminal justice system on helping the case manager."

Clients today are consumers of services, since many of them, under the auspices of managed care and other funding models, are able to choose where they will receive the services to which they are entitled. In other words, "it goes back to client choices . . . it is the consumer's choice to participate." Therefore, some agencies find themselves competing with other agencies for clients, so they spend time convincing clients to come to them for services.

Communication with other professionals is also viewed as a valuable skill. Networking is an important tool for finding resources for clients. Having a good relationship with other professionals and knowing which person or agency to call for help ultimately benefits clients. A second part of communication with other professionals involves relating to other staff, especially other team members. For most of these case managers, working with their colleagues is "pretty family oriented, everybody is really close." These professionals use each other for support so they don't feel "they are just hanging out there" alone. They are also working on problems that demand "getting a lot of heads together" to solve problems. For many case managers, the following quote summarizes the feelings they have about the people with whom they work: "I have survived these past two and one-half years . . . [by] establishing a relationship with my team."

Setting-Specific Knowledge

Case managers also mentioned setting-specific knowledge as critical to their job performance. This includes general skills, such as typing and computer usage, as well as more specialized knowledge (e.g., medical terminology, medications and their side effects, drug regimens). For others, a thorough understanding of human behavior, psychosocial issues, and various helping methods form a basis from which to work with clients, make assessments and recommendations, and develop plans. Also important is knowing how systems, such as Medicaid, probation, child welfare, and housing, help the case manager support the client's interaction with other agencies and services. Finally, many helpers believe that "street smarts" are essential for case management. Street sense allows case managers to provide realistic assistance to clients very different from themselves.

Ethical Decision Making

Another case management skill discussed by the case managers who participated in this study was **ethical decision making** (discussed in Chapter 4). They must be able to identify ethical issues (e.g., self-determination, confidentiality, and role conflict); to ask the questions that surround the issues; and to make appropriate professional responses. One case manager summarized the issue of self-determination this way: "People have the right to poor judgment. . . . We can educate them, but we can't take away their rights." The case managers interviewed gave many examples of incidents in which, in their opinion, clients made bad choices. Clients refused services, chose to retain independence rather than receive complete services, refused medications, ate poorly, returned to abusive situations, and violated parole agreements. Case managers often described their frustration when clients refused services or did not heed sound advice, but they were passionate about the rights that clients have to determine their own destinies.

One ethical dilemma concerns confidentiality. For many of the case managers we interviewed, there is often a question of what information goes into a report and what should be omitted. Another issue of confidentiality emerges as computers are more widely used. One case manager expressed a hope that computer security is receiving the attention that confidentiality demands. This concern was addressed in Chapter 4. Another issue for those whose agencies serve immigrants is the legal status of the individual. Is the person in the United States legally? If not, should you report it? One case manager simply doesn't ask.

One of the most difficult dilemmas that case managers face is role conflict. They describe it as "working both sides of the fence," "walking a thin line," or "protecting two sides." When role conflict occurs, case managers are asked to assume dual responsibilities that may be at variance with one another. One helper shared a role conflict she encountered as a parole officer. The parole officer's primary responsibility is to support the parolee's life outside the prison environment. If the parolee "messes up," the parole officer becomes the prosecutor. This type of conflict makes it difficult to maintain a supportive, trusting relationship with a parolee after the prosecution.

Boundaries

Surviving the intensity of helping was a concern to the case managers in this study. They felt it was essential for the case manager to establish **boundaries** between self and client. One helper explained, "You have to watch yourself because you get too attached to clients." Another elaborated: "You have to have an idea of your own boundaries and what your issues are. You don't want to get yourself confused about what is going on. . . . Sometimes it's helpful to step back and ask, 'Whose problem is this?'" Even though case managers work hard to establish boundaries between themselves and their clients, they still agonize over their clients and the difficult situations they face. One interviewee described it as "close detachment." One case manager described her reactions to client problems thus: "Sometimes I have had to go to the restroom to cry for a few minutes because it was a really hard case." Many workers even dream about their clients. These professionals are committed to handling boundary issues and recommend "staying realistic" and "leaving work at work."

Critical Thinking

The effective case manager is one who needs to think critically and clearly. One of the **critical thinking** skills needed is "seeing the whole in addition to individual, narrow parts." Because individual cases are so complex, there is a danger of focusing on the details and "not seeing the forest for the trees." For one case manager, critical thinking is "being able to procure and digest a large body of information from different people" and then determining the real issue. The existence of underlying issues means that case managers are "detectives" who are able to ask the right questions, not take things exactly as they are presented, conduct continual evaluations, and assess communication as it is happening. Helpers then "put it all together to get a better picture." Interviewees suggested that using their years of experience, tuning into their insight, "going in with a clear mind," and "performing a reality check—my fantasy versus reality" all enhance their critical thinking abilities.

Personal Qualities

The people we interviewed identified a number of personal qualities that enable a case manager to be effective, including *realistic, patient, flexible, and self-confident.* Two interviewees likened the desirable personal qualities to those of a fairy godmother and a chameleon.

Flexibility was a consistent recommendation. Constant interruptions to the plans for the day, unpredictable and emergency client needs, interviews conducted under unusual circumstances, and interactions with different clients and professionals who demand different styles are examples that illustrate the need for flexibility. Case managers also need to be "firm" in communications and "soft" at other times. One participant explained, "I just kind of roll with the punches. Whatever needs to be done, I just do it."

The ability to form good working relationships with clients is essential. For some case managers this is carried to an extreme of "somehow being reasonable to the point where you can no longer be reasonable." Others describe it as being tactful and respectful of others, "going out of the way to communicate with others," and "taking time with people." The case manager is in a people-oriented field, and whether working with clients or other professionals, it is important to "be able to talk with people and get along with them."

Another necessary quality for case managers is patience. One participant described her willingness to "take one step at a time" when working with her clients. According to these case managers, it is sometimes difficult to be patient. They reported that they remind themselves and their colleagues of the importance of "being able to let go and wait." Part of patience comes from being realistic, including having realistic expectations for clients. These helpers also realized that "when you are allowing clients to learn to help themselves, you cannot be in a hurry." A second factor in patience is persistence. For many professionals, "you just keep plugging." They acknowledge the difficulties and the resistance encountered from both clients and bureaucracies during the helping process.

Self-esteem provides the foundation for many of the difficulties encountered in case management. According to participants, "you must have self-confidence."

It helps to maintain a positive perspective when things do not go well: "You realize that you will get over it [failures] and things will move on." This confidence also fosters **assertiveness** when dealing with other professionals or resistant clients. One special challenge of working with other professionals is the bureaucratic hierarchy; many higher-ups do not acknowledge that case managers have important professional contributions to make. Self-esteem also helps them assume a leadership role when it is required and, as you read in Chapters 8 and 10, leadership is required more often today. These case managers recognized the need to assume authority in order to accomplish their goals.

Finally, according to participants, case managers need to have both a "sense of adventure" and "excitement" about their work. Terms like *anthropologist, private eye*, and *detective* described the challenges of case management: Human service providers must accurately identify problems, develop service plans to meet client needs, provide and seek out services, and evaluate the process. For many, the challenges are stimulating, not depressing; these people thrive on working in a demanding, fast-paced environment.

 # Case Study

The following case study illustrates many of the themes just discussed. It is based on several cases that occurred in the southeastern United States. This case crosses state lines, involves a number of professionals, and illustrates the realities of providing services to an ethnically diverse population. The case is complex as well as confusing, reflecting the reality of case management. Since it is told from the point of view of the case manager, Delores Fuentes, her narrative is in italics. As you read the case study, identify situations that illustrate the themes described in the previous section.

People Involved in the Case

Delores Fuentes	Case manager, community health and human services clinic
Mr. and Mrs. Ruiz	Parents of Juan
Dr. Hidalgo	Pediatrician
Ms. Brown	Nurse, Health and Rehabilitation Services, Georgia
Mrs. Marcos	Guardian in Minneapolis, Minnesota
Mexican Consulate	Atlanta, Georgia
Mr. Sanchez	Lawyer, Mexican Consulate, Georgia
Dr. Stapleton	Physician in Minneapolis
Mexican Consulate	Minneapolis
Police—first encounter	Minneapolis
Mr. Gluckey	Staff member, Mexican Consulate, Minneapolis
Police—second encounter	Minneapolis

My name is Delores Fuentes, and I am a case manager at a community health and human services clinic in rural Georgia. Since I work with a diverse group of clients, including migrant farm workers, my work is very challenging. One problem we face is the client's lack of understanding of the culture here, of the standards of living in this country, and of the laws that govern relationships, such as marriage and child custody. Let me share a recent case in which there was such a misunderstanding.

Mr. and Mrs. Ruiz came to the clinic to see the pediatrician, Dr. Hidalgo. Mr. Ruiz asked the doctor if he was willing to take care of his baby, Juan.

DR. HIDALGO:	I know you want me to see Juan. Where is he?
MR. RUIZ:	Oh, I need you to sign these papers saying that you are willing to take care of him so they can release him from the hospital in Minneapolis.
DR. HIDALGO:	Okay, but why is he there?
MR. RUIZ:	Oh, he was sick.
DR. HIDALGO:	So he is there in the hospital?
MR. RUIZ:	No, he is with a translator. The nurse from Health and Rehabilitation Services (HRS) needs to have a paper from you because if not, we will lose our baby.
DR. HIDALGO:	I don't understand this. Do you mind if I send you back to HRS?

Dr. Hidalgo asked me to look into this case so he could understand the situation better. Altogether it took me about two hours to figure out what was going on. The scrap of paper Mr. Ruiz gave the doctor had a phone number and a name on it—that was it. From there I started trying to figure out the case. First, I called the HRS in the state capital and said, "I have this family here and I need to speak to the director of nursing." Eventually, I spoke with a nurse, Ms. Brown.

MS. BROWN:	Don't even bother to take on the case. It is a lost cause. It was referred to me by a hospital in Minneapolis [where Mr. and Mrs. Ruiz had recently been living]. They wanted to make sure that the family has a suitable home for the baby. I made a visit; they live in a house with no electricity and no phone. There are three or four people living with them, so that is not a healthy environment for the kid. So I signed that Juan cannot be released into their care.

I asked her Juan's location now. She reported that he had already been released and was with a guardian. I needed the name of the guardian, and finally she gave it to me. I reported to the pediatrician, and he told me that he would call the guardian. He reported this conversation.

DR. HIDALGO: My name is Dr. Hidalgo and I am a pediatrician working at a community health clinic. I need some information about Juan Ruiz. Could you tell me what is going on with Juan?

MRS. MARCOS: Don't even bother asking questions of the parents. They don't have any more right to be trying to fight for him. I have custody of Juan.

I wondered how that could be. How is that possible? She should have been able to describe why she had custody. She could have said, "I have custody because the father mistreated the wife, and I was able to remove that kid from that unhealthy environment."

I called Mrs. Marcos back and told her that the pediatrician had spoken to me. I asked her what the issues are, and she told me not to assess the situation. She explained that she had observed the interaction between Mr. and Mrs. Ruiz. According to her, Mr. Ruiz is the only one who talks—his wife always looks down, she cannot face you. Mrs. Marcos concluded that it was a typical case of abuse.

In Mexican culture, it is a sign of respect for women to avoid eye contact while men answer the questions. I told the pediatrician what I had learned and that I would visit Mr. and Mrs. Ruiz to determine whether it was an abusive situation. I talked with them and their neighbors and friends and found no evidence of abuse.

I then called the Mexican Consulate for help, explaining that one of our clients had a problem and describing the details I had uncovered so far. I said that I thought the family needed a lawyer. They agreed. After investigating the case, the lawyer, Mr. Sanchez, called me.

MR. SANCHEZ: I found out that the couple signed custody papers for a Mrs. Marcos to take care of the baby, Juan, while he was in the hospital. Nothing was signed after that. The hospital released the baby without permission from the parents to this person who actually, at least initially, had the good will to help. She was a translator and spoke Spanish well, so the family thought she would help them. I think initially that was what she really intended. Later she became attached to Juan.

I called Mrs. Marcos back. I explained that the case is complicated and that although I do not have any legal background, I had found a lawyer for the family. I told her that she would receive a call from a lawyer working for the Mexican Consulate. He was exploring the issues and wanted her to tell him more about the situation. She said that Mr. Sanchez had already called her. I think she was becoming nervous because she called Dr. Hidalgo. She told him she wanted to take care of the baby and that Juan needs this and that. Put like that, it sounded like the baby was almost dead. Dr. Hidalgo asked me to contact the Minneapolis hospital for the name of the doctor. I talked with the doctor, Dr. Stapleton, who confirmed that nothing was wrong with the baby when he was released. I gave his name and phone number to the lawyer, Mr. Sanchez. Dr. Stapleton told Mr. Sanchez what he told me: "There is nothing wrong with the baby." Then Mr. Sanchez told me about his conversation with Mrs. Marcos.

| MR. SANCHEZ (to Mrs. Marcos): | The doctor said that the baby must go home to Georgia. |
| MRS. MARCOS: | Oh, no. I will not release the baby. |

Eventually, an agreement was reached that the baby, the guardian, the family, and the lawyer would all see Dr. Stapleton. Meanwhile, Mr. and Mrs. Ruiz learned that Mrs. Marcos had been seen putting some boxes in her car, and they called me at home at 2:00 a.m. I called Mr. Sanchez, who said that we needed to send the couple to Minneapolis immediately. In the meantime, he would call the police.

We sent the parents to Minneapolis in a van. While they were on the way, we made sure the family had a house in Georgia to return to, as well as work, a pediatrician for the baby, and a phone. They arrived in Minneapolis the next morning. The family reported that the translator, Mrs. Marcos, was very nice.

| MR. RUIZ: | She gave us a piece of cake and a coffee, and let us play with the baby. Then she disappeared. About ten minutes later the police came in, and the police literally chased us out, and said, "If you come back here, we will put you in jail." |

Mr. Ruiz called me from a gas station to report what had happened, and I told him to call me back in an hour. I called the Mexican Consulate to report on the situation. The Consulate here called the one in Minneapolis and asked them to have someone meet the family at the gas station. We called the police to tell them what was going on. The police agreed to check on it.

The police sergeant found a report with the correct time and date, stating that the parents were removed from the home of Mrs. Marcos, but he did not have the full report of what happened. So a member of the Mexican Consulate in Minneapolis and their lawyer, Mr. Gluckey, went to see Mrs. Marcos to get custody of Juan. All agreed to meet at the doctor's office the next morning, and if he said the baby could travel, then Mr. and Mrs. Ruiz could take Juan back to Georgia.

This case took three weeks to resolve.

Survival Skills

The Ruiz case illustrates the complexities of case management. It involves a set of parents (in Georgia), a pediatrician and a case manager (Georgia), a nurse (in the state capital), a translator turned guardian (Minneapolis), a doctor (Minneapolis), two Mexican consulates and their lawyers (Georgia and Minneapolis), two teams of police, and an indeterminate number of people involved in facilitating and obstructing the case management process. Many individuals were helpful; some were passive or indifferent. Several individuals did not trust the clients, the parents of the child. Some professionals believed that the child was the client, not the parents.

The activity in this case reflects many of the themes described earlier in this chapter, such as the performance of multiple roles, the ability to communicate, knowledge of ethical decision making, and the ability to be flexible and patient. The case also suggests how difficult performing the role of case manager can be. It emphasizes the stressful and emotional nature of the work, which that can sometime leads to burnout. Ms. Fuentes, the case manager, saw her work come to a positive end. It is one of her best successes. Often she works hundreds of hours with little reward for her efforts.

The skills of time management and assertiveness can provide the case manager with stability and support to counter many of the challenges of this difficult work. These are survival skills. The following section discusses the nature of burnout and describes how time management and assertiveness can prevent or alleviate it.

The Prevention of Burnout

A key to effectiveness as a helping professional is caring. The inability to care is the "most dangerous signal of burnout, ineffectiveness, and incompetence" (Skovholt, 2001, p. 12). Case management is only one facet of human services that illustrates the difficult nature of helping. Continually facing unsolvable problems, unmotivated clients, resistance, bureaucracy, psychological injury, and a lack of needed skills may lead to this inability to care. The losses that a case manager experiences daily may also contribute to this problem: clients who never return, ineffectiveness within a rigid bureaucracy, or few case closures. The term **burnout** has become a way to consider the exhaustion and disquiet on the job that often result from this difficult work.

There are many descriptions of burnout. Among them are fatigue, helplessness, hopelessness, exhaustion, and stress. Christine Maslach, a leading researcher of burnout, believes that it is a problem in the social environment in which people work, and she cites six sources: work overload, lack of control, insufficient reward, unfairness, breakdown of community, and value conflicts (Maslach & Leiter, 1997). In her view, the problem is with the organization, not the individual. The challenge for case managers is to find ways to feel they are succeeding—and to focus on what they can control. One case manager at a halfway house shelter in Brooklyn, New York, describes the intrinsic rewards in "having so much happiness when the women I work with, for the first time in their lives, have a home of their own. I love hearing them talk about their plans for the home, how they will decorate it and what they will cook there." Other strategies may include the development of supportive professional relationships, observation of minor client changes, and opportunities for professional development to develop additional skills or to further develop professional relationships. Self-care strategies that may also be helpful are developing a sense of humor, committing to a regular exercise program, participating in relaxing activities, enjoying recreation, and improving one's spiritual or religious self. A review of one's boundaries, both of time and emotions, may also be beneficial.

Professionals with case management responsibilities are particularly susceptible to burnout because many of the factors that contribute to burnout are integral to their work. Among those factors are the nature of the clients, the stresses of dealing with bureaucracy, and a personal tendency to react negatively in stressful situations.

For case managers who help individuals with more severe levels of trauma exposure, such as child abuse, violent crime, natural disasters, or torture, vicarious traumatization may result (Painter, 2012). What happens to helpers who work with victims of trauma? Researchers indicate that vicarious trauma may integrate into the case manager's worldview; the world no longer seems a safe and secure place (Harrison & Westwood, 2009). As with symptoms of burnout, vicarious trauma may also influence varying aspects of the case manager's life such as relationships, physical health, mental health, faith, and hope (Trippany et al., 2004). In addition, again similar to burnout, vicarious trauma may affect the helping abilities, resulting in decreased empathic abilities, defensive reactions, and a decreased concern for clients (Trippany et al.)

Let's look at how burnout and vicarious trauma relates to Delores Fuentes as she encounters these challenges in her work with the Ruiz family and other helping professionals.

Want to Know More?

There are increasing numbers of natural disasters occurring globally. Search the following terms for more information about natural disasters and how government agencies and not-for-profit agencies are responding to the expanding needs.

American Red Cross
World Bank
Tsunami
Flood
Natural disaster
Fire
Rape
Child abuse
Elder abuse

Delores Fuentes, who worked so hard to reclaim Juan for his parents, experienced many of the work characteristics that can lead to burnout. In this case, she did not react negatively. On the contrary, she was able to pursue her advocacy for Mr. and Mrs. Ruiz beyond the bounds of her agency's responsibilities and across state lines.

Much of the work of case management involves clients who have very complex and long-term difficulties: children who are at risk, adults who have disabilities or are elderly, and people with medical problems, such as AIDS or cancer. In fact, modern-day case management emerged as a methodology designed to handle the overwhelming stress of serving clients with multiple problems. Case managers daily come face-to-face with clients who have little hope and who often confront long-term illness and severe social difficulties. Other case managers work with clients who demand intense services over long periods. Often, no one asks if the case manager has the time or the resources to provide the services needed.

Delores Fuentes certainly experienced a situation that required her immediate and undivided attention. Her other work was left undone. In addition, she was working in "uncharted waters," with a variety of professionals from other states and international governmental agencies.

Case managers also struggle with the bureaucracy in which they work. One pervasive difficulty is the size of caseloads. Case managers often lack sufficient time to give each client quality treatment; they must settle for providing services that are adequate at best. The programs offered and the resources available often do not match clients' needs. Case managers find themselves in the middle—understanding clients' needs all too well, yet knowing the limitations of the system.

As Delores Fuentes struggled to provide a fair hearing for the parents, she found that the bureaucracy was ill equipped to handle a situation that involved non-U.S. citizens across state lines. Ms. Fuentes had to forge new alliances with agencies such as the Mexican Consulate.

Certain observable symptoms related to work habits, psychological and physical well-being, and relationships with clients can indicate that case managers are reacting negatively to stress. Absenteeism, tardiness, and high job turnover rates are three indicators of widespread burnout. A case manager might also have symptoms such as gastrointestinal problems, substance abuse, exhaustion, sleep disturbance, poor self-concept, inability to concentrate, and difficulties in personal relationships. Their attitudes and working patterns may change: They become disorganized, have difficulty working with colleagues, blame clients for their own problems, and depersonalize clients by referring to them with disparaging names. These changes soon affect job performance, and case managers can feel increasingly frustrated and worn down by their responsibilities. Burnout is a devastating problem for an agency, the case managers affected, and the clients served.

Acquiring time management and assertiveness skills can help case managers avoid burnout. Both skills allow them to be more effective case managers.

Managing Time

Time management is particularly important for case managers, but the nature of their work makes it especially difficult. Many types of activities are required

Voices from the Field: Helper Reactions to Working with Trauma Victims

NetCE (2012) provides a short case study that illustrates one helper's reaction to working with trauma victims. Other cases are available at http://www.netce.com/casestudies.php?courseid=744.

Mr. A is a psychotherapist who has worked at a community mental health clinic for the past ten years serving adults with a wide range of presenting problems. He has been treating Patient M for the past six months. Patient M is a highly educated, married woman, thirty-five years of age, from a country in Latin America. She has been in the United States for the past two years. She was referred to the mental health clinic by her primary care doctor because of her severe and frequent panic attacks, nightmares, experiences of seeing and hearing dead people talking to her, and severe depression. Over the course of the first several months of treatment, Patient M has shared bits and pieces of her story with Mr. A. He has learned that the patient was working as a teacher in her home community and was active in one of the opposition political groups in her country. She fled her country after soldiers killed opposition party supporters and their families in her town one night, including her husband and child. She was at a distant neighbor's house when the massacre took place, tending to a sick friend, and she believes that is why the soldiers did not find her. She tells Mr. A that she is too afraid to return to her country and is seeking asylum in the United States.

Mr. A finds himself flooded with many painful emotions in and after sessions with Patient M. He often feels horrified and has desires for revenge as she discusses her memories of finding her dead husband and child when she returned home that night. He feels terrified by the thought that Patient M may be deported to her native country, where her life may be in danger. Mr. A has not experienced much trauma in his own life and definitely does not identify with Patient M's experiences.

Mr. A finds it extremely difficult to tolerate the intensity of his feelings when working with this patient. In order to avoid the pain associated with these feelings, he unconsciously develops empathic withdrawal toward Patient M. . . . For example, Mr. A has unconsciously distanced himself from Patient M and often blankly stares at her when she brings up anything related to her traumas. Mr. A's reactions have led him to neglect to thoroughly assess the patient's traumatic experiences and the origins of her current symptoms.

SOURCE: Reprinted from Berthold SM. *Vicarious Trauma and Resilience*. Sacramento, CA: CME Resource. Copyright 2011, with permission from CME Resource.

of them: interviewing clients; writing reports; meeting with clients, team members, and others; arranging services; transporting clients; and monitoring services. Each of these mandates specific skills, a special orientation, and considerable time. In addition, this diversity of activities require interaction with many different individuals, including colleagues, clients, and families. Each contact takes time—making the initial contact, monitoring the work, and maintaining relationships with all those involved. Another facet of the job that makes time management difficult is the uniqueness of each case. At the beginning of the work with a client, it is hard to predict the full scope of the services that will be needed or the intensity of support the client will need.

Another difficulty is the lack of *consistent* demands on time. As a result, it is very difficult to establish a regular schedule and to plan and then perform certain tasks at appointed times. When crises interrupt this complicated work, the case manager must constantly weigh and shift priorities. Even the best time manager must set aside the plan for the day to assume additional responsibilities that emerge and demand attention.

Four questions facilitate thinking about time management (Grubbs, Cassell, & Mulkey, 2005):

- What personal characteristics, both strengths and weaknesses, influence how case managers manage time?
- Exactly how is time spent?
- What techniques are used to control time?
- What time-management system is used? What principles are followed?

Before case managers can manage time effectively, they must assess how they actually use time during the workday. They can use a time log to establish time spent and outcomes achieved. A time log records activity every 20 minutes or so, describes and summarizes interactions, and covers at least two weeks so as to capture variations in workflow. Once such a log is completed, the data are analyzed to reveal how the individual has allocated time.

There are several ways to analyze the data. One is to examine the distribution of time in various case management activities, such as intake interviewing, reading reports, team meetings, and making community visits. It is also helpful to know how much time is spent managing cases and providing services. A form like the one in **Figure 11.1** might be used for such an analysis.

Delores Fuentes spent three days working intensely for Mr. and Mrs. Ruiz and their baby, Juan. Because she managed a community health clinic and saw clients of her own, her other responsibilities had to be absorbed by her co-workers. At the same time, she had to be extremely organized to coordinate assessment, fact-finding, and the search for resources. Delores was the point person, receiving updates from all professionals working on this case. She made a plan each day, but since the events continued to change quickly, she was continually revising her plan.

	Number of Hours/Week	%
Managing		
Intake interviewing		
Report writing		
Team meetings		
Community visits		
Making referrals		
Monitoring the plan		
Community outreach		
Subtotal		
Direct service		
Direct service to clients		
Direct service to families		
Subtotal		
TOTAL		

Figure 11.1   Time allocated to case management activities (managing and direct service)

Time-Management Techniques

The time analysis of the tasks of one day, considered in light of the extent of control the case manager has, provides the basis for managing time. One component of time management is the planning and implementation of more effective use of time in several steps. It begins with setting goals and priorities and ends by assigning particular activities to times when they can best be completed.

Alan Laekin is the author of the time management classic *How to Get Control of Your Time and Your Life,* which was first published in 1976 and reissued in 1996. He recommends that individuals categorize desired goals into three time spans: the next five years, the next three months, and this day. The next step is to assign priorities to the goals. For example, an ABC scale can be used, on which A is the most important, B is moderately important, and C is least important. The goals can then be sorted into categories: short-term, intermediate, and long-range. How do case managers use this information in planning for the day? They focus on the most important activities and choose tasks that are ranked A (most important) from each of the categories—short-term, intermediate, and long-range.

The following guidelines also help determine how to plan each day.

THE RESPONSIBILITIES
- Pick the two most important goals and pursue them first.
- Address the toughest jobs first.
- Alternate difficult and easy tasks.
- Group similar tasks.

THE PLANNING OF TIME
- Plan some uninterrupted time.
- Plan time for the unexpected or for crises.
- Allot time according to deadlines.
- Assign time to plan.
- Assign time for paperwork.
- Have a list of quick tasks to use as filler.

TIME MANAGEMENT AND OTHERS
- Understand assignments.
- Delegate tasks that others can or should do.
- Develop a system for monitoring the work of others.
- Say no when the assignment is inappropriate or there is not enough time to complete the task.

Time management is one skill that can help the case manager establish priorities and assess the needs of a situation and the professional resources available. Assertiveness is another skill that can facilitate effective case management. Case manager Delores Fuentes used assertiveness skills as she worked with the parents, other professionals, and the guardian.

Assertiveness

All people have a right to express their thoughts, feelings, opinions, and preferences and to expect to be treated with respect and dignity. Assertiveness is a skill that case managers use to stand up for themselves, their clients, or both. It enables a person to speak for himself or herself or others, to act in one's own best interests, and to express one's self without denying the rights of others (Johnson, 2008).

Assertiveness is different from aggressive behavior, which is an attempt to hurt someone or destroy someone. Aggressive behavior violates the rights of others by treatment that is disrespectful and communication that is insulting, sarcastic, and hostile. Acting non-assertively is the opposite of aggressive behavior and involves passivity, hiding feelings, and allowing others to violate your personal right to be treated with respect and dignity. A balance between these two extremes is assertiveness which is an effective way to resolve conflicts and express opinions.

A key element in assertiveness training is the ability to distinguish among assertiveness, aggression, and non-assertiveness. The goal of assertiveness is to relate to others in ways that lead to maximum dialogue and a positive outcome for each participant. Other important elements are listening carefully when others talk, speaking in a respectful and nonjudgmental manner, and recognizing the equal rights of each person. Maintaining a positive attitude and valuing fairness also contribute to a satisfactory outcome. Developing these skills will enable case managers to represent themselves and their clients more effectively.

Delores Fuentes was assertive throughout this case. She assumed the role of advocate for Mr. and Mrs. Ruiz, tracking down information from Ms. Brown and pursuing the case when it was recommended that she drop it. She also contacted consulates and lawyers and engaged the help of the police. In fact, she did what she needed to do to help her clients. As she tells her story, it is clear that she understands the difference between passivity, assertiveness, and aggressiveness. She took action and she was assertive, but not aggressive, as she dealt with other professionals.

Assertive communication works in the case manager's favor because every facet of case management involves communicating with others. Whether you are initiating assertive communication or responding to passive or aggressive clients or colleagues, assertive communication is an important skill.

Communicating assertively also has much to do with the principles of case management discussed in Chapter 1. Both empowerment and advocacy benefit from assertive communication. *Advocacy* means standing up for the rights of those who cannot represent themselves, and when case managers engage in advocacy, they are acting assertively.

Empowerment is the activity of instilling the belief that clients have rights and abilities and can solve problems for themselves. Assertiveness promotes the self-confidence and self-respect that supports empowerment. At times, clients and their families do not want to be empowered, which creates a very difficult situation for case managers. Clients may be angry when they do not receive what they feel they are entitled to or when they are asked to help themselves.

Case managers can learn to be assertive and apply these skills in several ways. Sometimes case managers are asked by clients or colleagues to make judgments about particular matters, and sometimes they themselves feel the need to provide feedback. In both cases, there are ways of communicating that will be seen as helpful and thoughtful rather than critical or hurtful. Occasionally, the case manager must confront clients or other professionals. Assertiveness allows confrontation to occur in a way that respects the rights of both the speaker and the listener and promotes honest, fair, and positive outcomes.

It is often difficult for case managers to say no to requests from colleagues or clients. Their commitment to help others makes them inclined to say yes to any request. However, a request is unreasonable if the action would violate ethical or legal codes. Colleagues or clients may ask for action that causes case managers to be frustrated, anxious, or physically or emotionally threatened. To be reasonable, a request must take into consideration the rights of case managers, as well as those of clients and colleagues.

From time to time, everyone encounters people who communicate in an aggressive way. It is difficult to turn an aggressive encounter into an assertive one. One's reaction may be passive because the aggressiveness is so surprising, and sometimes aggressiveness is returned for aggressiveness. Either type of response leaves the feeling that rights have been violated and that communication is closed. Acting assertively helps the case manager to turn aggressiveness into reasonableness.

Professional Development: Focus on Supervision

One key element for professional growth and development as a case manager is supervision. As discussed in Chapter 10, in theory, formal and informal supervisors within an agency setting provide assistance and support to the new case manager. It is important to note that supervision is an essential component of all professional growth as a case manager. At times, individuals provide excellent supervision without your requesting it. On the other hand, case managers expect to seek supervision when it is not available. The following five guidelines help you obtain the supervision you need for your work as a case manager.

1. **Identify both formal and informal** supervisors with whom you work. Establish regular opportunities to talk about your work with these individuals. Also establish strategies for contacting supervisors for emergency consultation.
2. **Be aware of the ethical standards and** issues that guide and confront you as a case manager (as described in Chapter 4). Consult with other professionals regularly as ethical issues emerge in your work. Weigh options carefully, considering professional and legal dimensions, agency policy, and your own education and training.
3. **Prepare for supervision.** Establish an agenda for meetings with supervisors. Meetings may include the following:
   a. Questions about the agency or the work.
   b. Questions about clients.
   c. Questions about policies and regulations.
   d. Questions about paperwork.
   e. Samples of your work to share and receive feedback on.
4. **Be honest** in reporting difficulties or mistakes.
5. **Be open** to supervision and to positive and critical feedback.

Deepening Your Awareness: First-Person Accounts of Case Managers

Each of these case managers talks about how they "thrive and survive" in their role as case managers.

 ## Self Care

Jessica

I think it is very important for case managers to practice humility. This means always seek supervision or advice from other professionals if you don't know what direction to go in with a client. I started this job when I was twenty-three, and it took a lot of learning to develop the skill set that I have now. So beginning case managers need to rely on people that have knowledge and experience. Case managers also need passion and compassion. They need to understand that it is really easy to get burned out when you are not making any money and you are doing so much and you are giving so much of yourself, but you have to find a way to balance your life because reactions to work are going to seep in. It's not the kind of job where you go there at 9 o'clock in the morning and you leave at 5 and everything stays there. If you are the kind of person who gets into human services, you are not the kind of person who can leave the kinds of things we hear at the office. So self-care is huge.

—Permission granted from Jessica Brothers Brock, 2012, text from unpublished interview

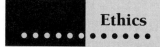 ## Ethics

Sara

I think an ethical challenge for me has always been my caseload, the number of kids I have on the caseload and being able to serve them properly. I worry about being able to get them quality care. I think there have been times when we do not provide the best care we can. I think it is because we never have enough staff. We have high turnover with staff and so a lot of the work falls on the few staff members remaining. We do work as a team and if I need to cover the floor I need to cover the floor. But my boss needs to know that that means I'm still going to do what I'm supposed to do. I'm still going to see my kids four times a week and provide all of the documentation I need to. And when new staff members begin, there is some relief, but not full-time relief. It takes a while for them to learn their job.

I really want staffing decisions to be good for the kids. I understand that we are a business and I understand that to have these kids we have to make money and we have to. . . I understand all of those things but it's really about the kids to me. I want them to get the best care that they can get. I know sometimes those lines can get blurred. I think this is because as the case managers and clinicians in the agency, we see things one way and we see the administrative and the business side too. So it can definitely be a challenge to fight for the kids and the agency.

—Permission granted from Sara Bergeron, 2012, text from unpublished interview

Boundaries

Katie

Boundaries. Now that's something that I've had to work on. I'm not very good at letting things go. I really let things affect me a lot. My husband tells me that I'm too empathetic. He says it is a bad thing. But I just take too many things personally that I don't have any control over. So I have a hard time letting things go. Recently I have tried; I talk to my supervisor about the job, so that helps. Obviously I can talk to her about my kids, so I talk to her weekly about what is going on. I've tried to exercise but that really does not work. I'm not very consistent with that. Recently what I've started doing is writing things down, like things that I'm worried about. It is almost like a prayer journal. I think it helps me let things go like what goes on when kids are at home and I'm not there, I know I've done everything that I can in a certain situation. So I just write what worries me about that kid. This helps me process my feelings and thoughts, so it's helped me let it go.

And it's the ones that I can't help, where I know bad things happen at their house and I can't do anything about it, that are difficult. I can't fix it and I can't change it. Or the frustrating ones where the parents say, "I need help and nobody's helping me." That's what I'm there for, to help them. But they don't take any suggestions, they don't want to change anything, so what's going on with their kid is not going to change.

—Permission granted from Katie Ferrell, 2012, text from unpublished interview

Deepening Your Awareness Discussion Questions

Answer the following questions:

1. What dilemmas does each of the case managers confront?
2. What advice does each have?
3. What did you learn from each about thriving and surviving as a case manager?

Chapter Summary

This chapter provides a summary of eight themes derived from interviews with case managers who talked about their work in human services. These reality-based themes illustrate the concepts presented in the previous chapters. These case managers emphasized the importance of performing multiple roles when working with their clients. For example, sometimes they are brokers, linking their clients to services, and other times they must advocate for new services that will help them meet their clients' needs. Other themes are survival skills, such as organization, communication, and critical thinking, which are important in effective case management practice. Organizational abilities allow case managers to handle large caseloads, work with clients who have multiple needs, maintain contacts and make referrals to various agencies and organizations, and handle the paperwork. Communication pervades the case management process and is essential in working with both clients and other professionals. Written communication, report writing, and case summaries all have their place in the case management process. Critical thinking skills allow the case manager to process information and make decisions by being clear, logical, thoughtful, attentive to the facts, and open to alternatives.

Ethical decision making and boundaries are directly related to the dilemmas and challenging situations that confront case managers each day. Both concerns require a recognition that problems exist and a willingness to work to resolve these problems. The personal qualities that are necessary to be effective in the job include being realistic, patient, flexible, and self-confident. The case manager must function in a variety of situations, under intense pressure, with very difficult and challenging clients.

In this chapter, Delores Fuentes, a case manager in a community health and human services clinic, presents the case of Juan. The case illustrates just how intense and complex the work of case management can be. It affirms and demonstrates the themes the case managers discussed in the interviews, emphasizing the complications that can occur when working with people from other cultures.

As Delores Fuentes's case demonstrates, case managers can have very stressful jobs. Too many clients, too few resources, unhelpful bureaucracies, and challenging clients are some of the factors that contribute to stress. Too much stress often results in burnout and, at times, vicarious traumatization, which both are accompanied by an inability to work with clients and others effectively. S may influence the case manager personally. Two skills help counter burnout or prevent it. Time management is particularly important for case managers, but the nature of their work makes it difficult. Case managers can use guidelines and planning techniques to coordinate their priorities and goals in the time available.

Assertiveness is based on the principle of equal rights for speaker and listener. Often, the case manager must be assertive to procure services for a client and when working with a difficult client. The goal of assertive communication is to relate to others in ways that lead to maximum dialogue and a positive outcome for each participant.

Finally supervision—receiving or seeking it—becomes important for the case manager's professional growth and development. Identifying formal and informal supervisors provides a way in which case managers can engage with other colleagues for help and support.

Chapter Review

Key Terms

Jack or Jill of all trades
Organizational skills
Communication skills
Ethical decision making
Boundaries

Critical thinking
Assertiveness
Burnout
Time management
Empowerment

Reviewing the Chapter

1. Why are there times when case managers have to perform multiple roles?
2. What are the challenges in case management that require organizational abilities?
3. What are the ways that case managers can communicate with their clients and other professionals?
4. What are the ethical decisions that confront case managers?
5. Why are boundaries an issue for case managers?
6. Define critical thinking.
7. Describe the personal qualities that case managers need.
8. Describe the complexities of the case of Juan Ruiz.
9. Define burnout.
10. Why do case managers burn out?
11. What are the symptoms of burnout?
12. Describe how the time-management guidelines might have helped Delores Fuentes with Juan's situation.
13. Define assertiveness.
14. What are ways in which case managers may be assertive?
15. Describe how assertiveness principles might have helped Delores Fuentes in the case of Juan.

Questions for Discussion

1. Describe five themes relevant to case management today. Provide an example for each theme.
2. Why is burnout an issue for case managers? Discuss how you would prevent burnout.
3. Are you a good time manager? Illustrate your self-assessment with three examples from your own life. What skills could you improve?
4. Why is assertiveness an important characteristic for the case manager? Describe a situation in which you would find it difficult to be assertive.

References ....

Grubbs, L. A., Cassell, J., & Mulkey, W. (2005). *Rehabilitation caseload management.* (2nd ed.). Knoxville, TN: University of Tennessee Regional Rehabilitation Continuing Education Program.

Harrison, R. L., & Westwood, M. J. (2009). Preventing vicarious traumatization of mental health therapists: Identifying protective practices. *Psychotherapy Theory, Research, Practice, Training, 46*(2), 203–219.

Johnson, D. W. (2008). *Reaching out: Interpersonal effectiveness and self-actualization (10th ed.).* Boston: Allyn and Bacon.

Laekin, A. (1976). *How to get control of your time and your life.* New York: Signet.

Maslach, C., & Leiter, M. P. (1997). *The truth about burnout.* San Francisco: Jossey-Bass.

Painter, E. R. (2012). *Trauma work and the impact of vicarious traumatization.* Unpublished paper, Knoxville, Tennessee.

Skovholt, T. M. (2001). *The resilient practitioner: Burnout prevention and self-care strategies for counselors, therapists, teachers, and health professionals.* Boston: Allyn and Bacon.

Trippany, R. L., White Kress, V. E., & Wilcoxon, S. A. (2004). Preventing vicarious trauma: What counselors should know when working with trauma survivors. *Journal of Counseling and Development, 82*, 31–37.

Woodside, M., McClam, T., Diambra, J., & Varga, M. (2012). The meaning of time for human service professionals. *Human Service Education.*

Glossary

Accountability   Responsibility to another person for one's actions and use of resources

Achievement test   A test for evaluating an individual's present level of functioning

Active listening   Attending to both the verbal and nonverbal messages, as well as the thoughts and feelings of the person communicating

Advance directive   A general term for a person's instructions about future medical care in the event he or she is unable to speak for him- or herself

Advocacy   Speaking or writing in defense of a person or cause, pleading his or her case, or standing up for his or her rights

Advocate   A person who speaks on behalf of clients when they are unable to speak for themselves or when they speak and no one listens

Ageism   Discrimination or expression of bias against someone based on the person's age

Applicant   An individual who requests services from a human service agency or who is referred to such an agency for services

Aptitude test   A test that gives an indication of an individual's potential for learning or acquiring a skill

Assertiveness   Expressing oneself with confidence and conviction, while respecting the rights of others

Assessment   A phase of case management that involves evaluating a need or request for services and determining eligibility for services

Assessment interview   An interaction that provides information for the evaluation of an applicant for services

Attending behavior   Ways in which a person communicates interest and attention; examples are eye contact, attentive body language, and vocal qualities

Authority   The ability to control resources and actions

Autonomy   The client's right to make choices

Broker   A role in which the case manager acts as a go-between, linking those who seek services and those who provide services

Budget   A numerical expression of an agency's expected income and planned expenditures for an upcoming period

Burnout   Emotional exhaustion resulting from the stress of interpersonal contact

Case history interview   A comprehensive interview that includes open-ended questions and questions that require factual answers. It may include family history and a chronology of major life events

Case management   A creative and collaborative method of service delivery, involving skills in assessment, consulting, teaching, modeling, and advocacy, that is intended to enhance the functioning of the client

Case manager   A helping professional who provides direct services to clients, links clients to services, and monitors the process

Case notes   A written account of each visit, contact, or interaction a case manager has with or about a client

Case review   A periodic examination of a client's case

Chain of command   The flow of authority in an agency or organization, from the position with the most authority to the position with the least authority

Client   An applicant whose request for services has been approved by an agency

Client empowerment   Developing the client's self-sufficiency to enable him or her to manage life without total dependence on the human service delivery system

Closed questions   Questions that encourage clients to provide a minimal response

Communication skills   An important skill set for a case manager; includes listening, questioning, and persuading

Collaboration   The process of working with others

Confidentiality   A guarantee to the client that information disclosed during the helping process will be kept in confidence

Consultant   A professional who has expertise to support case management functions

Continuity of care   Comprehensive care provided during and after service delivery

Continuous improvement   A commitment and subsequent action whose purposes are improving service delivery

Contracts   Agreements that secure funding from governmental agencies or private corporations

Coordinator   A role in which the case manager works with other professionals and staff to integrate services

Counselor   Case manager role of therapist

Critical thinking   Process of seeing the whole as well as the individual parts—a valuable skill for the case manager

Deinstitutionalization   Moving clients from self-contained settings to various community-based settings, such as halfway houses

Departmental team   A small number of professionals who have similar job responsibilities and support each other's work

Documentation   Written presentation of data, observations, interviews, and services

DSM-IV-TR   The *Diagnostic and Statistical Manual of Mental Disorders* (4th edition, text revision), published by the American Psychological Association, which classifies types of mental disorders Soon to be revised as DSM-V

Duty to warn   A situation in which a helping professional must violate the confidentiality promised to a client in order to warn others that the client is a threat to self or to others

Effective communication   Verbal and nonverbal messages (delivered through greeting, eye contact, and responses) that let the client know that the interviewer is attentive

Empowerment   The belief that all individuals, regardless of their needs, have integrity and worth. Because of this belief, the case manager places the client in a central role in case management

End-of-life care   Attention to decisions concerning terminally ill family members or clients

Equal access to services   Nondiscrimination in granting services. Because of the commitment to equal access, the case manager assumes the role of advocate and develops ways of providing and extending services

Ethical decision making   The ability to identify issues, such as self-determination and confidentiality, and make appropriate professional responses

Ethnocentrism   The belief that one's own group has desirable characteristics and others outside the group are substandard

Feedback logs   Records that provide feedback to the agencies that deliver services to help ensure quality information and referral services

Fee-for-service   Payment for cost of service provided

For-profit agency   An agency that, when it makes money, distributes its profits to its owners

Goal   A statement describing a broad intent or a desirable condition

Halo effect   A favorable or unfavorable early impression that biases the judgment of the observer or interviewer

HMO (health maintenance organization)   A managed care model that is very structured and controlled and emphasizes positive health promotion

Implementation   The third phase of case management, in which service delivery occurs; the case manager either provides the services or oversees their delivery

Informal structure   The everyday way an organization does business; determined by personal relationships and influence, it varies from traditional organizational structure and formal lines of communication

Information and referral system   Source of information about what services are locally available and how to gain access to those services

Intake interview   A structured discussion, usually occurring when a person applies for services, which is guided by a set of questions in the form of an application

Intake summary   A type of documentation, written at some point during the assessment phase, which summarizes initial observations about the client

Integration of services   A coordinated effort on the part of many agencies and professions to bring together services to help a client

Intelligence test   A test that measures mental ability according to the criteria the test creators use to define the term

Interdisciplinary team   A team that includes professionals from various disciplines, each of which represents a service the client might receive

Interview   A face-to-face meeting between the case manager and the applicant. It may have a number of purposes, including getting or giving information, providing therapy, resolving a disagreement, or considering a joint undertaking

Jack or Jill of all trades   Describes an individual with multiple skills who performs a wide range of roles; a necessary trait for a case manager

Job description   A general guideline that describes the work for which an employee is held responsible

Living will   An individual's written instructions about medical treatment or intervention at the end of life

Managed care   An agreement with providers of physical and mental health care for provision of services

Maximum performance test   A test that asks examinees to do their best at something

Medical consultation   An appointment with a physician to interpret available medical data, determine any medical and vocational implications for health and employment, and recommend further medical care if needed

Medical diagnosis   An appraisal of an individual's general health status to establish whether a physical or mental impairment is present

Medical power of attorney   A signed, legally binding document that identifies an individual who can make decisions for the signer about medical care if the signer is unable to do so

Mental status examination   A structured interview consisting of questions designed to evaluate an individual's current mental status, considering factors such as appearance, behavior, and general intellectual process

Mission statement   A summary of the guiding principles of an agency

Mobilizer   One who works with other community members to obtain new resources or services for clients and communities

Monitoring services   Reviewing the services the client receives, the conditions that may have changed since planning, and progress toward the goals and objectives of the plan

Not-for-profit agency   An agency that uses its profits to meet the organization's objectives

Objective   An intended result of service provision

Open inquiries   Questions that elicit broad answers, allowing the client to express thoughts, feelings, and ideas

Organization-based case management   A model of case management that focuses on ways of configuring services that are comprehensive and meet the needs of clients who have multiple problems

Organizational chart   A symbolic representation of authority, accountability, and information flow within an organization or agency

Organizational climate   Those conditions of the work environment that affect how people experience their work

Physical examination   Inspection, palpation (feeling), percussion (sounding out), and auscultation (listening) by a physician. Typically, this is conducted from the skin inward through various orifices and from the top of the head to the toes

Plan A document written before service delivery that sets forth goals, objectives, and activities

Plan development A process that includes setting goals, writing objectives, and planning specific interventions

Planner The role of the case manager when preparing for the service or treatment the client is to receive

Planning The second phase of case management, in which the case manager and the client develop a service plan

POS (point of service) A managed care option in which clients are encouraged to use providers in the managed care system but do not lose all their benefits if they choose medical care outside the system

PPO (preferred provider organization) A type of health care plan that falls between the traditional HMO and the standard indemnity health insurance plan

Pretreatment review The review and approval by a managed care professional of a treatment plan before service delivery

Privileged communication A legal concept referring to the right of clients not to have their communications with a professional used in court without their consent

Problem solver The case manager role that assists in client self-sufficiency, decision making, and problem solving

Problem solving The process of identifying challenges and resolving difficulties

Process recording A narrative account of an interaction with another individual

Psychological evaluation The measurement of characteristics pertaining to an individual's behavior and mental capacity; used to understand an individual, psychological evaluation is an integral part of the client study process

Psychological report A document that reports on the evaluation of behavioral characteristics and mental capacity

Psychological test A device for measuring characteristics of human beings that pertain to behavior

Public agency An agency that exists by public mandate and receives funding from federal, state, regional, county, or municipal governments

Quality assurance Program that focuses on developing standards of care

Quality care An emphasis on providing the best services to the client in terms of effectiveness and efficiency

Questioning An important technique for soliciting information, especially in the intake interview

Racism Discrimination or unfavorable opinions based on an individual's race

Record Any type of information related to a client's case, including history, observations, examinations, diagnosis, consultations, and financial and social information

Recordkeeper   One of the primary roles of the case manager; involves collecting information and adding information to the client's file

Referral   Connecting the client to a resource within the agency or at another agency

Reliability   The degree of consistency with which a test measures whatever it is measuring

Resource selection   The process of choosing an individual, program, or agency to meet the client's needs

Responsibility-based case management   A model of service delivery in which case management can be conducted by the family, supportive care network, volunteers, or the client

Role-based case management   A model of case management that centers on the roles and responsibilities the case manager is expected to perform

Second-opinion mandate   The requirement that a second professional be consulted before a treatment plan can be approved

Service coordination   The process of locating, arranging, and supporting the client's use of community resources

Sexism   Discrimination or unfavorable opinions based on an individual's gender

Social diagnosis   A systematic way in which helping professionals gather information and study the nature of client problems

Social history   The telling of the client's story in his or her own words, with guidance from the helper; reflects the client's life and individual characteristics

Social service directory   A catalog listing the problems handled and services delivered by other agencies

Sources of error   Potential biases that affect an interview's reliability and validity

Staff notes   Comments written at the time of each visit, contact, or interaction that any helping professional has with a client

Structured clinical interview   An interview consisting of specific questions asked in a designated order

Structured interviews   Direct and focused interviews, usually guided by a form or a set of questions that elicit specific information

Summary recording   An organized, condensed presentation of facts; the form of recording preferred by most human service agencies

Teamwork   Professionals sharing responsibility for clients

Termination   The final step in case management, when the client and the case manager review the problem, goals, plan, services, and outcomes

Test   A measurement device

Time management   A systematic way of planning your work in order to see your clients, complete paperwork, and make allowance for crises

Treatment team   A group of professionals who meet to review client problems, evaluate information, and make recommendations about priorities, goals, and expected outcomes

Typical performance test   A test, such as an interest or personality test, that gives an idea of what an examinee is like on a day-to-day basis

Unstructured interview   An interaction that consists of a sequence of questions that follow from what has been said

Validity   The extent to which a test or interview measures what you wish to measure

Verbal following   Minimal verbal responses that let the client know you are listening

Violence   Any physical or verbally assaultive behaviors, including harm to self or others or the physical destruction or damaging of property

Whole person   A holistic view of the individual

Word root   The main part or stem of a word

Index